Question Bank on Statistics

NIPA® GENX ELECTRONIC RESOURCES & SOLUTIONS P. LTD.
New Delhi-110 034

About the Author

Rakesh Goel, M.Phil (Statistics), Assistant Professor, Department of Animal Genetics and Breeding, College of Veterinary Science and Animal Husbandry, U.P. Pandit Deen Dayal Upadhyaya Pashu Chikitsa Vigyan Vishwavidyalaya Evam Go-Anusandhan Sansthan, Mathura, has more than 30 years of teaching and research experience in the field of Computer & Biostatistics. He is associated with the research activities of the students, teachers and research scholars of Veterinary and Animal Sciences, Medical Sciences, Agricultural Sciences and Animal Biotechnology. He has also organized several computer trainings.. He is also the author of "Statistics and Computers for Animal and Veterinary Sciences" written to meet the requirements of the undergraduate students of Veterinary and Animal Science, Animal Biotechnology and other related fields.

Question Bank on Statistics

Rakesh Goel
Assistant Professor
Department of Animal Genetics and Breeding
College of Veterinary Science and Animal Husbandry
U.P. Pandit Deen Dayal Upadhyaya Pashu Chikitsa Vigyan Vishwavidyalaya
Evam Go-Anusandhan Sansthan
Mathura, Uttar Pradesh

NIPA® GENX ELECTRONIC RESOURCES & SOLUTIONS P. LTD.
New Delhi-110 034

NIPA® GENX ELECTRONIC RESOURCES & SOLUTIONS P. LTD.

101,103, Vikas Surya Plaza, CU Block
L.S.C.Market, Pitam Pura, New Delhi-110 034
Ph : +91 11 27341616, 27341717, 27341718
E-mail:newindiapublishingagency@gmail.com
www: www.nipabooks.com

For customer assistance, please contact
Phone: + 91-11-27 34 17 17 Fax: + 91-11- 27 34 16 16
E-Mail: feedbacks@nipabooks.com

ISBN: 978-81-19002-23-8

Composed and Designed by NIPA®.

उ.प्र.पं. दीनदयाल उपाध्याय पशु चिकित्सा विज्ञान विश्वविद्यालय
एवं गो अनुसंधान संस्थान, मथुरा-281001 (उ.प्र)
U.P. Pandit Deen Dayal Upadhyaya Pashu Chikitsa Vigyan Vishwavidyalaya Evam Go Anusandhan Sansthan (DUVASU)
Mathura-281001 (U.P.)

प्रो.(डॉ.) ऐ.के. श्रीवास्तव
कुलपति
Prof. (Dr.) A.K. Srivastava
Vice-Chancellor

Foreword

The present book entitled **"Question Bank on Statistics"** will meet the requirements of the under graduate students of Veterinary and Animal Sciences, Animal Biotechnology, Business Management, Agriculture, Medical and other related fields. The book is particularly written for the benefit of the students, preparing for various competitive exams like **JRF, SRF, NET, State level,** and other examinations.

I am sure that the book in its present form would also be of great help to researchers and teachers, particularly related to Agriculture, Biotechnology and Animal Sciences.

I appreciate the efforts made by the author in bringing out the book in present form.

(A.K. Srivastava)

Tel. : 0565-2470199; Fax: 0565-2470819; Mob. 9466592661; Email: duvasuvc@gmail.com
https://www.upvetuniv.edu.in

Preface

It's my pleasure to present the book, its being published with an aim to spread the knowledge attained by the undersigned in the vast teaching career as Statistician amongst the interested persons. There are manifold purposes of writing this book on the subject. Basically, it caters to the needs of the candidates aspiring for competitive examinations, and for the beginners to understand the intricacies of the subject.

It is observed that the very name of the subject, statistics evokes fear in the minds of the students, through this book an effort has been made to dispel that fear and students should be able to approach the subject confidently. It should be learning with pleasure for the students and they should face the competitions any time and at any place with confidence. The latest trend of education is the teaching through multiple choice questions. The MCQ's in this book are intended to enable students to prioritize and plan their learning through regular practicing.

Statistics has assumed a lot of importance in recent times. It is imperative for every student to study and gain knowledge of statistics, its uses as well as its applications. The format and contents of this book has been carefully drafted, the Chapters have been written with a view to explain the concepts clearly and examples have been given in the text to help the students to understand the application of these concepts in real life situations. MCQs at the end of every unit will help the students to make a self-assessment of the knowledge assimilated by going through the Chapters. Through this book, the author has made an effort to provide rationale for the solutions. The book, therefore, meets the expectations of the students as it answers the demand and the quest in their mind and give rise to real learning which would stand in good stead for the student's career and his life.

The quality of the MCQs presented in this book stands out from the other books available in the market. Speciality of this book is that there are varieties of MCQs which comprehend all the related questions on the subjects as well the contents have been so designed as to include in the true spirit of the syllabus of basics of statistics.

Entire syllabus of statistics is divided into Chapters, and another innovation in this book is that each Chapter is divided into various sections, sub-sections and headings. The book is user-friendly and provides information in a well structured manner. It provides comprehensive and critical study of the various concepts of the subject matter.

It is felt that the contents should be crystal clear and efforts are made to ensure this however a word or suggestion from reader's side may add another feather to the cap of the subject matter of the book. The author looks forward to the comments, suggestions and criticism from the readers. Constructive suggestions and feedback from users would be highly appreciated, gratefully acknowledged and suitably incorporated.

The author is highly obliged to his wife Mrs. Namita Goel and daughter Miss Priya Goel for keeping patience and moral support.

Author would like to take this opportunity to thank Prof. (Dr.) A.K. Srivastava, the Honorable Vice Chancellor, Uttar Pradesh Pandit Deen Dayal Upadhyaya Pashu Chikitsa Vigyan Vishwavidyalaya Evam Go-Anusandhan Sansthan, Mathura along with Prof. (Dr.) P. K. Shukla, Dean, Veterinary College and other faculty members of the University.

With best wishes,

Author

Contents

1

Statistics: An Introduction

Introduction to Statistics

The word statistics has been derived from Latin word "**status**", the German word "**statistik**", the Italian word "**statistica**" or the French word "**statistique**". During early periods, it was used by emperors and kings only for collecting the information about state and other information which was needed about their people, their number, land, status of people, revenue of the state etc. Gradually, it meaning and uses extended and there onwards its nature changed.

Some scientist/ researchers were given definition of statistics as per time and place.

- An ancient date, **A.L. Bowley** gives some definitions according their use and place. He seems to be inadequate because these do not include all aspects of statistics. As per time, place or requirement, he gave three definitions of statistics:
  - Statistics may be called the science of counting
  - Statistics may rightly be called the science of average
  - Statistics as the science of counting and average
- In the words of **Boddington,** "Statistics is science of estimates and probabilities".
- According to **Wallis and Roberts**, "Statistics is regard as a body of methods of decision making in the face of uncertainty".
- According to **Croxton and Cowden**, "Statistics is the collection, presentation, analysis and interpretation of numerical data".
- As all these definitions are incomplete and restricted the scope of science. **Prof. A. Saligman** gave the most comprehensive definition of statistics is "Statistics is the science which deals with the application and methods of collection, classification, comparison, presentation, analysis and interpretation of data in any spare of enquiry".

There have been many definitions of the term ‘statistics’ which can be broadly classified in two senses:

- Statistics in plural sense, and
- Statistics in singular sense

Statistics in plural sense: It is used for denoting numerical facts systematically collected with some definite object in view, in any field of enquiry, whatsoever of observation, measurement or experiment. ***For example:*** Statistics of the population of a country, males and females, births and deaths, heights and weights, income and expenditure, food production, etc.

Statistics in singular sense: It is a science which investigates the statistical methods and deals with scientific methods of collection, organization, presentation, analysis and interpretation of data obtained by conducting a survey or an experimental study.

Some basic concepts

- **Population:** It is a well defined set, group or aggregate of observations relating to a phenomenon under statistical investigation. It can be **finite** or **infinite** according to the set contains a finite or infinite number of observations. The infinite populations are **hypothetical** in nature. i.e., their existence can only be imagined or conceptualized.
- **Sample:** It is a small part or a fraction of a population drawn by a suitable method.
- **Statistic:** If we have data of any series, like: houses with different numbers of rooms, families with different numbers of children, goal record in football matches, and so on, we can calculate the average number of rooms in a house, number of children per family, the greatest score in football matches, etc. Any such number is called a “**statistic**”.
- **Constant:** A variable whose domain contains only one value is called a **constant**.
- **Attribute:** The qualitative character of the items chosen are called **attribute**. **For example:** Breed of an animal, etc.
- **Variate:** Any single value in the domain of the variable is called a **variate**. **For example:** The number of inches or centimeters that represents the height of an individual is a **variate**.
- **Variable:** A **variable** is a quantity which may take up any value from a given assemblage of values, called its **domain**. Thus, sex of an animal is a variable of which the domain consists of only two values, *i.e.*, male and female. Variables may be of three types:

- **Quantitative variable:** Whenever the measurement of a variable is possible on a scale in some appropriate unit, it is called a **quantitative** variable. **For example:** height, weight, age, speed, shape, number of children in the family, etc.
- **Qualitative variable:** It is un-measureable on a scale and, as such, it is un-expressible in magnitudes. **For example:** sex, literacy, honesty, colour of flowers, etc.
- **Random variable:** When the values obtained as a chance factors, the variable is called a **random variable**. These are also two types:
- **Continuous random variable:** If a variable can take all values from one given number to another, it is called **continuous random variable. For example:** Body weight of the student.
- **Discrete random variable:** If the values of the variable differ from one another by definite amounts. i.e., they varies only by finite 'jumps' or 'breaks' and can assume finite values in a finite interval. **For example:** Number of students in a class, marks obtained by students, etc.

- **Parameter:** A constant which changes its values from one situation to another is often called a **parameter**.
- **Biostatistics:** It deals with the application of statistics to the biological data. It is also called **'Biometry'**. **Francis Galton** (1822-1911) is called the **"Father of Biostatistics"**. **W.F.R. Weldon** (1860-1906) coined the term **'Biometry'**. Biostatistics covers application and contribution not only from health, biotechnology, medicine and nutrition but also from fields such as agriculture, genetics, biology, biochemistry and many others.
- **Statistical Inference:** It is a conclusion about a population on the basis of information contained in a sample. **For example:** Inferences about daily milk recording in a dairy farm in the study are statistical inferences.
- **Accuracy:** It shows the closeness of a calculate value or measured value to the true value.
- **Precision:** It shows the closeness of two or more measurements are to each other. Usually, accuracy and precision are used synonymously equal, but in statistics they are used different. **For example:** Suppose, a deep refrigerator thermometer is read 5 times, i.e. 20.1, 19.8, 19.9, 20.2, 20.1. However, the real temperature inside the de-refrigerator is -20^{0}C. The thermometer isn't accurate, but all the numbers are close to -20^{0}C. In this case the accurate temperature is -20^{0}C, but these fluctuations shows that the temperature of deep refrigerator almost -20^{0}C.

- **Parametric statistics:** It is based on the assumption that the distribution of population from which the sample was taken.
- **Nonparametric statistics**: It is not based on assumptions, that is, the data can be collected from a sample that does not follow a specific distribution.

MCQ

Q.1. The word 'Statistik" in German means:

(a) Calculation (b) **Government**

(c) Math's (d) Classification

Q2. The application of statistical methods in biology is called:

(a) Statistics in biology (b) Statistics in vivo

(c) **Biostatistics** (d) All of them

Q3. Biostatistics is also called as:

(a) **Bionumerology** (b) Bionemerology

(c) Statistics in biology (d) None of these

Q4. Who is regarded as father of biostatistics:

(a) Fisher (b) Karl Pearson

(c) **Francis Galton** (d) W.S.Gossett

Q5. The term 'biometry' was coined by:

(a) R.A. Fisher (b) Karl Pearson

(c) Francis Galton (d) **Walter Weldon**

Q6. In biostatistics, group of individuals taken for study is called as:

(a) **Population** (b) Block

(c) Group (d) Flock

Q7. Branches of statistics includes

(a) Applied statistics (b) Mathematical statistics

(c) Industry statistics **(d) Both (a) and (b)**

Q8. The word 'statistics' is used as:

(a) Singular (b) Plural

(c) **Both (a) and (b)** (d) None of the above

Bold means correct answer

Q9. Sample statistics are denoted by the

(a) Upper case Greek letter (b) Associated roman alphabets

(c) **Roman letters** (d) Lower case Greek letter

Q10. Which of the following represents data?

(a) Only a single value (b) Two values in a set

(c) **Group of values in a set** (d) None of these

Q11. In statistics, a population consists of:

(a) All people living in a country

(b) **All people living in the area under study**

(c) All subjects or objects whose characteristics are being studied

(d) None of the above

Q12. Branch of statistics which deals with findings of solution in field of medicine, education and economics is classified as

(a) Economic statistics (b) **Applied statistics**

(c) Mathematical statistics (d) Industry statistics

Q13. Branch of statistics which deals with development of particular statistical methods is classified as:

(a) Economic statistics (b) Applied statistics

(c) **Mathematical statistics** (d) Industry statistics

Q14. Branch of statistics in which data is collected according to ordinal scale or nominal scale is classified as:

(a) Distribution statistics (b) Sampling statistics

(c) Parametric statistics (d) **Non-parametric statistics**

Q15. Statistics deals with:

(a) **Quantitative information** (b) Qualitative information

(c) Both (a) and (b) (d) None of (a) and (b)

Q16. Statistical results are:

(a) Absolutely correct (b) **Not absolutely correct**

(c) Always incorrect (d) Misleading

Q17. Statistics can:

(a) Prove anything

(b) Disprove anything

(c) **Neither prove nor disprove anything, is just a tool**

(d) None of the above

Q18. Statistics can best be considered as:

(a) An art
(b) Science
(c) **Both art and science**
(d) Neither art nor science

Q19. In the development of statistical methods, the greatest contribution is that of:

(a) Businessmen
(b) Scientists
(c) **Mathematicians**
(d) Economists

Q20. In its present usage, the word statistics is:

(a) **A century old**
(b) Two centuries old
(c) Fifty centuries old
(d) A decade old

Q21. How do statistics inform the reader of the central massage?

(a) **Summarizing and creating meaning from aggregated data**

(b) By saying what is true

(c) They don't inform the reader

(d) None of these

Q22. The branch of biostatistics that deals with methods of collection, organization and presentation of data is called as:

(a) Inferential biostatistics
(b) **Descriptive biostatistics**
(c) Comparative biostatistics
(d) None of these

Q23. The branch of biostatistics that deals with testing of hypothesis, making predictions using data collected is called as:

(a) **Inferential biostatistics**
(b) Descriptive biostatistics
(c) Comparative biostatistics
(d) None of these

Q24. Variables whose values can be expressed numerically are called:

(a) Qualitative variable
(b) **Quantitative variable**
(c) Continuous variable
(d) Absolute variable

Q25. The characteristics or quantity that may vary from one individual to another is called:

(a) **Variable**
(b) Static group
(c) Dynamic group
(d) Dynamism

Q26. Flower color is:

(a) **Qualitative variable**
(b) Quantitative variable
(c) Continuous variable
(d) Absolute variable

Q27. Height of students in a class is:

(a) Qualitative variable
(b) Quantitative variable
(c) **Continuous variable**
(d) Absolute variable

Q28. Quantitative variables that have only fixed or finite values are called:

(a) Qualitative variable
(b) Quantitative variable
(c) Continuous variable
(d) **Discrete variable**

Q29. Quantitative variables that have any numerical values are called:

(a) Qualitative variable
(b) Quantitative variable
(c) **Continuous variable**
(d) Discrete variable

Q30. Number of fruits in a tress is:

(a) Qualitative variable
(b) Quantitative variable
(c) Continuous variable
(d) **Discrete variable**

Q31. In every phenomenon, process of thought that focus on identifying, controlling and reduction of variations in data is classified as__________ thinking:

(a) Parallel
(b) Serial
(c) **Statistical**
(d) Managerial

Q32. Process of converting inputs into outputs in presence of repeatedly same conditions is classified as

(a) Sampler (b) Parameters

(c) **Process** (d) Mixer

Q33. One of category of statistical method is:

(a) Managerial statistics (b) Decision science

(c) **Inferential statistics** (d) Industry statistics

Q34. Analytical study of breakeven point and decisions related to investments in uncertain conditions is an example of:

(a) Statistics in personnel management

(b) **Statistics in finance**

(c) Statistics in marketing

(d) Statistics in production

Q35. Science and art which is used to present, analyze and interpret collected observations is classified as:

(a) Serial analysis (b) **Statistics**

(c) Management (d) Accounting

Q36. Use of statistics in analysis of time when to order and quantity required for in sourcing or outsourcing of materials is an example of usage of statistics in:

(a) Marketing (b) **Production**

(c) Personnel management (d) Finance

Q37. Statistical methods which facilitate decision making process for population sample results are classified as:

(a) Decisional procedure (b) **Inferential statistics**

(c) Elementary statistics (d) Social methodology

Q38. Group of observations for which better understanding is needed and group is related to specific phenomenon is classified as:

(a) **Population** (b) Population statistic

(c) Phenomenon statistic (d) Parameter statistic

Q39. Who stated that statistics is a branch of applied mathematics which specializes in data?

(a) L.R.Connor (b) A.L.Bowley

(c) **R.A.Fisher** (d) Tippet

Q40. The statement, 'Statistics is both a science and an art', was given by:

(a) L.R.Connor (b) A.L.Bowley

(c) R.A.Fisher (d) **Tippet**

Q41. Who originally gave the formula for the estimation of errors?

(a) L.R.Connor (b) A.L.Bowley

(c) R.A.Fisher (d) **A. L. Boddington**

Q42. Type of questions included in questionnaire to record responses in which respondents have set of alternative are classified as:

(a) Open ended questions (b) **Close ended questions**

(c) Multiple choices (d) Itemized question

Q43. Branch of statistics which deals with findings of solution in field of medicine, education and economics is classified as

(a) Economic statistics (b) **Applied statistics**

(c) Mathematical statistics (d) Industry statistics

Q44. Type of questions for questionnaire includes

(a) Multiple choices (b) Open ended

(c) Dichotomous (d) **All of above**

Q45. Statistical techniques that summarize and organize the data are classified as:

(a) Population statistics (b) Sample statistics

(c) **Descriptive statistics** (d) Inferential statistics

Q46. A researcher is curious about the IQ of students at the Utrecht University. The entire group of students is an example of a:

(a) Parameter (b) Statistic

(c) **Population** (d) Sample

Q47. A researcher studies the factors that determine the number of children future couples decide to have. The variable 'number of children' is a:

(a) **Discrete variable** (b) Continuous variable

(c) Categorical variable (d) Ordinal variable

Q48. The analysis of labor turnover rates, performance appraisal, training programs and planning of incentives are examples of role of:

(a) Statistics in marketing (b) **Statistics in personnel management**

(c) Statistics in finance (d) Statistics in production

Q49. Collection of observations for all the variables related to some research or findings is classified as

(a) **Data** (b) Population

(c) Statistical process (d) Mesokurtic

Q50. When statistical inference is made on basis of sample results about characteristics of population then this is classified as

(a) Inferential statistics (b) **Sample statistics**

(c) Population statistics (d) Population variability

Q51. The branch of statistics which considers the ratio scale and interval scale is considered as:

(a) Sampling statistics (b) Distribution statistics

(c) **Parametric statistics** (d) Non-parametric statistics

Q52. Characterization, collection and presentation of particular set of data in organized way is classified as

(a) **Descriptive statistics** (b) Education statistics

(c) Social statistics (d) Business statistics

Q53. Scale used in statistics which provides difference of proportions as well as magnitude of differences is considered as

(a) Satisfactory scale (b) **Ratio scale**

(c) Goodness scale (d) Exponential scale

Q54. Procedures of descriptive statistics and control charts which are used to improve process are classified as

(a) **Statistical tools** (b) Parallel tools

(c) Serial tools (d) Behavioral tools

Q55. Examples of variables in statistical phenomenon consists

(a) Job satisfaction (b) Consumer behaviors

(c) Leadership ability (d) **All of above**

Q56. Branches of statistics includes

(a) Applied statistics (b) Mathematical statistics

(c) Industry statistics (d) **Both (a) and (b)**

Q57. One of category of statistical method is

(a) Managerial statistics (b) Decision science

(c) **Inferential statistics** (d) Industry statistics

Q58. The questions which have different answers for its subparts are considered as:

(a) Single barreled questions (b) **Double barreled questions**

(c) Triple barreled questions (d) Multiple barreled questions

Q59. A population value or characteristic that is of interest to us and that we would like to estimate is called:

(a) Hypothesis (b) Statistic

(c) Population (d) **Parameter**

Q60. A summary measure that is computed from a sample to describe a characteristic of the population is called

(a) Parameter (b) **Statistic**

(c) Population (d) Inferential statistics

Q61. For quality improvements, the philosophy which acts like a guide and provides solid foundation is classified as:

(a) **Management philosophy** (b) Statistical philosophy

(c) Serial philosophy (d) Parallel philosophy

Q62. In every phenomenon, the process of thought that focus on identifying, controlling and reduction of variations in data is classified as

(a) Parallel thinking
(b) Serial thinking
(c) **Statistical thinking**
(d) Managerial thinking

Q63. Which of the following is not the goal of descriptive statistics?

(a) Summarizing data
(b) Displaying aspects of the collected data
(c) Reporting numerical findings
(d) **Estimating characteristics of the population**

Q64. Statistical inference means:

(a) The process of drawing conclusions about a sample based on population data
(b) The process of drawing conclusions about a statistic based on a parameter
(c) The process of drawing conclusions about a population based on a parameter
(d) **The process of drawing conclusions about a population based on sample data**

Q65. The science of collecting, organizing, presenting, analyzing and interpreting data to assist in making more effective decisions is called:

(a) **Statistic**
(b) Parameter
(c) Population
(d) Statistics

Q66. Methods of organizing, summarizing, and presenting data in an informative way are called:

(a) **Descriptive statistics**
(b) Inferential statistics
(c) Theoretical statistics
(d) Applied statistics

Q67. The methods used to determine something about a population on the basis of a sample is called:

(a) **Inferential statistics**
(b) Descriptive statistics
(c) Applied statistics
(d) Theoretical statistics

Q68. When the characteristic being studied is nonnumeric, it is called:

(a) Quantitative variable (b) **Qualitative variable**

(c) Discrete variable (d) Continuous variable

Q69. When the variable studied can be reported numerically, the variable is called:

(a) **Quantitative variable** (b) Qualitative variable

(c) Independent variable (d) Dependent variable

Q70. A specific characteristic of a population is called:

(a) Statistic (b) **Parameter**

(c) Variable (d) Sample

Q71. A specific characteristic of a sample is called:

(a) Variable (b) Constant

(c) Parameter (d) **Statistic**

Q72. A set of all units of interest in a study is called:

(a) Sample (b) **Population**

(c) Parameter (d) Statistic

Q73. A part of the population selected for study is called:

(a) Variable (b) Data

(c) **Sample** (d) Parameter

Q74. Listing of the data in order of numerical magnitude is called:

(a) Raw data (b) **Arrayed data**

(c) Discrete data (d) Continuous data

Q75. Listings of the data in the form in which these are collected are known as:

(a) Secondary data (b) **Raw data**

(c) Arrayed data (d) Qualitative data

Q76. Data that are collected by anybody for some specific purpose and use are called:

(a) Qualitative data (b) **Primary data**

(c) Secondary data (d) Continuous data

Q77. The data which have under gone any treatment previously is called:

(a) Primary data (b) **Secondary data**

(c) Symmetric data (d) Skewed data

Q78. The data obtained by conducting a survey is called:

(a) **Primary data** (b) Secondary data

(c) Continuous data (d) Qualitative data

Q79. The data collected from published reports is known as:

(a) Discrete data (b) Arrayed data

(c) **Secondary data** (d) Primary data

Q80. A survey in which information is collected from each and every individual of the population is known as:

(a) Sample survey (b) Pilot survey

(c) Biased survey (d) **Census survey**

Q81. Data used by an agency that originally collected them are:

(a) **Primary data** (b) Raw data

(c) Secondary data (d) Grouped data

Q82. Registration is the source of:

(a) Primary data (b) **Secondary data**

(c) Qualitative data (d) Continuous data

Q83. Data in the population census reports are:

(a) Ungrouped data (b) Secondary data

(c) **Primary data** (d) Arrayed data

Q84. Issuing a national identity card is an example of:

(a) Sampling (b) Statistic

(c) Census (d) **Registration**

Q85. A variable that assumes only some selected values in a range is called:

(a) Continuous variable (b) Quantitative variable

(c) **Discrete variable** (d) Qualitative variable

Q86. A variable that assumes any value within a range is called:

(a) Discrete variable (b) **Continuous variable**
(c) Independent variable (d) Dependent variable

Q87. A variable that provides the basis for estimation is called:

(a) Dependent variable (b) **Independent variable**
(c) Continuous variable (d) Qualitative variable

Q88. The variable that is being predicted or estimated is called:

(a) **Dependent variable** (b) Independent variable
(c) Discrete variable (d) Continuous variable

Q89. Monthly rainfall in a city during the last ten years is an example of:

(a) Discrete variable (b) **Continuous variable**
(c) Qualitative variable (d) Independent variable

Q90. The proportion of females in a sample of 50 accounts officers is an example of:

(a) Parameter (b) **Statistic**
(c) Array (d) Variable

Q91. Number of family members in different families in a town is an example of:

(a) **Discrete variable** (b) Continuous variable
(c) Dependent variable (d) Qualitative variable

Q92. Colours of flowers are an example of:

(a) Quantitative variable (b) **Qualitative variable**
(c) Skewed variable (d) Symmetric variable

Q93. If each measurement in a data set falls into one and only one of a set of categories, the data set is called:

(a) Quantitative (b) **Qualitative**
(c) Continuous (d) Constant

Q94. Any phenomenon which is not measurable is called:

(a) Variable (b) Constant
(c) Parameter (d) **Attribute**

Q95. A constant can assume values:

(a) Zero (b) One

(c) **Fixed** (d) Not fixed

Q96. A value which does not change from one individual to another individual is called:

(a) Variable (b) Statistic

(c) **Constant** (d) Array

Q97. In the plural sense, statistics means:

(a) **Numerical data** (b) Methods

(c) Population data (d) Sample data

Q98. In the singular sense, statistics means:

(a) **Methods** (b) Numerical data

(c) Sample data (d) Population data

Q99. Weight of earth is:

(a) Discrete variable (b) Qualitative variable

(c) **Continuous variable** (d) Difficult to tell

Q100.Weights of students in a class marks are:

(a) Discrete data (b) **Continuous data**

(c) Qualitative data (d) Constant data

Q101.Life of a electric tube light is:

(a) Discrete variable (b) **Continuous variable**

(c) Qualitative variable (d) Constant

Q102.Questionnaire method is used in collecting:

(a) **Primary data** (b) Secondary data

(c) Published data. (d) True data

Q103. Census returns are:

(a) **Primary data** (b) Secondary data

(c) Qualitative data (d) True data

Q104.Students divided into different groups according to their intelligence and gender will generate:

(a) Quantitative data (b) **Qualitative data**

(c) Continuous data (d) Constant

Q105.Statistics are:

(a) **Aggregate of facts and figures** (b) Always true

(c) Always continuous (d) Always qualitative

Q106.Statistics results are:

(a) Randomly true (b) Always true

(c) Not true (d) **True on average**

Q107.Statistics does not study:

(a) Constant (b) Statistic

(c) Parameter (d) **Individual**

Q108.A statistical population may consist of:

(a) Finite number of values (b) Infinite number of values

(c) **Either of (a) and (b)** (d) None of (a) and (b)

Q109.The only continuous variable here is:

(a) **Rain fall on different days in a city**

(b) Number of customers entering a store on different days

(c) Number of flights landing on an airport on different days

(d) None of them

Q110.Example of descriptive statistics is:

(a) **60% people in India live in rural areas.**

(b) 50% people are likely to vote in the national election

(c) 20% of the bulbs produced in a factory will be defective

(d) Difficult to tell.

Q111.Example of inferential statistics is:

(a) Percentage of smokers in India

(b) Percentage of skilled workers in a factory.

(c) **Estimate of increase in prices in the next year**

(d) None of the above

Q112.Statistics are always:

(a) Exact
(b) **Estimated values**
(c) Constant
(d) Population values

Q113.Statistics must be:

(a) **Comparable**
(b) Not comparable
(c) Discrete in nature
(d) Qualitative in nature

Q114.Given 6 quantities, X_1 through X_6, the correct notation for adding quantities 3 through 6 is:

(a) $\sum_{i=6} X_i$
(b) $\sum_{i=1}^{6} X_i$
(c) $\sum_{i=2}^{N} X_i$
(d) $\sum_{i=3}^{6} \mathbf{X}_i$

Q115.Given $X_1 = 12, X_2 = 19, X_3 = 10, X_4 = 7$, $\sum_{i=2}^{4} X_i$ equals:

(a) 29
(b) **36**
(c) 41
(d) 48

Q116.The symbolic notation $\sum_{i=3}^{6} X_i$ tells us to:

(a) Add all quantities from Y1 through Yn

(b) Add all quantities from Y=2 through Yn

(c) Add all quantities from Y=2 through Y=n

(d) **Add all quantities from Y2 through Yn**

Q117.Analytical study of breakeven point and decisions related to investments in uncertain conditions is an example of statistics in:

(a) Personnel management
(b) **Finance**
(c) Marketing
(d) Production

Q118. Which of the following is a branch of statistics?

(a) Descriptive statistics
(b) Inferential statistics
(c) Industry statistics
(d) **Both (a) and (b)**

Q119. What are the variables whose calculation is done according to the weight, height, and length known as?

(a) Flowchart variables
(b) Discrete variables
(c) **Continuous variables**
(d) Measuring variables

Q120. A population value or characteristic that is of interest to us and that we would like to estimate is called:

(a) a statistic
(b) a hypothesis
(c) **a parameter**
(d) a population

Q121. The time frame to complete a transaction in bank is classified as:

(a) **Process**
(b) Sampler
(c) Mixer
(d) Parameter

Q122. Which of the following is not the goal of descriptive statistics?

(a) Reporting numerical findings
(b) Summarizing data
(c) **Estimating characteristics of the population**
(d) None of these

2

Classification and Tabulation

Statistical Series: When observations of any characteristic present in the individual of a group are recorded and arranged one after the other in a systematic order, they form a series, known as **statistical series**. **For example:** milk yield of 20 cows, height of the students in a class in a systematic order, etc. These are some important types of series:

- **Discrete series:** In this series, observations can take up only exact values and not any fractional values, is called **discrete series**. That means, this type of series has whole number of counting. **For example:** number of the students in a class, members of a family, number of births in a certain year, etc
- **Continuous series:** A series which can take up any numerical value (integral/ fractional) within a certain range is called **continuous series**. **For example:** milk yield of 20 cows, height, weight, temperature, etc.
- **Array series:** When we arranged the observations of a series either in ascending order of their magnitude or descending order of their magnitude, then it becomes an **array series**.
- **Spatial series:** When items of a series are given according to space or place or time, then such type of series is known as **spatial series**. **For example:** month wise milk yield of a herd of cows, month wise income or expenditure of an industry, population of a country according to states, etc.

Data: The facts and figures collected, analyzed, and summarized for presentation and interpretation, or we can sat, a large number of related figures are known as **data**. It may include numbers of individuals that make up the census of a city, ages of students in a certain class, morning and evening milk recording in the dairy farm, etc.

Collection of data

Statistical data are classified under two categories:

- **Primary data:** On those which form raw material of an enquiry. It is also known as original data or raw data. The data which are recorded at the source is known as original data for the **primary data**. Actually primary data are those which have never be utilized before. **For Example,** Daily milk yield recorded at the dairy farm, Commodities of ration issue to animals, etc

- **Secondary data:** When primary data through any statistical method, it becomes the **secondary data**. In other words, when any statistical method is applied on primary data, then it becomes the **secondary data. For Example,** Total of monthly milk yield in the dairy farm, total expenditure, total income, etc.

Classification of data

The arrangement of dividing the huge data in different groups or classes according to resemblances and similarities is known as **Classification of data**. Statistical data are classified in respect of their characteristics. Broadly, there are four basic types of classification namely-

- **Chronological Classification:** Collected data are arranged according to the order of time expressed in years, months, weeks, etc. The data is generally classified in ascending order of time.
- **Geographical or Special Classification:** Classification of the data is classified according to geographical region or place. ***e.g.***, District wise milk production in Uttar Pradesh.
- **Qualitative Classification:** Classification of the data is classified on the basis of some attributes or qualities like: sex, literacy, breed, etc.
- **Quantitative Classification:** Classification of data are grouped with reference to characteristic which can be measured and numerically described such as- height, weight, age, etc.

Frequency: It may be defined as 'the number of observations falling in a class interval.

Frequency distribution

Croxton and Cowden defined frequency distribution as "a statistical table which shows the sets of all distinct values of the variable arranged in order of magnitude, either individually or in groups, with their corresponding frequencies side by side". In other words, a frequency distribution is defined as the way in which the original data are divided into un-differential class intervals and it contains in first column the class intervals and in the second column, corresponding frequencies.

Tabulation of data

The presentation of classified data in a suitable tabular form is known as "**Tabulation of data**". In other words, **tabulation** may be defined as the arrangement of data in different rows and columns. The tabulation is always done after classification only and more over tabulation is not a statistical process but it is simply a mechanical process. In other words, tabulation may be defined as the arrangement of data in different rows and columns. The tabulation is classified as follows:

- **Single tabulation:** When sub-division of total data is done on the basis of **one** factor, then it is known as **single tabulation**. It can provide answer of one question only. **For example:** In India, division of population according to different **states**.

- **Double tabulation:** When sub-division of total data is done on the basis of **two** factors, then it is known as **double tabulation**. It can provide answer of two questions simultaneously. **For example:** Division of Indian population **first** on the basis of **states** then further sub-division on the basis of **sex**.

- **Triple tabulation:** When sub-division of total data is done on the basis of **three** factors, then it is known as **triple tabulation**. It can provide answer of **three** questions simultaneously. **For example:** Division of Indian population **first** on the basis of **states, second** on the basis of **sex** and **third** on the basis of **literacy**.

- **Manifold tabulation:** When sub-division of total data is done on the basis of **three** factors, then it is known as **triple tabulation**. When sub-division of total data is done on the basis of **more than three** factors, then it is known as **manifold tabulation**. It can provide answer of **more than three** questions simultaneously. **For example:** Division of Indian population **first** on the basis of **states, second** on the basis of **sex, third** on the basis of **literacy** and forth on the basis of **age**.

MCQ

Q.1. Which of the following is not an example for a primary data?

(a) Mailed questionnaire

(b) Local correspondents

(c) Indirect oral investigation

(d) **Survey reports in newspapers, journals**

Q2. Procedures of descriptive statistics and control charts which are used to improve process are classified as:

(a) **Statistical tools** (b) Parallel tools

(c) Serial tools (d) Behavioral tools

Q3. Scale used in statistics which provides difference of proportions as well as magnitude of differences is considered as:

(a) Satisfactory scale (b) **Ratio scale**

(c) Goodness scale (d) Exponential scale

Bold means correct answer

Q4. Branch of statistics which considers ratio scale and interval scale is considered as:

(a) **Parametric statistics** (b) Non-parametric statistics

(c) Distribution statistics (d) Sampling statistics

Q5. Variables whose measurement is done in terms such as weight, height and length are classified as:

(a) **Continuous variables** (b) Measuring variables

(c) Flowchart variables (d) Discrete variables

Q6. Type of variable which can take fixed integer values is classified as

(a) Flowchart variable (b) **Discrete variable**

(c) Continuous variable (d) Measuring variables

Q7. Type of variable which can take any numerical figure for calculation is classified as:

(a) **Continuous variable** (b) Measuring variable

(c) Flowchart variable (d) Discrete variable

Q8. Examples of variables in statistical phenomenon consists

(a) Job satisfaction (b) Consumer behaviors

(c) Leadership ability (d) **All of above**

Q9. Number of employees according to human resource manager is an example of

(a) Flowchart variable (b) **Discrete variable**

(c) Continuous variable (d) Measuring variable

Q10. A numerical value used as a summary measure for a sample, such as sample mean, is known as a:

(a) Population parameter (b) Sample parameter

(c) **Sample statistic** (d) Population mean

Q11. Since the population size is always larger than the sample size, then the sample statistic:

(a) Can never be larger than the population parameter

(b) Can never be equal to the population parameter

(c) Can never be smaller than the population parameter

(d) **None of the above**

Q12. The sum of the percent frequencies for all classes will always equal:

(a) One | (b) Number of classes

(c) Number of items in the study | (d) **100**

Q13. Analysis of labor turnover rates, performance appraisal, training programs and planning of incentives are examples of role of statistics in:

(a) **Personnel management** | (b) Finance

(c) Marketing | (d) Production

Q14. Focus groups, individual respondents and panels of respondents are classified as:

(a) Pointed data sources | (b) Itemized data sources

(c) Secondary data sources | (d) **Primary data sources**

Q15. Variables whose measurement is done in terms such as weight, height and length are classified as:

(a) **Continuous variables** | (b) Measuring variables

(c) Flowchart variables | (d) Discrete variables

Q16. Technique used to analyze unemployment rate, inflation rate anticipation and capacity utilization to manufacture goods is classified as__________ technique:

(a) Data supplying | (b) **Forecasting**

(c) Data importing | (d) Data exporting

Q17. Numerical methods and graphical methods are specialized procedures used in:

(a) Social statistics | (b) Business statistics

(c) **Descriptive statistics** | (d) Education statistics

Q18. Measure of how well is a technique, concept or process is considered as:

(a) Continuity of variables | (b) Goodness of variables

(c) **Validity** | (d) Reliability

Q19. Simple classification and manifold classification are types of

(a) **Qualitative classification** (b) Quantitative classification

(c) Open end classification (d) Time series classification

Q20. Reports on quality control, production and financial accounts issued by companies are considered as________________________ data sources:

(a) External secondary (b) **Internal secondary**

(c) External primary (d) Internal primary

Q21. Type of classification in which class is subdivided into subclasses and one attribute is assigned for statistical study is considered as

(a) Rational classification (b) Reflected classification

(c) **Simple classification** (d) Manifold classification

Q22. Scales which categorizes events in collectively exhaustive manner and mutually exclusive manner is classified as __________ scale.

(a) Discrete (b) Continuous

(c) Valid (d) **Nominal**

Q23. Type of rating scale which allows respondents to choose most relevant option out of other stated options is classified as______________ rating scale.

(a) Marking (b) Graphical

(c) **Itemized** (d) Pointed

Q24. Government and non-government publications are considered as ___________ data sources.

(a) **External secondary** (b) Internal secondary

(c) External primary (d) Internal primary

Q25. Type of variable which can take fixed integer values is classified as:

(a) Flowchart variable (b) **Discrete variable**

(c) Continuous variable (d) Measuring variables

Q26. Data which is generated within company such as routine business activities is classified as ________________________ data sources.

(a) External primary (b) Internal primary

(c) External secondary (d) **Internal secondary**

Q27. Questions which have different answers for its subparts are considered as:

(a) **Double barreled questions** (b) Single barreled questions

(c) Multiple barreled questions (d) Dichotomous questions

Q28. Analytical study of relationship between output commodity and its price is classified as

A. **Demand analysis** B. Supply analysis

C. Imports analysis D. Export analysis

Q29. Group of observations for which better understanding is needed and group is related to specific phenomenon is classified as

(a) **Population** (b) Population statistic

(c) Phenomenon statistic (d) Parameter statistic

Q30. Successful life of product, time, weight and height are classified as

(a) **Continuous random variable**

(b) Discrete random variable

(c) Continuous waiting time variable

(d) Continuous hyper geometric variable

Q31. Type of variable which can take any numerical figure for calculation is classified as:

(a) **Continuous variable** (b) Measuring variable

(c) Flowchart variable (d) Discrete variable

Q32. Tools such decision making by nominal groups, brain storming and term buildings are all considered as:

(a) Serial tools (b) **Behavioral tools**

(c) Statistical tools (d) Parallel tools

Q33. Less than type› cumulative frequency distribution is considered as correspondent to

(a) **Upper limit of class intervals**

(b) Lower limit of class intervals

(c) Upper limit of range

(d) Lower limit of range

Q34. More than type› cumulative frequency distribution is considered as correspondent to

(a) Upper limit of class intervals

(b) **Lower limit of class intervals**

(c) Upper limit of range

(d) Lower limit of range

Q35. Time frame to complete a transaction in bank is classified as:

(a) Parameters (b) **Process**

(c) Mixer (d) Sampler

Q36. Type of rating scale which represents response of respondents by marking at appropriate point is classified as

(a) **Graphic rating scale** (b) Responsive scale

(c) Pointed scale (d) Marking scale

Q37. First step in constructing frequency distribution is to

(a) **Select appropriate class intervals**

(b) Determine class intervals

(c) Determine class limits

(d) Determine midpoints of classes

Q38. Scale which is used to determine ratios equality is considered as

(a) Satisfactory scale (b) **Ratio scale**

(c) Goodness scale (d) Exponential scale

Q39. An example of variables in statistical phenomenon consists:

(a) Job satisfaction (b) Consumer behaviors

(c) Leadership ability (d) **All of above**

Q40. Procedures of descriptive statistics and control charts which are used to improve process are classified as:

(a) **Statistical tools** (b) Parallel tools

(c) Serial tools (d) Behavioral tools

Q41. Class interval classification method which ensures data continuity is classified as

(a) Midpoint method
(b) Ratio method
(c) **Exclusive method**
(d) Inclusive method

Q42. Measurement scale which allows researchers and statisticians to perform certain operations on data collected from respondents is classified as:

(a) **Interval scale**
(b) Flow measuring scale
(c) Validity scale
(d) Reliability scale

Q43. Type of questions included in questionnaire to record responses in which respondent can answer in any way are classified as

(a) Multiple choices
(b) Itemized question
(c) **Open ended questions**
(d) Close ended questions

Q44. Numerical or descriptive measure which is associated with variable to describe entire population of statistical phenomenon is classified as:

(a) Mixer
(b) Sampler
(c) **Parameter**
(d) Process

Q45. Model which consists of management philosophy, behavioral tools and statistical methods as key steps towards improvement is considered as___________ improvement process model.

(a) Serial
(b) Behavioral
(c) **Quality**
(d) Statistics

Q46. Collection of all elements such as group of variables for research is classified as

(a) Statistical process
(b) Marginal error
(c) Data
(d) **Population**

Q47. According to types of questionnaires, structured undisguised questionnaire is one in which purpose of study is:

(a) Non-interval respondents
(b) Disclosed to respondents
(c) **Not disclosed to respondents**
(d) None of above

Q48. Characterization, collection and presentation of particular set of data in organized way is classified as:

(a) **Descriptive statistics** (b) Education statistics

(c) Social statistics (d) Business statistics

Q49. Data measurement which arises from a specific measuring process is classified as

(a) Reliable data (b) Valid data

(c) Continuous data (d) **Discrete data**

Q50. Measurement scale which allows ranking of numbers rather than arithmetic operations on data is classified as

(a) Valid scale (b) Discrete scale

(c) **Ordinal scale** (d) Continuous scale

Q51. Scale used in statistics which provides difference of proportions as well as magnitude of differences is considered as

(a) Satisfactory scale (b) **Ratio scale**

(c) Goodness scale (d) Exponential scale

Q52. Measurement scale in which values are categorized to represent qualitative differences and ranked in meaningful manner is classified as

(a) Valid scale (b) Discrete scale

(c) **Ordinal scale** (d) Continuous scale

Q53. Measurement scale which allows determination of differences in intervals is classified as

(a) **Interval scale** (b) Flow measuring scale

(c) Validity scale (d) Reliability scale

Q54. Use of statistics in analysis of sales territories, advertising strategies and routing of sales personnel are classified as uses of statistics in:

(a) Personnel management (b) Finance

(c) **Marketing** (d) Production

Q55. Types of structured questions does not include

(a) Nominal questions (b) Interval questions

(c) Ratio questions (d) **Non-disguised questions**

Q56. In response questionnaires, questions which require reflections from respondent side are called:

(a) Reflection questions (b) Responsive questions

(c) **Opinion questions** (d) Reliability questions

Q57. Analysis based on study of price fluctuations, production of commodities and deposits in banks is classified as

(a) Sample series analysis (b) **Time series analysis**

(c) Numerical analysis (d) Experimental analysis

Q58. Bias which occurs when randomly drawn sample from population fails to represent whole population is classified as

(a) Populated bias (b) Random sample bias

(c) **Under coverage bias** (d) Coverage bias

Q59. Reports published by International Labor Organization and International Monetary Fund are considered as

(a) External primary data sources

(b) Internal primary data sources

(c) **External secondary data sources**

(d) Internal secondary data sources

Q60. Type of questions for questionnaire includes

(a) Multiple choices (b) Open ended

(c) Dichotomous (d) **All of above**

Q61. Measurement of how well particular concept and technique measures variables is classified as:

(a) **Reliability** (b) Validity

(c) Continuity of variables (d) Goodness of variables

Q62. Data measurement which arises from a specific process of counting is classified as

(a) Continuous data (b) **Discrete data**

(c) Reliable data (d) Valid data

Q63. Structured undisguised questionnaire is one in which purpose of study is

(a) Not disclosed to respondents (b) **Disclosed to respondents**

(c) Interval respondents (d) Ratio respondents

Q64. The .... process would be required to ensure that the data is complete and as required:

(a) Analysis (b) Tabulation

(c) **Editing** (d) Ordering

Q65. Subset of selected population is called

(a) Descriptive portion (b) Elementary portion

(c) Inferential portion (d) **Sample**

Q66. A variable is any characteristic which can assume .... values:

(a) Fixed (b) **Different**

(c) Assumed (d) Similar

Q67. Scale which is used in determination of categorical information is called

(a) Discrete scale (b) Continuous scale

(c) Valid scale (d) **Nominal scale**

Q68. Collection of observations for all variables related to some research or findings is classified as:

(a) **Data** (b) Population

(c) Statistical process (d) None of these

Q69. For quality improvements, philosophy which acts like a guide and provides solid foundation is classified as:

(a) **Management philosophy** (b) Statistical philosophy

(c) Serial philosophy (d) Parallel philosophy

Q70. Structured undisguised questionnaire is one in which purpose of study is

(a) Not disclosed to respondents (b) **Disclosed to respondents**

(c) Interval respondents (d) Ratio respondents

Q71. All individual elements of sample are considered as

(a) **Experimental units** (b) Population units

(c) Statistical units (d) Process units

Q72. Stem and leaf displaying technique is used to present data in

(a) Descriptive data analysis (b) **Exploratory data analysis**

(c) Nominal data analysis (d) Ordinal data analysis

Q73. Number of employees according to human resource manager is an example of:

(a) Flowchart variable (b) **Discrete variable**

(c) Continuous variable (d) Measuring variable

Q74. Characteristics that are intended to be analyzed and investigated for a given population are classified as:

(a) Exponents (b) Constants

(c) **Variables** (d) Exponential base

Q75. Uses of index numbers include:

(a) Commodity prices (b) Cost of living

(c) Exports and imports (d) **All of above**

Q76. Types of questions included in questionnaire to record responses in which respondents have set of alternatives are classified as:

(a) Open ended questions (b) **Close ended questions**

(c) Multiple choices (d) Itemized question

Q77. Data measurement which arises from a specific process of counting is classified as

(a) Continuous data (b) **Discrete data**

(c) Reliable data (d) Valid data

Q78. Measurement of how well particular concept and technique measures variables is classified as:

(a) **Reliability** (b) Validity

(c) Continuity of variables (d) Goodness of variables

Q79. Scale used in statistics which provides difference of proportions as well as magnitude of differences is considered as

(a) Satisfactory scale (b) **Ratio scale**

(c) Goodness scale (d) Exponential scale

Q80. In the case of the questionnaire method of gathering data, it should be made certain that all the questions have been:

(a) Interpreted (b) Read

(c) **Answered** (d) All of them

Q81. Numerical data presented in descriptive form are called:

(a) Classified presentation (b) Tabular presentation

(c) Graphical presentation (d) **Textual presentation**

Q82. In an exclusive type distribution, the limits excluded are:

(a) Upper limits (b) Lower limits

(c) **Either (a) or (b)** (d) Both (a) and (b)

Q83. Whether classification is done first or tabulation?

(a) **Classification precedes tabulation**

(b) Classification follows tabulation

(c) Both are done simultaneously

(d) No criterion

Q84. In a grouped data, the numbers of classes preferred are:

(a) Maximum possible (b) Minimum possible

(c) **Adequate** (d) Any arbitrary number

Q85. If the lower and upper limits of a class are 5 and 45, respectively, the mid-point of the class is:

(a) 15.0 (b) 20.0

(c) **25.0** (d) 30.0

Q86. A series showing the sets of all distinct values individually with their frequencies is known as:

(a) **Simple frequency distribution**

(b) Grouped frequency distribution

(c) Cumulative frequency distribution

(d) None of the above

Q87. A series showing the sets of all values in classes with their corresponding frequencies is known as:

(a) Simple frequency distribution

(b) **Grouped frequency distribution**

(c) Cumulative frequency distribution

(d) None of the above

Q88. Class of variable which can accept only values from set of integers is classified as

(a) **Discrete random variable**

(b) Continuous random variable

(c) Posterior random variable

(d) Interior random variable

Q90. Which of the following represents data?

(a) Single value (b) Only two values in a set

(c) **Group of values in a set** (d) None of the above

Q91. Types of structured questions does not include

(a) Nominal questions (b) Interval questions

(c) Ratio questions (d) **Non-disguised questions**

Q92. If 'A' is the actual value and 'B' is its estimated value, the absolute error is:

(a) (A – B) (b) **|A – B|**

(c) A / B (d) B / A

Q93. If 'A' is the actual value and 'B' is its estimated value, the relative error is:

(a) (A – B)/B (b) (A – B)/B

(c) |A – B| /A (d) **|A – B| /B**

Q94. Relative error is always:

(a) Positive (b) Negative

(c) Zero (d) **Both (a) and (b)**

Q95. If the estimated value of an item is 20 and its actual value is 40, the absolute error is:

(a) 0.5 (b) 1.0

(c) – 20 (d) **20**

Q96. If the estimated value of an item is 20 and its actual value is 40, the relative error is:

(a) 0.5 (b) **1.0**

(c) 2.0 (d) 20

Q97. Mailed questionnaire method of enquiry can be adopted if respondents:

(a) Are known (b) **Are educated**

(c) Have high income (d) Lives in cities

Q98. Statistical data are collected for:

(a) Any purpose

(b) **Given purpose**

(c) Collecting data without any purpose

(d) None of the above

Q99. Method of complete enumeration is applicable for:

(a) Knowing the production

(b) Knowing the population

(c) Knowing the quantum of export and import

(d) **All of them**

Q100.A statistical population may consist of:

(a) A finite number of items (b) An infinite number of items

(c) **Either (a) or (b)** (d) Both (a) and (b)

Q101.A study based on complete enumeration is known as:

(a) Pilot survey
(b) Sample survey
(c) **Census survey**
(d) None of the above

Q102.Which of the following can be classified as hypothetical population?

(a) Male population of a country
(b) Female population of a country
(c) **Population of real numbers between 0 and 100**
(d) Students of the world

Q103.Measurement scale in which values are categorized to represent qualitative differences and ranked in meaningful manner is classified as

(a) Valid scale
(b) Discrete scale
(c) **Ordinal scale**
(d) Continuous scale

Q104.If each value of frequency distribution is divided by total number of recorded observations in distribution then resultant value is called

(a) Interval frequency distribution
(b) **Percentage frequency distribution**
(c) Nominal frequency distribution
(d) Ordinal frequency distribution

Q105.Table in which data represented is extracted from some other data table is classified as

(a) Classification tables
(b) Cumulative tables
(c) **Derived table**
(d) Dispersion tables

Q106.Type of table in which study variables provides large number of information with interrelated characteristics is classified as

(a) Lower order table
(b) Manifold table
(c) Higher order table
(d) **Both (b) and (c)**

Q107.Measurement scale which allows ranking of numbers rather than arithmetic operations on data is classified as

(a) Valid scale
(b) Discrete scale
(c) **Ordinal scale**
(d) Continuous scale

Q108.Data measurement which arises from a specific measuring process is classified as

(a) Reliable data (b) Valid data

(c) Continuous data (d) **Discrete data**

Q109.According to types of questionnaires, structured undisguised questionnaire is one in which purpose of study is

(a) Non-interval respondents (b) Disclosed to respondents

(c) **Not disclosed to respondents** (d) None of above

Q110.Smallest numerical value is subtracted from largest numerical value and then divided to number of class desired to calculate:

(a) Simple class interval (b) **Width of class interval**

(c) Number of classes (d) Manifold class intervals

Q111.Process of arranging data on basis of certain properties in classes or groups is classified as

(a) **Classification of data** (b) Reflection of data

(c) Sample of population (d) Sample observations

Q112.Original tables used to represent data are considered as

(a) Dispersion tables (b) **Classification tables**

(c) Cumulative tables (d) Distributed tables

Q113.Class frequency is divided by number of observations in frequency distribution to convert it into

(a) Relative margin distribution

(b) Relative variable distribution

(c) **Relative frequency distribution**

(d) Relative width distribution

Q114.Classification on basis of time order is called as

(a) Disclosed classification (b) **Chronological classification**

(c) External classification (d) Internal classification

Q115.Halfway point between lower class limits and upper class limits is classified as

(a) Nominal mid-point (b) **Class mid-point**

(c) Interval mid-point (d) Ordinal mid-point

Q116.Set of raw data arranged in ascending or descending order is called

(a) **Ordered array** (b) Nominal array

(c) Ordinal array (d) Interval array

Q117.Branch of statistics in which data is collected according to ordinal scale or nominal scale is classified as

(a) Distribution statistics (b) Sampling statistics

(c) Parametric statistics (d) **Non-parametric statistics**

Q118.Classification of data on geographical basis is also called

(a) Reflected classification (b) Populated classification

(c) Sampling classification (d) **Spatial classification**

Q119.Type of variable which can take any numerical figure for calculation is classified as

(a) **Continuous variable** (b) Measuring variable

(c) Flowchart variable (d) Discrete variable

Q120.Frequency distribution which is result of cross classification is called

(a) **Bivariate frequency distribution**

(b) Univariate frequency distribution

(c) Multi-variables distribution

(d) Close ended distribution

Q121.Data table which is presented in tabular form on basis of single characteristics is classified as

(a) **Simple table** (b) Complex table

(c) Percentage table (d) Interval table

Q122.Type of classification in which class is subdivided into subclasses and subclasses are divided into more classes is considered as

(a) Simple classification (b) **Manifold classification**

(c) Rational classification (d) Reflected classification

Q123.Data table which is presented in tabular form on basis of two or more simultaneous characteristics is classified as:

(a) Percentage table (b) Interval table

(c) Simple table (d) **Complex table**

Q124.Data classification which is based on variables such as production, cost, height and weight is considered as

(a) Qualitative classification (b) **Quantitative classification**

(c) Open end classification (d) Time series classification

Q125.Distribution which shows cumulative figure of all observations placed below upper limit of classes in distribution is considered as:

(a) **Cumulative frequency distribution**

(b) Upper limit distribution

(c) Class distribution

(d) Cumulative class distribution

Q126.Third step in constructing frequency distribution is to

(a) Select appropriate class intervals

(b) Determine class intervals

(c) **Determine class limits**

(d) Determine midpoints of classes

Q127.The goal of ___________ is to focus on summarizing and explaining a specific set of data.

(a) Inferential statistics (b) **Descriptive statistics**

(c) None of the above (d) All of the above

Q128.To interpret the relationship between two categorical variables, a contingency table should be constructed with either column or row percentages, and ----.

(a) If the percentages are calculated down the columns, then comparisons should be made across the rows

(b) If the percentages are calculated across the rows, comparisons should be made down the columns

(c) **Both (a) and (b) are correct**

(d) Neither (a) nor (b) is correct

Q129.A statistical table consists of at least:

(a) Two parts (b) Three parts

(c) **Four parts** (d) Five parts

Q130.In a statistical table, prefatory note is shown:

(a) Below the body (b) Box head

(c) Foot note (d) **Below the title**

Q131. A source note in a statistical table is given:

(a) **At the end of a table** (b) In the beginning of a table

(c) In the middle of a table (d) Below the body of a table

Q132. In a statistical table, column captions are called:

(a) **Box head** (b) Stub

(c) Body (d) Title

Q133. In a statistical table, row captions are called:

(a) Box head (b) **Stub**

(c) Body (d) Title

Q134.The headings of the rows of a table are called:

(a) Prefatory notes (b) Titles

(c) **Stubs** (d) Captions

Q135.The headings of the columns of a table are called:

(a) Stubs (b) **Captions**

(c) Footnotes (d) Source notes

Q136.The following frequency distribution:

Classes	Less than 10	Less than 20	Less than 30	Less than 40	Less than 50
Frequency	2	3	16	19	20

is classified as:

(a) **Less than cumulative frequency distribution**

(b) More than cumulative frequency distribution

(c) Discrete frequency distribution

(d) Cumulative percentage frequency distribution

Q137. The following frequency distribution

Classes	50 - 55	55 - 60	60 - 65	65 - 70	70 - 75
Frequency	40	36	30	16	4

is classified as:

(a) Relative frequency distribution

(b) Less than cumulative frequency distribution

(c) **More than cumulative frequency distribution**

(d) Bivariate frequency distribution

Q138.A frequency distribution formed considering two variables at a time is called:

(a) Univariate frequency distribution

(b) **Bivariate frequency distribution**

(c) Trivariate frequency distribution

(d) Bimodal distribution

Q139. The sum of rows or sum of columns, of a bivariate, frequency distribution is equal to:

(a) $\sum X$ (b) $\sum fX$

(c) $\sum (f+X)$ (d) $\sum \mathbf{f}$

Q140. The arrangement of data in rows and columns is called:

(a) Classification

(b) **Tabulation**

(c) Frequency distribution

(d) Cumulative frequency distribution

Q141.When the qualitative or quantitative raw data are classified according to one characteristic, the tabulation of different groups is called:

(a) Dichotomy (b) Manifold-division

(c) Bivariate (d) **One-way**

Q142.In a relative frequency distribution, the total of the relative frequencies is:

(a) 100 (b) **One**

(c) $\sum f$ (d) $\sum X$

Q143. In a percentage frequency distribution, the total of the percentage frequencies is always equal to:

(a) 1 (b) $\sum f$

(c) **100%** (d) $\sum X$

Q144. The cumulative frequency of first group in more than cumulative frequency distribution is always equal to:

(a) 1 (b) 100

(c) $\boldsymbol{\sum f}$ (d) $\sum X$

Q145. The cumulative frequency of last class in less than cumulative frequency distribution is always equal to:

(a) $\boldsymbol{\sum f}$ (b) $\sum X$

(c) 1 (d) 100

Q146.The class frequency divided by the total number of observations is called:

(a) Percentage frequency (b) **Relative frequency**

(c) Cumulative frequency (d) Bivariate frequency

Q147.The relative frequency multiplied by 100 is called:

(a) **Percentage frequency** (b) Cumulative frequency

(c) Bivariate frequency (d) Simple frequency

Q148.The following frequency distribution

Classes	10 - 20	20 - 30	30 - 40	40 - 50	50 - 60
Frequency	2	4	6	4	2

is classified as:

(a) **Exclusive classification** (b) Inclusive classification

(c) Geographical classification (d) Two-way classification

Q149.The following frequency distribution

Classes	0 - 4	5 - 9	10 - 14	15 - 19	20 - 24
Frequency	2	3	7	5	3

is classified as:

(a) Multiple classifications (b) Qualitative classification

(c) **Inclusive classification** (d) Exclusive classification

Q150. The following frequency distribution

Classes	More than 2	More than 4	More than 6	More than 8	More than 10
Frequency	2	6	16	19	20

is classified as:

(a) Geographical classification

(b) Chronological classification

(c) Inclusive classification

(d) **Exclusive classification**

Q151.When the classes are 40 - 44, 45 - 49, 50 - 54, ... the class interval is:

(a) 4 (b) 2

(c) 10 (d) **5**

Q152.A grouping of data into mutually exclusive classes showing the number of observations in each class is called:

(a) Frequency polygon (b) Relative frequency

(c) **Frequency distribution** (d) Cumulative frequency

Q153.The following frequency distribution

Classes	Less than 2	Less than 4	Less than 6	Less than 8	Less than 10
Frequency	2	6	16	19	20

is classified as:

(a) Inclusive classification (b) **Exclusive classification**

(c) Discrete classification (d) Cross classification

Q154.A frequency distribution that contains a class with limits of "10 and under 20" would have a midpoint:

(a) 10 (b) 14.9

(c) **15** (d) 20

Q155. If the number of workers in a factory is 128 and maximum and minimum hourly wages are 100 and 20 respectively. For the frequency distribution of hourly wages, the class interval is:

(a) 8 (b) 9

(c) **10** (d) 80

Q156.Width of interval h is equal to:

(a) $\frac{\text{Largest number - 20}}{\text{Number of classes}}$

(b) $\frac{\text{Largest number + Smallest number}}{\text{Number of classes}}$

(c) $\frac{\textbf{Largest number + Smallest number}}{\textbf{Number of classes}}$

(d) $\frac{\text{Number of classes}}{\text{Range}}$

Q157. Length of interval is calculated as:

(a) **The difference between upper limit and lower limit**

(b) The sum of upper limit and lower limit

(c) Half of the difference between upper limit and lower limit

(d) Half of the sum of upper limit and lower limit

Q158. The class marks are given below: 10,12,14,16,18. The first class of the distribution is:

(a) 9 - 12 (b) 10.5 - 12.5

(c) **9 - 11** (d) 10 - 12

Q159.If the midpoints are 10, 15, 20, 25 and 30. The last class boundary of the distribution is:

(a) 25 - 30
(b) **27.5 - 32.5**
(c) 20 - 35
(d) 30 - 35

Q160. The number of classes depends upon:

(a) Class marks
(b) Frequency
(c) **Class interval**
(d) Class boundary

Q161.The class interval is the difference between:

(a) Two extreme values
(b) Two successive frequencies
(c) **Two successive upper limits**
(d) Two largest values

Q162.Class boundaries are also called:

(a) **Mathematical limits**
(b) Arithmetic limits
(c) Geometric limits
(d) Qualitative limits

Q163.The average of lower and upper class limits is called:

(a) Class boundary
(b) Class frequency
(c) **Class mark**
(d) Class limit

Q164.The lower and upper class limits are 20 and 30, the midpoint of the class is:

(a) 20
(b) **25**
(c) 30
(d) 50

Q165.The suitable formula for computing the number of classes is:

(a) 3.322 logN
(b) 0.322 logN
(c) **1+3.322 logN**
(d) 1- 3.322 logN

Q166. The number of classes in a frequency distribution is obtained by dividing the range of variable by the:

(a) Total frequency
(b) **Class interval**
(c) Mid-point
(d) Relative frequency

Q167. If the number of workers in a factory is 256, the number of classes will be:

(a) 8 (b) **9**

(c) 10 (d) 12

Q168. The largest and the smallest values of any given class of a frequency distribution are called:

(a) Class Intervals (b) Class marks

(c) Class boundaries (d) **Class limits**

Q169.If there are no gaps between consecutive classes, the limits are called:

(a) Class limits (b) **Class boundaries**

(c) Class intervals (d) Class marks

Q170. The extreme values used to describe the different classes in a frequency distribution are called:

(a) Class intervals (b) Class boundaries

(c) **Class limits** (d) Cumulative frequency

Q171.If in a frequency table, either the lower limit of first class or the upper limit of last class is not a fixed number, then classes are called:

(a) One-way classes (b) Two-way classes

(c) Discrete classes (d) **Open-end classes**

Q172.The class boundaries can be taken when the nature of variable is:

(a) Discrete (b) **Continuous**

(c) Both (a) and (b) (d) Qualitative

Q173.In an exclusive classification, the limits excluded are:

(a) Upper limits

(b) Lower limits

(c) Both lower and upper limits

(d) **Either lower or upper limits**

Q174. The series

Weights(pounds)	15 - 20	20 - 25	25 - 30	30 - 35	35 - 40
No. of items	10	15	30	10	5

is categorized as:

(a) **Continuous series** (b) Discrete series

(c) Time series (d) Geometric series

Q175. The series

Year	2010	2011	2012	2013	2014
Profit (000 Rs.)	10	15	30	10	5

will be called as:

(a) **Time series** (b) Discrete series

(c) Continuous series (d) Individual series

Q176.Frequency distribution is often constructed with the help of:

(a) Entry table (b) Tally sheet

(c) **Both (a) and (b)** (d) Neither (a) and (b)

Q177.The data given as 3, 5, 15, 35, 70, 84 and 96 will be called as:

(a) **Individual series** (b) Discrete series

(c) Continuous series (d) Time series

Q178. Frequency of a variable is always in:

(a) Fraction form (b) Percentage form

(c) Less than form (d) **Integer form**

Q179. Data arranged in ascending or descending order of magnitude is called:

(a) Ungrouped data (b) Grouped data

(c) Discrete frequency distribution (d) **Arrayed data**

Q180. The grouped data are called:

(a) Primary data (b) **Secondary data**

(c) Raw data (d) Difficult to tell

Q181.A series of data with exclusive classes along with the corresponding frequencies is called:

(a) Discrete frequency distribution

(b) **Continuous frequency distribution**

(c) Percentage frequency distribution

(d) Cumulative frequency distribution

Q182.The number of tally sheet count for each value or a group is called:

(a) Class limit
(b) Class width
(c) Class boundary
(d) **Frequency**

Q183. The frequency distribution according to individual variate values is called:

(a) **Discrete frequency distribution**

(b) Cumulative frequency distribution

(c) Percentage frequency distribution

(d) Continuous frequency distribution

Q184. A series arranged according to each and every item is known as:

(a) Discrete series
(b) Continuous series
(c) **Individual series**
(d) Time series

Q185. A frequency distribution can be:

(a) Qualitative
(b) Discrete
(c) Continuous
(d) **Both (b) and (c)**

Q186. The following frequency distribution:

X	5	15	38	47	68
f	2	4	9	3	1

is classified:

(a) Relative frequency distribution
(b) Continuous distribution
(c) Percentage frequency distribution
(d) **Discrete distribution**

Q187.Classification is applicable in case of:

(a) Normal characters
(b) Quantitative characters
(c) Qualitative characters
(d) **Both (b) and (c)**

Q188. In classification, the data are arranged according to:

(a) **Similarities**
(b) Differences
(c) Percentages
(d) Ratios

Q189. When data are arranged at regular interval of time, the classification is called:

(a) Qualitative
(b) Quantitative
(c) **Chronological**
(d) Geographical

Q190. When an attribute has more than three levels it is called:

(a) **Manifold-division**
(b) Dichotomy
(c) One-way
(d) Bivariate

Q191. The series

Country	Pakistan	India	Holland	Egypt	Japan
Birth rate	45	40	10	35	10

is of the type:

(a) Discrete
(b) Continuous
(c) **Individual**
(d) Time series

Q192. The series

Country	Pakistan	India	Holland	Egypt	Japan
Death rate	15	16	10	12	10

is of the type:

(a) Inclusive
(b) Exclusive
(c) **Geographical**
(d) Time series

Q193. In an array, the data are:

(a) In ascending order
(b) In descending order
(c) **Either (a) or (b)**
(d) Neither (a) or (b)

Q194. When data are classified according to a single characteristic, it is called:

(a) Quantitative classification
(b) Qualitative classification
(c) Area classification
(d) **Simple classification**

Q195. Classification of data by attributes is called:

(a) Quantitative classification
(b) Chronological classification
(c) **Qualitative classification**
(d) Geographical classification

Q196. Classification of data according to location or areas is called:

(a) Qualitative classification

(b) Quantitative classification

(c) **Geographical classification**

(d) Chronological classification

Q197. You ask five of your classmates about their height. On the basis of this information you conclude that the average height of all students is 170 cm. This is an example of:

(a) Descriptive statistics (b) **Statistical inference**

(c) A parameter (d) A population

Q198. The relative frequency for a class in a frequency distribution is calculated by:

(a) Dividing the frequency of the class by the number of classes

(b) Dividing the frequency of the class by the class width

(c) **Dividing the frequency of the class by the total number of observations in the data set**

(d) Subtracting the lower limit of the class from the upper limit and multiplying the difference by the number of classes

Q199. The sum of the relative frequencies for all classes in a frequency distribution will always equal to:

(a) The number of classes

(b) The class width

(c) The total number of observations in the data set

(d) **One**

Q200. Consider these two variables: (i) whether you are a SA citizen (ii) your marital status. In the order given, these variables are:

(a) Nominal, Interval (b) Interval, Nominal

(c) Interval, Interval (d) **Nominal, Nominal**

Q201.The process of systematic arrangement of data in rows and columns is called:

(a) Array (b) **Tabulation**

(c) Classification (d) None of these

Q202.Total Relative Frequency is always:

(a) Quarter (b) Half

(c) **One** (d) Two

Q203.Cumulative Frequency is ________________ frequency:

(a) **Increasing** (b) Decreasing

(c) Fixed (d) None of these

Q204.Dividing the upper and lower limits of a particular class we get:

(a) Class interval (b) Class frequency

(c) **Class marks** (d) Class boundary

Q205.The average value of the lower and upper limit of a class is called:

(a) **Mid-point** (b) Class boundary

(c) Class interval (d) Class frequency

Q206.While constructing Frequency Distribution, the number of classes used depends upon:

(a) Number of observation (b) **Size of class**

(c) Range of data (d) None of these

Q207.The data which have already been collected by someone are called:

(a) Raw data (b) Array data

(c) **Secondary data** (d) Fictitious data

Q208.How many methods are used for the collection of data?

(a) 4 (b) 3

(c) **2** (d) 1

Q209. Classification method in which upper limit of interval is same as of lower limit class interval is called:

(a) **Exclusive method** (b) Inclusive method

(c) Mid-point method (d) Ratio method

Q210. Type of cumulative frequency distribution in which class intervals are added in top to bottom order is classified as:

(a) Variation distribution

(b) **Less than type distribution**

(c) More than type distribution

(d) Marginal distribution

Q211. Data based on workers salary is as 2500, 2700, 2600, 2800, 2200, 2100, 2000, 2900, 3000, 2800, 2200, 2500, 2700, 2800, 2600 and number of classes desired is 10 then width of class interval is

(a) 400 (b) 300

(c) **100** (d) 200

Q212. Largest value is 60 and smallest value is 40 and number of classes desired is 5 then class interval is:

(a) 20 (b) **4**

(c) 25 (d) 15

Q213. Summary and presentation of data in tabular form with several non-overlapping classes is referred as:

(a) Nominal distribution (b) Ordinal distribution

(c) Chronological distribution (d) **Frequency distribution**

Q214. Classification method in which upper and lower limits of interval is also in class interval itself is called:

(a) Exclusive method (b) **Inclusive method**

(c) Mid-point method (d) Ratio method

Q215. General tables of data used to show data in orderly manner is called as

(a) Single characteristics tables (b) **Repository tables**

(c) Manifold tables (d) Double characteristic table

Q216.Frequencies of all specific values of x and y variables with total calculated frequencies are classified as:

(a) Variate frequencies
(b) Unconditional frequencies
(c) Conditional frequencies
(d) **Marginal frequencies**

Q217.Type of cumulative frequency distribution in which class intervals are added in bottom to top order is classified as:

(a) **More than type distribution**
(b) Marginal distribution
(c) Variation distribution
(d) Less than type distribution

Q218.'Less than type distribution' and 'More than type distribution' are types of

(a) Class distribution
(b) Cumulative class distribution
(c) **Cumulative frequency distribution**
(d) Upper limit distribution

Q219.Exclusive method and inclusive method are ways of classifying data on basis of

(a) Manifold classes
(b) Rational intervals
(c) Class width
(d) **Class intervals**

Q220.Third step in constructing frequency distribution is to

(a) Select appropriate class intervals
(b) Determine class intervals
(c) **Determine class limits**
(d) Determine midpoints of classes

Q221.Distribution which shows cumulative figure of all observations placed below upper limit of classes in distribution is considered as:

(a) **Cumulative frequency distribution**
(b) Upper limit distribution
(c) Class distribution
(d) Cumulative class distribution

Q222.Data classification which is based on variables such as production, cost, height and weight is considered as

(a) Qualitative classification
(b) **Quantitative classification**
(c) Open end classification
(d) Time series classification

Q223.Type of classification in which class is subdivided into subclasses and subclasses are divided into more classes is considered as:

(a) Simple classification
(b) **Manifold classification**
(c) Rational classification
(d) Reflected classification

Q224.Data table which is presented in tabular form on basis of two or more simultaneous characteristics is classified as:

(a) Percentage table
(b) Interval table
(c) Simple table
(d) **Complex table**

Q225.Data table which is presented in tabular form on basis of single characteristics is classified as

(a) **Simple table**
(b) Complex table
(c) Percentage table
(d) Interval table

Q226.Frequency distribution which is result of cross classification is called

(a) **Bivariate frequency distribution**
(b) Univariate frequency distribution
(c) Multi-variables distribution
(d) Close ended distribution

Q227.Classification of data on geographical basis is also called

(a) Reflected classification
(b) Populated classification
(c) Sampling classification
(d) **Spatial classification**

Q228.Set of raw data arranged in ascending or descending order is called

(a) **Ordered array**
(b) Nominal array
(c) Ordinal array
(d) Interval array

Q229.Halfway point between lower class limits and upper class limits is classified as

(a) Nominal mid-point
(b) **Class mid-point**
(c) Interval mid-point
(d) Ordinal mid-point

Q230.Classification on basis of time order is called as

(a) Disclosed classification
(b) **Chronological classification**
(c) External classification
(d) Internal classification

Q231.Class frequency is divided by number of observations in frequency distribution to convert it into

(a) Relative margin distribution
(b) Relative variable distribution
(c) **Relative frequency distribution**
(d) Relative width distribution

Q232.Original tables used to represent data are considered as

(a) Dispersion tables
(b) **Classification tables**
(c) Cumulative tables
(d) Distributed tables

Q233. If classification of collected data is based on characteristics such as religion, education and caste, then this is considered as:

(a) Open end classification
(b) Time series classification
(c) **Qualitative classification**
(d) Quantitative classification

Q234.Process of arranging data on basis of certain properties in classes or groups is classified as

(a) **Classification of data**
(b) Reflection of data
(c) Sample of population
(d) Sample observations

Q235.Smallest numerical value is subtracted from largest numerical value and then divided to number of class desired to calculate:

(a) Simple class interval
(b) **Width of class interval**
(c) Number of classes
(d) Manifold class intervals

Q236.Type of table in which study variables provides large number of information with interrelated characteristics is classified as:

(a) Lower order table
(b) Manifold table
(c) Higher order table
(d) **Both (b) and (c)**

Q237.Table in which data represented is extracted from some other data table is classified as

(a) Classification tables
(b) Cumulative tables
(c) **Derived table**
(d) Dispersion tables

Q238.If each value of frequency distribution is divided by total number of recorded observations in distribution then resultant value is called:

(a) Interval frequency distribution
(b) **Percentage frequency distribution**
(c) Nominal frequency distribution
(d) Ordinal frequency distribution

Q239.Record of daily shipment is 34, 35, 41, 30, 55, 45, 30, 34, 32, 52, 42, 40, 60, 36, 38, 48, 56, 53, 34, 33, 32, 41, 55, 59, 34, 51, 54, 53, 36, then range of values to calculate class interval is:

(a) 2
(b) **30**
(c) 90
(d) 180

Q240.Largest numerical value is 85 and smallest numerical value is 65 and classes desired are 8 then width of class interval is

(a) 18.75
(b) 14.75
(c) 13.75
(d) **2.5**

Q241.Stem and leaf displaying technique is used to present data in

(a) Descriptive data analysis
(b) **Exploratory data analysis**
(c) Nominal data analysis
(d) Ordinal data analysis

Q242.Class interval classification method which ensures data continuity is classified as

(a) Midpoint method
(b) Ratio method
(c) **Exclusive method**
(d) Inclusive method

Q243.First step in constructing frequency distribution is to

(a) **Select appropriate class intervals**

(b) Determine class intervals

(c) Determine class limits

(d) Determine midpoints of classes

Q244.Distribution which requires inclusion of open ended classes is considered as

(a) Inclusive distribution
(b) Midpoint distribution
(c) Close ended distribution
(d) **Open ended distribution**

Q245.Curve of cumulative frequency is also known as

(a) **Ogive**
(b) A-give
(c) C-give
(d) B-give

Q246.Type of classification in which class is subdivided into subclasses and one attribute is assigned for statistical study is considered as:

(a) Rational classification
(b) Reflected classification
(c) **Simple classification**
(d) Manifold classification

Q247.Cumulative frequency distribution which is 'greater than' type is correspondent to

(a) Upper limit of range

(b) Lower limit of range

(c) Upper limit of class intervals

(d) **Lower limit of class intervals**

Q248.Simple classification and manifold classification are types of

(a) **Qualitative classification**
(b) Quantitative classification
(c) Open end classification
(d) Time series classification

Q249.Second step in constructing frequency distribution is to

(a) Determine class limits

(b) Determine midpoints of classes

(c) Select appropriate class intervals

(d) **Determine width of class intervals**

Q250.Complex types of table in which variables to be studied are subdivided with interrelated characteristics is called as:

(a) **Two way table**
(b) One way table
(c) Subparts of table
(d) Order level table

Q251.Largest numerical value is 45 and smallest numerical value is 25 and classes desired are 4 then width of class interval is:

(a) 65
(b) 45
(c) 17.5
(d) **5**

Q252.For 'more than ogive' the x-axis represents:

(a) Frequency
(b) **Lower limits of class interval**
(c) Mid-values of class-interval
(d) Upper limits of class-interval

Q253.Class mark of a class is obtained by using

(a) (Upper limit – lower limit) / 2
(b) **(Upper limit + lower limit) / 2**
(c) Upper limit – lower limit
(d) Upper limit + lower limit

Q254.Weights of 50 eggs were recorded as given below:

Weight (in gms)	40-42	43-45	46-48	49-51	52-54
No. of eggs	5	12	18	9	6

The lower limit of modal class is:

(a) 40
(b) 42.5
(c) **45.5**
(d) 46

Q255.Table which shows frequency of each score is called:

(a) Polygon
(b) Pie chart
(c) Histogram
(d) **Frequency distribution table**

Q256.Total numbers of observations, which are below a certain value, are known as:

(a) Class boundaries
(b) Class marks
(c) **Cumulative frequency**
(d) None of these

Q257.Number of times each value appears is called values:

(a) Range
(b) **Frequency**
(c) Average
(d) None of these

Q258.The sum of the percent frequencies for all classes will always equal to:

(a) Zero
(b) One
(c) **Hundred**
(d) None of these

Q259.Discrete variables and continuous variables are two types of

(a) Open end classification
(b) Time series classification
(c) Qualitative classification
(d) **Quantitative classification**

Q260.Classification method in which the upper limit of interval is same as of lower limit class interval is called

(a) **Exclusive method**
(b) Inclusive method
(c) Mid-point method
(d) Ratio method

Q261.Summary and presentation of data in tabular form with several non-overlapping classes is referred as

(a) Nominal distribution
(b) Ordinal distribution
(c) Chronological distribution
(d) **Frequency distribution**

Q262.Classification method in which upper and lower limits of interval is also in class interval itself is called

(a) Exclusive method
(b) **Inclusive method**
(c) Mid-point method
(d) Ratio method

Q263.General tables of data used to show data in orderly manner is called as

(a) Single characteristics tables
(b) **Repository tables**
(c) Manifold tables
(d) Double characteristic table

Q264.Data table which is presented in tabular form on the basis of single characteristics is classified as

(a) **Simple table** (b) Complex table

(c) Percentage table (d) Interval table

Q265.Type of table in which study variables provides large number of information with interrelated characteristics is classified as:

(a) Lower order table (b) Manifold table

(c) Higher order table (d) **Both (b) and (c)**

Q266.Table in which the data represented is extracted from some other data table is classified as

(a) Classification tables (b) Cumulative tables

(c) **Derived table** (d) Dispersion tables

Q267.Constructing a frequency distribution:

(a) Is one of the most common means of summarizing data

(b) Begins by recording the number of times a particular value occurs

(c) Is the basis for construction of a percentage distribution

(d) **All of the above**

Q268.Largest numerical value is 40 and smallest numerical value is 20 and classes desired are 4 then width of class interval is:

(a) **5** (b) 10

(c) 15 (d) 20

Q269.Body weights of students in a class were recorded as given below:

Weight (in kg)	40 - 45	45 - 50	50 - 55	55 - 60	60 - 65
No. of Students	5	12	18	9	6

The number of students less than 55 kg is:

(a) 17 (b) **35**

(c) 44 (d) 51

Q270. While computing mean of grouped data, we assume that the frequencies are [NCERT Exemplar Problems]

(a) Evenly distributed over all the classes

(b) **Centered at the class-marks of the classes**

(c) Centered at the upper limits of the classes

(d) Centered at the lower limits of the classes

Q271. In the given data:

Class Interval	5 - 15	15 - 25	25 - 35	35 - 45	45 - 55
Frequency	5	12	18	9	6

The difference of the upper limit of the median class and the lower limit of modal class is:

(a) 5 (b) **10**

(c) 25 (d) 35

Q272. Which of the following is not found in a frequency distribution?

(a) Class intervals

(b) Class limits

(c) **Individual observations within each class**

(d) None of these

Q273. The sum of the relative frequencies for all classes in a frequency distribution will always equal:

(a) The class width

(b) The number of classes

(c) The total number of observations in a data set

(d) **One**

3

Sampling Theory

Population: The word **population** or **universe** in statistics means a set of observations relating to a phenomenon under investigation.

Sample: It is a fraction or a subset of population drawn through a valid statistical procedure so that it can be regarded as representative of the entire population.

Sampling: The statistical procedure of drawing a sample from the population is called **Sampling**.

Census enquiry or complete enumeration: When each and every unit of the population is investigated for the characteristic under study, known as **census enquiry** or **complete enquiry**.

Sample enquiry or sample enumeration: It is a part of the population is investigated for the characteristic under study, known as **sample enquiry or sample enumeration.**

Necessity of sampling: Since, the process of collecting information from all the elements of a large population may be, in general, expensive, time consuming and difficult. We help to sampling and generalise the properties of the sample to the **entire population**. The statistical procedures which are used for drawing inferences or conclusions about the population from sample data are covered under **Inferential Statistics** or **Statistical inference**. Thus, sampling theory is the basis of statistical inference in which we wish to obtain maximum information about the population with minimum effort and maximum precision.

Sampling techniques

The choice of sampling procedure greatly depends on the purpose of enquiry and the nature of population. There are two types of sampling techniques –

- Random or probability sampling
- Non-Random or non-probability sampling
- **Random or probability sampling:** Any type of sampling in which every unit of the population has a definite, pre-assigned probability of being selected in the sample is called **random or probability sampling**. In other

words, a random or probability sample is a sample selected in such a way that every unit in the population has equal chance of being included. Some of the random or probability sampling techniques are –

(i) Simple random sampling

(ii) Stratified random sampling

(iii) Systematic sampling

(iv) Cluster sampling

(v) Multistage sampling

(vi) Sequential sampling

(i) Simple random sampling: In **simple random sampling** each unit of the population has an equal and independent chance of being included in the sample. It may be with or without replacement according as a unit selected is not or is excluded from further selection. It eliminates the possibility of any bias selection. There are two methods of selecting a simple random sample –

- **Lottery method:** In this method every unit of the population is numbered or named on separate slips of paper which are identical in size and shape. These slips are then well mixed in a container or a drum and then one draws the required number of slips to constitute the desired size.
- **Table of random numbers:** In practice, tables of random number can also be used for selecting a **simple random sample**. Several standard tables of random numbers are available of which some in common use are –
  - Tippet's random number tables; and
  - Fisher and Yates tables.

For using these tables, the population units should be numbered from 1 to N, which makes it possible to determine the range of numbers to be selected. If N is less than 100, two digit numbers are selected; if N is less than 1000, three digits numbers are selected and so on.

(ii) Stratified random sampling: When a population is heterogeneous in nature, then the entire heterogeneous population is divided into a number of homogeneous groups, usually called **strata** or **sub-populations**. Then units are sampled at random from each of this stratum. The sample, which is the set of all the sampling units drawn from each stratum, is called a **stratified sample** and the technique of drawing the sample is termed as **stratified random sampling**. The stratified sampling method adds to the precision of the sample estimates when:

- The units within each stratum are more or less homogeneous.
- The units in different strata are unlike or heterogeneous.

(iii) Systematic sampling: This is a convenient method when complete list of sampling units, called sample frame, is readily available or can be easily prepared. The sampling scheme consists of selecting only the first unit at random and the rest are then automatically selected according to some predetermined pattern. A systematic sample gives more precise estimates compared to simple random sample if the units selected periodicity in the list; we expect the selection of more or less homogeneous units in the sample resulting in biased estimates.

(iv) Cluster sampling: A group of elementary units in the population is called a **cluster**. When a cluster is taken as a sampling unit, the procedure of sampling is called **cluster sampling**. It is useful in situations when we observe an unequal concentration of individual units in the population. In such cases, certain blocks or clusters of higher concentration are randomly selected for complete enquiry.

(v) Multistage sampling: Sometimes sampling is done in stages to reduce the cost of the survey. In this sampling method, the population is divided into first stage sampling units also called **primary units**. Then a random sample of first stage units is made. Further division is made of the first stage sampling units selected, and a random sample is then taken from these second stage sampling units. The process can be obtained for a number of stages. The process of sampling is called **Multi-stage sampling**.

(vi) Sequential sampling: It consists of a sequence of samples of size one drawn random from the population. i.e., observations are sequentially recorded one by one. Such sampling is generally adopted in statistical quality control. In this sampling, a test may be determined at any stage of experimentation. Here we randomly draw sample observations one by one and make decisions on the basis of the combined evidence of already drawn observations. In this regard let us consider an example where a lot quality is to be judged by sequential sampling of items from the lot. If the quality of a first few is good enough to accept the lot then we stop testing and terminate the test. Similarly if the quality of already drawn items at any stage is such that it is bad enough to reject the lot, than we again terminate further sampling and reject the lot quality. Thus, in **sequential sampling**, the sample space is divided into three zones, namely-

- The region of acceptance
- The region of rejection
- The zone of indifference in which the sampling is continued and observations are drawn from the population one by one.

- **Non-Random or non-probability sampling:** In non-random sampling method, the units in the population are selected on subjective basis and no probability law is applied in the selection of a sample from the population. Some of such sampling methods are –

(i) Subjective or judgment or purposive sampling

(ii) Quota sampling

(iii) Convenience sampling

(i) **Subjective or judgment or purposive sampling:** Any type of sampling in which the sample selected depends on personal discretion or judgment of the investigator is called a **subjective or judgment or purposive sampling**. This type of sampling is used with a definite purpose in view and as such is not used for general purposes. In this method, the choice of sample items depends exclusively on the judgement of investigator.

(ii) **Quota sampling:** This is a restricted type of subjective or judgment or purposive sampling. This sampling method is adopted when stratification is difficult and still we wish to derive the benefits of stratification without using probability sampling. Quotas may be fixed for each stratum and then required size of samples is drawn by judgment sampling. This method is usually adopted for opinion-pole and for socio-economic investigations.

(iii) **Convenience sampling:** The method of convenience sampling is also called the **chunk**. A chunk refers to that fraction of the population being investigated which is selected neither by probability nor by judgement but by convenience. A sample obtain from readily available lists such as automobile registrations, telephone directories, etc, is a **convenience sample** and not a random sample even if the sample is drawn at random from the lists. However, **convenience sampling** is often used for making pilot studies.

MCQ

Q.1. Subset of selected population is called

(a) Descriptive portion (b) Elementary portion

(c) Inferential portion (d) **Sample**

Q2. Sample is subset of:

(a) Data (b) Set

(c) **Population** (d) Distribution

Q3. Any population constant is called a:

(a) Statistic (b) **Parameter**

(c) Estimate (d) Estimator

Bold means correct answer

Q4. Any calculation on the sampling data is called:

(a) Parameter (b) **Static**

(c) Mean (d) Error

Q5. List of all the units of the population is called:

(a) Bias (b) Random sampling

(c) **Sampling frame** (d) Probability sampling

Q6. Any measure of the population is called:

(a) Static (b) **Parameter**

(c) Mean (d) Random error

Q7. If all the units of a population are surveyed, it is called:

(a) Random sample (b) Indirect sampling

(c) Non-random sampling (d) **Complete enumeration**

Q8. Probability distribution of a statistics is called:

(a) Sampling (b) Parameter

(c) Data (d) **Sampling distribution**

Q9. The difference between a statistic and the parameter is called:

(a) Probability (b) **Sampling error**

(c) Random (d) Non-random

Q10. The sum of the frequencies of the frequency distribution of a statistic is equal to:

(a) Sample size (b) Population size

(c) **Possible samples** (d) Sum of X values

Q11. Standard deviation of sampling distribution of a statistic is called:

(a) Serious error (b) Dispersion

(c) **Standard error** (d) Difference

Q12. If we obtain a point estimate for a population mean μ, the difference between and μ is:

(a) Standard error (b) Bias

(c) **Error of estimation** (d) Difficult to tell

Q13. A distribution formed by all possible values of a statistics is called:

(a) Binomial distribution (b) Poisson distribution

(c) Normal distribution (d) **Sampling distribution**

Q14. A population about which we want to get some information is called:

(a) Finite population (b) Infinite population

(c) Sampling population (d) **Target population**

Q15. The population consists of the results of repeated trials is named as:

(a) Finite population (b) Infinite population

(c) Real population (d) **Hypothetical population**

Q16. A population consisting of the items which are all present physically is called:

(a) Finite population (b) Infinite population

(c) **Real population** (d) Hypothetical population

Q17. Study of population is called:

(a) Parameter (b) Statistic

(c) Standard error (d) **Census**

Q18. For making voters list in India we need:

(a) Sampling error (b) Standard error

(c) **Census** (d) Simple random sampling

Q19. Sampling based upon equal probability is called:

(a) Probability sampling (b) Systematic sampling

(c) **Simple random sampling** (d) Stratified random sampling

Q20. In sampling with replacement, an element can be chosen:

(a) Less than once (b) **More than once**

(c) Only once (d) Difficult to tell

Q21. Standard deviation of sample mean without replacement__________ standard deviation of sample mean with replacement:

(a) Less than (b) More than

(c) Two times (d) Equal to

Q22. In sampling without replacement, an element can be chosen:

(a) Less than once
(b) More than once
(c) **Only once**
(d) Difficult to tell

Q23. In sampling with replacement, the following is always true:

(a) $n = N$
(b) $n < N$
(c) $n > N$
(d) **All of the above**

Q24. Which of the following statement is true?

(a) Standard error is always one
(b) Standard error is always zero
(c) Standard error is always negative
(d) **Standard error is always positive**

Q25. Random sampling is also called:

(a) **Probability sampling**
(b) Non-probability sampling
(c) Sampling error
(d) Random error

Q26. Non-random sampling is also called:

(a) Biased sampling
(b) **Non-probability sampling**
(c) Random sampling
(d) Representative sample

Q27. Sampling error can be reducing by:

(a) Non-probability sampling
(b) Increasing the population
(c) Decreasing the sample size
(d) **Increasing the sample size**

Q28. The selection of cricket team for the world cup is called:

(a) Random sampling
(b) Systematic sampling
(c) **Purposive sampling**
(d) Cluster sampling

Q29. A complete list of all the sapling units is called:

(a) Sampling design
(b) **Sampling frame**
(c) Population frame
(d) Cluster

Q30. A Plan for obtaining a sample from a population is called:

(a) Population design (b) **Sampling design**

(c) Sampling frame (d) Sampling distribution

Q31. If a survey is conducted by a sampling design is called:

(a) **Sample survey** (b) Population survey

(c) Systematic survey (d) None of these

Q32. The difference between the expected value of a statistic and the value of the parameter being estimated is called a:

(a) Sampling error (b) Non-sampling error

(c) Standard error (d) **Bias**

Q33. The standard deviation of any sampling distribution is called:

(a) **Standard error** (b) Non-sampling error

(c) Type- I error (d) Type II-error

Q34. Which of the following statement is not true?

(a) S.E() $\neq 0$ (b) S.E() $\neq 1$

(c) S.E() = -2 (d) **All of the above**

Q35. The standard error increases when sample size is:

(a) Increase (b) **Decrease**

(c) Fixed (d) More than 30

Q36. The mean of sampling distribution of mean is equal to:

(a) $\overline{X}$ (b) μ

(c) p (d) None of the above

Q37. The mean of the sample means is exactly equal to the:

(a) Sample mean (b) **Population mean**

(c) Weighted mean (d) Combined mean

Q38. Sum of all sample means / Total number of samples, is equal to:

(a) $E(\overline{X})$ (b) μ

(c) **Both (a) and (b)** (d) None of the above

Q39. A sample which is free from bias is called:

(a) Biased
(b) **Unbiased**
(c) Positively biased
(d) Negatively biased

Q40. If $E(\overline{X}) = \mu$ then bias is:

(a) Positive
(b) Negative
(c) **Zero**
(d) 100%

Q41. The sampling procedure in which the population is first divided into homogenous groups and then a sample is drawn from each group is called:

(a) Probability sampling
(b) Simple random sampling
(c) **Stratified random sampling**
(d) Sampling with replacement

Q42. When a random sample is drawn from each stratum, it is known as:

(a) Simple random sampling
(b) **Stratified random sampling**
(c) Probability sampling
(d) Purposive sampling

Q43. When the procedure of selecting the elements from the population is not based on probability is known as:

(a) Purposive sampling
(b) Judgment sampling
(c) Subjective sampling
(d) **All of the above**

Q44. Suppose a finite population has 6 items and 2 items are selected at random without replacement, then all possible samples will be:

(a) 6
(b) 12
(c) **15**
(d) 36

Q45. Suppose a finite population contains 7 items and 3 items are selected at random without replacement, then all possible samples will be:

(a) 7
(b) 14
(c) 21
(d) **35**

Q46. A population contain N item and all possible sample of size n are selected without replacement. The possible number of sample will be:

(a) N (b) n^N

(c) N^n (d) **^NC_n**

Q47. Suppose a finite population contains 4 items and 2 items are selected at random with replacement, then all possible samples will be:

(a) 4 (b) 6

(c) 8 (d) **16**

Q48. A population contains 2 items and 4 items are selected at random with replacement, then all possible samples will be:

(a) 4 (b) 6

(c) 8 (d) **16**

Q49. Suppose a population has N items and n items are selected with replacement. Number of all possible samples will be:

(a) ^NC_n (b) **N^n**

(c) N ; (d) n

Q50. In random sampling, the probability of selecting an item from the population is:

(a) Unknown (b) **Known**

(c) Un-decided (d) One

Q51. If N is the size of the population and n is size of the sample, then sampling fraction is:

(a) n^N (b) N^n

(c) **n/N** (d) ^NC_n

Q52. If $\overline{X} = 10$ and $\mu = 12$ then sampling error is equal to:

(a) 22 (b) 12

(c) 10 (d) **2**

Q53: Listing of elements in population with identifiable number is classified as

(a) Regularity experimental frame (b) Indirect experiment frame

(c) Direct experimental frame (d) **Frame for experiment**

Q54: Type of sampling In which each element of population has equally likely chance of occurrence in a random sample is classified as

(a) **regular and irregular sampling**

(b) error free sampling

(c) Inertia sampling

(d) Simple random sampling

Q55: Quota sampling, judgment sampling and convenience sampling are classified as types of

(a) Random sampling (b) **Non random sampling**

(c) Direct sampling (d) Indirect sampling

Q56: Unknown or exact value that represents whole population is classified as

(a) **Parameters** (b) Estimators

(c) Absolute statistics (d) Coverage estimator

Q57: Listing of elements in population with identifiable number is classified as

(a) Regularity experimental frame (b) Indirect experiment frame

(c) Direct experimental frame (d) **Frame for experiment**

Q58: In statistical analysis, sample size is considered large if

(a) **n > or = 30** (b) n < or = 30

(c) n > or = 50 (d) n < or = 50

Q59: If standard deviation of population is known then μ must be equal to

(a) Absolute value of estimator (b) Error free mean

(c) **Expected value of mean** (d) Inferential value of mean

Q60: Methods in statistics that uses sample statistics to estimate parameters of population are considered as:

(a) **Inferential statistics** (b) Absolute statistics

(c) Coverage statistics (d) Random sample statistics

Q61: Interviewing all members of a given population is called a:

(a) Sample (b) Gallup poll

(c) **Census** (d) Nielsen audit

Q62. Which ONE of the following classifications contains divisions with the labels A, B, C1, C2, D and E?

(a) Social class.
(b) Socio-economic groupings
(c) Socio-economic classes
(d) **Social grade.**

Q63. Sampling means following a sequence of stages. Which ONE of the following stages should come before the others?

(a) Proceed with the fieldwork
(b) Find suitable source for the population members
(c) Define the people of interest
(d) **Examine the objective of the study**

Q64. Which ONE of these sampling methods is a probability method?

(a) Quota
(b) Judgement
(c) Convenience
(d) **Simple random**

Q65. For sampling, which ONE of the following should be up-to-date, complete and affordable?

(a) A census
(b) A CAPI machine
(c) **A sampling frame**
(d) A respondent.

Q66. Which ONE of these sample sources will be the most incomplete for a sample of nationally representative households?

(a) The Electoral Register
(b) The postcode address file (PAF)
(c) Telephone directories
(d) **A customer database**

Q67. Which ONE of the following is the benefit of using simple random sampling?

(a) **We can calculate the accuracy of the results**
(b) The results are always representative
(c) Interviewers can choose respondents freely
(d) Informants can refuse to participate

Q68. Which ONE of the following is the main problem with using non-probability sampling techniques?

(a) The expense

(b) The results are never representative

(c) **Human judgement error**

(d) Informants can refuse to participate.

Q69. Which ONE of the following is the best - but an often unused - way to decide on sample size?

(a) By using industry standards (b) **By calculation**

(c) By 'building blocks' (d) By budget available.

Q70. Which ONE of the following methods is generally used in qualitative sampling?

(a) Random digit dialing (b) **Quota**

(c) Stratified random (d) Simple random

Q71. The number of possible samples of size n from a population of size N units with replacement is:

(a) n^N (b) $\mathbf{N^n}$

(c) ${}^NC_n$ (d) N!

Q72. The variance of the set of the values is zero if the observations are:

(a) **Same** (b) Different

(c) Both (A) and (B) (d) None

Q73. Which of the following property of the estimator deals with the behaviour of the estimator when sample size tends to infinity?

(a) Unbiasedness (b) **Consistency**

(c) Sufficiency (d) Efficiency

Q74. In proportional allocation the sample size in the hth stratum is given by:

(a) $n_h = W_h / n$ (b) $\mathbf{n_h = n.W_h}$

(c) $n_h = n/W_h$ (d) None of the above

Q75. Stratification is the division of the population into certain number of subpopulations, which are:

(a) Heterogeneous within stratum

(b) **Homogeneous within stratum**

(c) Both (A) and (B)

(d) None of the above

Q76. Circular systematic sampling is used when

(a) **N is not divisible by n** (b) N is whole number

(c) N is a multiple of n (d) None of the above

Q77. In ratio and regression methods of estimation sample mean is:

(a) Unbiased estimator of population mean

(b) Sufficient estimator of mean

(c) **Biased estimator of population mean**

(d) None of the above

Q78. Ratio estimators are more efficient when the regression line passes through:

(a) Means (b) **Origin**

(c) Intercept (d) None of these

Q79. In complete enumeration the kinds of errors present are:

(a) Sampling errors

(b) **Non-sampling errors**

(c) Both sampling and non-sampling errors

(d) Systematic errors

Q80. Which of the following allocations provides self-weighing sampling in stratified random sampling?

(a) Equal allocation (b) **Proportional allocation**

(c) Neyman allocation (d) Optimum allocation

Q81. Which of the following sampling scheme is known as two phase sampling?

(a) Stratified Sampling (b) Two stage sampling

(c) **Double sampling** (d) Systematic sampling

Q82. The number of all possible samples of size n from a population of size N in circular systematic sampling:

(a) **N**
(b) n
(c) ${}^{N}C_{n}$
(d) N^{n}

Q83. When frame of elementary unit is not available, then which of the following sampling design can be used for sampling?

(a) Double sampling
(b) Multi stage sampling
(c) Cluster sampling
(d) **Both (b) and (c)**

Q84. Which of the following estimator can be used if the study variables and an auxiliary variable related to it are negatively correlated?

(a) Ratio estimator
(b) Regression estimator
(c) Product estimator
(d) **Both (b) and (c)**

Q85. Estimate of population mean of the same population at different points of time is obtained by employing:

(a) Systematic sampling
(b) Double sampling
(c) Two-stage sampling
(d) **Successive sampling**

Q86. The difference between statistic and parameter is called:

(a) Sampling error
(b) Random error
(c) Standard error
(d) **Error**

Q87. In random sampling, the probability of selecting an item from the population is:

(a) **Known**
(b) Unknown
(c) Zero
(d) One

Q88. The number of possible samples of size n out of N population units with replacement is:

(a) N^2
(b) n^2
(c) N!
(d) **∞ (Infinite)**

Q89. The number of possible samples of size n out of N population units without replacement is:

(a) N^2
(b) n^2
(c) N!
(d) **${}^{N}C_{n}$**

Q90. A sample consists of:

(a) 5 percent units of the population

(b) 15 percent units of the population

(c) **Any fraction of the population**

(d) All units of the population

Q91. Probability of drawing a unit at each selection remains same is:

(a) **Simple random sampling with replacement**

(b) Simple random sampling without replacement

(c) Both (a) and (b)

(d) None of the above

Q92. An unordered sample of size n can occur is:

(a) One way

(b) n ways

(c) **n! ways**

(d) n^2 ways

Q93. Probability of any one sample of size n being drawn out of N units is:

(a) 1/N

(b) n/N

(c) 1/n!

(d) **$1/{}^{N}C_{n}$**

Q94. Probability of including a specified unit in a sample of size n selected out of N units is:

(a) **1/N**

(b) n/N

(c) 1/n!

(d) $1/{}^{N}C_{n}$

Q95. When an investigator wants a sample containing n units which possess a rare attribute, the appropriate sampling procedure is:

(a) Purposive sampling

(b) **Inverse sampling**

(c) Stratified sampling

(d) Judgement sampling

Q96. Sampling frame is a term used for:

(a) A list of population units

(b) **A list of sampling units of a population**

(c) A list of random numbers

(d) A list of voters

Q97. In simple random sampling with replacement, the same sampling unit may be included in the sample:

(a) Only once (b) Only twice

(c) **More than once** (d) None of the above

Q98. If the items are destroyed under investigation, we have to go for:

(a) **Sampling studies** (b) Census studies

(c) Both (a) and (b) (d) None of these

Q99. Cluster sampling, stratified sampling and systematic sampling are types of:

(a) Direct sampling (b) Indirect sampling

(c) **Random sampling** (d) Non-random sampling

Q100. Which of the following is NOT part of the sampling design process?

(a) Determining the relevant sample frame

(b) **Refining the research question**

(c) Defining of the population of the study

(d) Specifying the sampling unit

Q101. Which of the following in NOT true of probability sampling?

(a) **The results will always be more than non-probability sampling**

(b) Sampling units are selected by chance as opposed to the judgement of the researcher

(c) Estimates are statistically projectable to the population

(d) The number of elements to be included in the sample set can be pre-specified

Q102. What is the least expensive and least time-consuming of all sampling techniques?

(a) Simple random sampling (b) Stratified sampling

(c) **Convenience sampling** (d) Judgement sampling

Q103. What are the distinguishing features of simple random sampling?

(a) Random numbers determine which elements are included in the sample

(b) A sampling frame must be compiled in which each element has a unique identification number

(c) Each element in the population has a known and equal probability of selection

(d) **All of the above**

Q104. Which of the following is NOT a characteristic of quota sampling?

(a) It is a relatively fast and cheap way of finding out about public opinions

(b) The researchers choose who to approach and so might bias the sample

(c) **The random selection of units makes it possible to calculate the standard error**

(d) The sample is subdivided into groups from which the researcher must obtain as certain number of responses

Q105. Snowball sampling can help the researcher to:

(a) Collect data cost effectively

(b) Access difficult or hidden populations

(c) Overcome the problem of not having an accessible sampling frame

(d) **All of the above**

Q106. Which of the following is NOT a type of non-probability sampling?

(a) Judgement sampling

(b) Quota sampling

(c) Convenience sampling

(d) **Cluster sampling**

Q107. Which one of these is a self-administered questionnaire?

(a) Personal questionnaire

(b) **Postal questionnaire**

(c) Telephone questionnaire

(d) Face-to-face questionnaire

Q108. Which one of these is an interviewer-administered questionnaire?

(a) On-line questionnaire

(b) Postal questionnaire

(c) **Telephone questionnaire**

(d) Delivery and collection questionnaire

Q109.In situations where not all respondents are sufficiently informed to answer a question:

(a) **Filter questions should be used**

(b) Open-ended questions should be used

(c) Multiple questionnaires should be designed

(d) Quota sampling should be used

Q110.What is a mistake that researchers who write questionnaires should avoid?

(a) Writing questions with filters and routes

(b) **Writing questions respondents do not understand**

(c) Writing questions respondents are willing to answer

(d) Writing questions respondents are able to answer

Q111.Under equal allocation in stratified sampling, the sample from each stratum is:

(a) Proportional to stratum size

(b) **Same size from each stratum**

(c) In proportion to the per unit cost of survey of the stratum

(d) All of the them

Q112.Under proportional allocation, the size of the sample from each stratum depends on:

(a) Total sample size

(b) Size of the stratum

(c) Population size

(d) **All of them**

Q113.Under proportional allocation one get:

(a) **A self-weighing sample**

(b) An optimum sample

(c) Both (a) and (b)

(d) None of these

Q114.How many types of optimum allocation is in common use?

(a) One

(b) Two

(c) **Three**

(d) Four

Q115.If a sample is drawn from each stratum minimizing the probabilities of non-preferred samples, it is known as:

(a) Selection with proportional allocation

(b) Haphazard selection

(c) **Controlled selection**

(d) None of these

Q116.Systematic sampling means:

(a) Selection of n middle units in a sequence

(b) Selection of n largest units

(c) **Selection of n units situated at equal distances**

(d) Selection of n contiguous units

Q117.If the number of population units N is an integral multiple of sampling size n, the systematic sampling is called:

(a) Circular systematic sampling

(b) **Linear systematic sampling**

(c) Random systematic sampling

(d) None of the above

Q118.Circular systematic sampling is used when:

(a) N is a multiple of n

(b) N is whole number

(c) **N is not divisible by n**

(d) None of the above

Q119.Linear and circular systematic sampling methods are equivalent if and only as:

(a) N is a whole number

(b) n is a whole number

(c) N = n

(d) **None of the above**

Q120.Selected units of a systematic sample are:

(a) Not easily locatable

(b) **Easily locatable**

(c) Not representing the whole population

(d) All the above

Q121.In what situation(s) a systematic sample is more preferred than others?

(a) When the data are on cards

(b) When the items are in row

(c) When the items situated at equal distances are uncorrelated

(d) **All the above**

Q122.A systematic sample does not yield good results if:

(a) Units at regular intervals are correlated

(b) Variation in units is periodic

(c) **Both (a) and (b)**

(d) None of (a) and (b)

Q123.Greatest drawback of systematic sampling is that:

(a) Data are not easily accessible

(b) **No single reliable formula for standard error of mean is available**

(c) One requires a large sample

(d) None of the above

Q124.Double sampling is also known as:

(a) **Two phase sampling** (b) Two stage sampling

(c) Two directional sampling (d) All of them

Q125.Double sampling has its utility in:

(a) Stratified sampling

(b) Regression method of estimation

(c) Ratio method of estimation

(d) **All of the above**

Q126.Cluster sampling is appropriate:

(a) When sampling frame is not available

(b) When all elementary units are not easily identified

(c) When the units are situated for apart

(d) **All the above**

Q127.Cluster sampling is more efficient:

(a) Choosing clusters having least variation between clusters

(b) Choosing clusters having largest variation between clusters

(c) By taking clusters of small size

(d) **All the above**

Q128.In what situation two stage sampling is better than single stage sampling?

(a) When the elements in the same stage are uncorrelated

(b) When the elements in the same stage are positively correlated

(c) **When the elements in the same stage are negatively correlated**

(d) None of the above

Q129.Problem of non-response:

(a) **Can be solved**

(b) Has no solution

(c) Is meaningless

(d) None of the above

Q130.The errors emerging out of faulty planning of surveys are categorized as:

(a) Non-response errors

(b) **Non-sampling errors**

(c) Sampling errors

(d) Absolute error

Q131.Two stage sampling design is more efficient than single stage sampling if the correlation between units in the first stage is:

(a) **Positive**

(b) Negative

(c) Zero

(d) None of the above

Q132.What sampling design is most appropriate for cluster sampling?

(a) Simple random sampling with replacement

(b) **Simple random sampling without replacement**

(c) Quota sampling

(d) Stratified random sampling

Q133.The discrepancy between estimates and population parameters is known as:

(a) **Sampling error**

(b) Human error

(c) Formula error

(d) Enumeration error

Q134.To meet requirement of the principle of validity of sampling methods, one must adopt:

(a) **Probability sampling**
(b) Non-probability sampling
(c) Restricted sampling
(d) Purposive sampling

Q135.Stratified sampling belongs to the category of:

(a) Subjective sampling
(b) Non-random sampling
(c) Judgement sampling
(d) **Controlled sampling**

Q136.The magnitude of the standard error of an estimate is an index of its:

(a) Efficiency
(b) Accuracy
(c) **Precision**
(d) All of the above

Q137.Sampling error can be reduced by:

(a) Using a suitable formula for estimation
(b) Choosing a proper probability sampling
(c) Selecting a sample of adequate size
(d) **All of them**

Q138. Method of random sampling which is also called area sampling method is classified as

(a) Statistical sub cluster sampling
(b) Sub cluster sampling
(c) Proportional inertia sampling
(d) **Cluster sampling**

Q139.Procedure in which numbers of elements in stratum are not in proportional to number of elements in population is classified as:

(a) Indirect strata procedure
(b) Direct strata procedure
(c) **Non proportional procedure**
(d) Proportional procedure

Q140.Method of sampling in which random sampling will not be possible because population is widely spread is classified as:

(a) Secondary stage sampling
(b) **Multistage sampling**
(c) Primary stage sampling
(d) Sub stage sampling

Q141.In probability sampling, probability of selecting an item from the population is known and is:

(a) Equal to zero
(b) **Non zero**
(c) Equal to one
(d) All of the above

Q142.Type of sampling In which each element of population has equally likely chance of occurrence in a random sample is classified as

(a) **Regular and irregular sampling**

(b) Error free sampling

(c) Inertia sampling

(d) Simple random sampling

Q143.Type of sampling in which desired and useful information is gathered from best position holder is classified as:

(a) Quota sampling
(b) Convenience sampling
(c) Purposive sampling
(d) **Judgment sampling**

Q144. Use of statistics in analysis of time when to order and quantity required for in sourcing or outsourcing of materials is an example of usage of statistics in:

(a) Marketing
(b) **Production**
(c) Personnel management
(d) Finance

Q145.Method in which sample statistic is used to estimate value of parameters of population is classified as

(a) **Estimation**
(b) Valuation
(c) Probability calculation
(d) Limited theorem estimation

Q146.Important principles to determine valid statistical inference must includes

(a) Principle of sampling error

(b) Principle of statistical regularity

(c) Principle of inertia

(d) **Both (b) and (c)**

Q147.Considering sample size, sampling distribution standard error decreases when the

(a) **Size of sample increases**
(b) Size of sample decreases
(c) Margin of error increases
(d) Margin of error decreases

Q148.In stratified sampling, sample drawn randomly from strata is classified as

(a) Sub strata (b) **Sub sample**

(c) Direct sub group (d) Indirect sub group

Q149.Method of sampling in which population is divided in to mutual exclusive groups that have useful context in statistical research is classified as:

(a) **Stratified sampling** (b) Regular group sampling

(c) Irregular group sampling (d) Direct group sampling

Q150.In cluster sampling, elements of selected clusters are classified as:

(a) **Elementary units** (b) Primary units

(c) Secondary units (d) Proportional units

Q151.In systematic sampling, population is 200 and selected sample size is 50 then sampling interval is:

(a) 0.25 (b) **4**

(c) 40 (d) 200

Q152.The sampling procedure that includes all units of the population in the study is called:

(a) Complete coverage (b) Saturation survey

(c) **Both (a) & (b)** (d) Neither (a) nor (b)

Q153.As the sample size increases, the:

(a) Standard deviation of the population decreases

(b) Population mean increases

(c) **Standard error of the mean decreases**

(d) Standard error of the mean increases

Q154.If we consider the simple random sampling process as an experiment, the sample mean is

(a) Always zero

(b) Always smaller than the population mean

(c) **A random variable**

(d) Exactly equal to the population mean

Q155. Two of the methods of unit selection in simple random sampling are

(a) Stratified sampling and quota sampling

(b) Multi-stage sampling and purposive sampling

(c) Cluster sampling and probability sampling

(d) **Lottery method and random numbers method**

Q156. Theoretical sampling means that

(a) Sampling procedures follow theoretical principles

(b) Sampling is geared towards developing or testing a theory

(c) Each sampling procedure applies its own theory

(d) Each sampling procedure applies its own theory

Q157. A researcher entered a large restaurant and interviewed very briefly the oldest person sitting at every second table. This type of sampling is:

(a) Systematic probability sampling

(b) Stratified sampling

(c) **Spatial sampling**

(d) Area sampling

Q158. Which of the following is NOT a random sampling technique?

(a) **Purposive sampling**

(b) Stratified sampling

(c) Cluster sampling

(d) Systematic sampling

Q159. Dean Student Welfare (DSW) is an investigating student's attitude toward year-round performance. He is particularly interested in describing the attitudes to the teachers from first year, internship, and PG students. Which sampling procedure should be used by DSW to ensure his sample is representative of these types of students?

(a) Simple random sampling

(b) **Stratified random sampling**

(c) Systematic sampling

(d) Convenience sampling

Q160. Which of the following is a basic step in selecting a sample from a population regardless of the specific technique being used?

(a) Identify the population
(b) Define the sample size
(c) Select a sample
(d) **All of these**

Q161. Which type of sampling procedure should be use to draw a sample that is not biased?

(a) Non-probability
(b) Concurrent
(c) **Random**
(d) Convenience

Q162. The process of selecting a number of participants for a study in such a way that they represent the larger group from which they were selected is known as:

(a) Research design
(b) **Sampling**
(c) Data collection
(d) Random assignment

Q163. In the systematic sampling, the value of k is classified as:

(a) **Sampling interval**
(b) Sub stage interval
(c) Secondary stage interval
(d) Multistage interval

Q164. In systematic sampling, the population is 200 and the selected sample size is 50 then the sampling interval is:

(a) 250
(b) 0.25
(c) 4
(d) 40

Q165. In cluster sampling, the elements of selected clusters are classified as

(a) **Elementary units**
(b) Primary units
(c) Secondary units
(d) Proportional units

Q166. Type of sampling in which the desired and useful information is gathered from the best position holder is classified as:

(a) Quota sampling
(b) Convenience sampling
(c) Purposive sampling
(d) **Judgment sampling**

Q167. Method of sampling in which the random sampling will not be possible because the population is widely spread is classified as:

(a) Secondary stage sampling
(b) **Multistage sampling**
(c) Primary stage sampling
(d) Sub stage sampling

Q168.Type of stratified proportion sampling in which the information is gathered on convenience basis from different groups of population is classified as

(a) Purposive sampling (b) Judgment sampling

(c) **Quota sampling** (d) Convenience sampling

Q169.Method of sampling in which the population is divided in to mutual exclusive groups that have useful context in statistical research is classified as

(a) **Stratified sampling** (b) Regular group sampling

(c) Irregular group sampling (d) Direct group sampling

Q170.In stratified sampling, the sample drawn randomly from the strata is classified as

(a) Sub strata (b) **Sub sample**

(c) Direct sub group (d) Indirect sub group

Q171.Procedure in which the number of elements in stratum is not in proportional to the number of elements in population is classified as:

(a) Indirect strata procedure (b) Direct strata procedure

(c) **Non proportional procedure** (d) Proportional procedure

Q172.Stratified samples may be distinguished from quota samples because with a stratified sample, sample elements are selected:

(a) **Randomly** (b) Judgmentally

(c) Proportionately (d) Sequentially

Q173.A researcher divides the population of product users into three groups based on degree of use. If the researcher then draws a random sample from each user group independently, she has created a ______ sample:

(a) Random (b) **Stratified**

(c) Judgment (d) quota

Q174.Name a sampling procedure in which initial respondents are selected by probability methods, and then additional respondents are obtained from information provided by initial respondents.

(a) Semi-random (b) Initial

(c) Quota (d) **Snowball**

Q175.Random sampling error

(a) Is the difference between a survey that includes only those who responded and a survey that also includes those who failed to respond

(b) Does not occur in non probability samples

(c) Results from the nature of a study's design and the inappropriate or random administration of the sampling process

(d) **Is a function of sample size**

Q176.In systematic sampling, population is 240 and selected sample size is 60 then sampling interval is ______

(a) 0.25 (b) **4**

(c) 60 (d) 240

Hint: (b); Sampling interval is defined as the interval in which the population is divided. The sampling interval is given as the population/ sample size = 240/60 = 4.

Q177.If the mean of population is 27 then the mean of sampling distribution is:

(a) 24 (b) 25

(c) 26 **(d) 27**

Hint: (d); In a sampling distribution the mean of the population is equal to the mean of the sampling distribution. Hence mean of population=29. Hence mean of sampling distribution=27.

Q178.The method of selecting a desirable portion from a population which describes the characteristics of whole population is called as :

(a) **Sampling** (b) Segregating

(c) Dividing (d) Implanting

Q179.The cluster sampling, stratified sampling or systematic samplings are types of:

(a) Direct sampling (b) Indirect sampling

(c) **Random sampling** (d) Non random sampling

Q180.The selected clusters in a clustering sampling are known as:

a) **Elementary units** b) Primary units

c) Secondary units d) Proportional units

Q181.A population has N items. Samples of size n are selected without replacement. Find the number of possible samples.

(a) $^{N}C_{n}$ (b) $^{n}C_{N}$

(c) 2^{n} (d) 2^{N}

Q182.Find the number of all possible samples from a population containing 8 items from which 2 items are selected at random without replacement.

(a) 56 (b) **28**

(c) 66 (d) 38

Hint: (b); The number of ways of selecting n samples from a population containing n items is $^{N}C_{n}$. The population is N = 8 and sample size is n = 2. Therefore the number of possible samples are $^{8}C_{2} = 28$.

Q183.The sampling technique in which every element of the population has an equal, non-zero probability of being selected in a sample is called:

(a) Quota sampling (b) Probability sampling

(c) **Purposive sampling** (d) Stratified sampling

Q184.A magazine conducts a survey and asks its readers to cut the questionnaire from the magazine, fill it and send it via mail. It is a type of:

(a) Purposive sampling (b) Probability sampling

(c) **Convenience sampling** (d) Stratified sampling

Q185.The sample is a subset of:

(a) Group (b) Distribution

(c) Data (d) **Population**

Q186.In sampling with replacement, a sampling unit can be selected:

(a) Less than one time (b) **More than one time**

(c) Only once (d) None of these

Q187. The sampling error is:

(a) Always negative
(b) **Always positive**
(c) Equal to the population mean
(d) Population parameter

Q188. The sampling technique in which every element of the population has an equal, non-zero probability of being selected in a sample is called:

(a) Stratified sampling
(b) Quota sampling
(c) **Purposive sampling**
(d) Probability sampling

Q189. The sampling procedure that includes all units of the population in the study is called:

(a) Saturation survey
(b) Survey population
(c) **Complete coverage**
(d) Both (b) and (c)

Q190. In random sampling, the probability of selecting an item from the population is:

(a) One
(b) Zero
(c) **Known**
(d) Unknown

4

Diagrammatic and Graphical Presentation of Data, Skewness, Moments and Kurtosis

Diagrammatic Presentation of Data

The quantitative data when classify are converted into averages and if the number of average is large then it is not possible to draw a conclusion from their large numbers of averages. Moreover quantitative data are dull, confusing, boring for many of us. Therefore, a statistical device to make such confusing data more attractive was developed i.e. if such complex data are presented by geometric figures, pictures, graphs or curves then it will become more attractive and more impressive, is called as **diagrammatic representation of data.** These diagrams are nothing but the use of geometrical figures to improve the overall presentation and offer visual assistance for the reader.

Types of Diagrams

Generally, diagrams are made for a special series or a time series and they are divided into five different categories:

1) One dimensional diagrams
2) Two dimensional diagrams
3) Three dimensional diagrams
4) Pictograms
5) Cartograms

1) One dimensional diagrams: If the ratio between smallest and the greatest quantities is nearly **1:10**, then we make **one dimensional diagram**. It consists of thick vertical bars though their thickness is not taken into account. Only their heights are considered and their heights are taken proportional to the given quantities.

Types of One Dimensional Diagram

This is further sub-divided into three types-

- **Simple bar diagram:** When single quantities in a spatial series are given, then we make **simple bar diagrams**. In this case, we take different points of the special series on a common base line of equal distances. On each point we make bar having high proportional to the given quantities against that point in the series. The widths of all the bars are taken equal. The bars can be drawn horizontally or vertically.
- **Multiple bar diagram:** It can represent set of two or more interrelated phenomenon. The method of drawing these diagrams is similar to simple bar diagrams. The only difference is that a separate bar of different shade, colour, etc. represents each characteristic of phenomenon in a particular set. These diagrams are suitable for comparison of the various related phenomenon. When the number of quantities are given against different points of a special series and these quantities cannot be totaled i.e. their total has no meaning. In other words, their quantities belong to different independent characteristics then on each point on the common base line are make a set of bars is equal to the number of quantities against point of a special series. These bars are made touching to each other. There widths are equal but heights are taken proportional to the given quantities. There will be equal space between different sets on the base line. In a set different bar are given different colours representing different quantities. There sequence should remain same in all the sets.
- **Sub-divided bar diagram:** The totals of these component quantities at different points are taken as the heights of different bars at different points on equal intervals on a common horizontal line *i.e.* with the help of this total a simple bar diagram is made. Then given magnitude can be broken into parts of which it is composed or if these are independent quantities constituting sub-division of totals, then the bars may be sub-divided to show the realization of the parts to the whole. In these diagrams, the various components or quantities in each bar should be sub-divided in the same order otherwise proper comparison will not be possible. The components can be distinguished in the bar by different colours or shades.

2) Two dimensional diagrams: If the ratio between smallest and the greatest quantities is nearly **1:100**, then we make **two dimensional diagram**. In these diagrams instead of length as well as the width of the bars is considered. These are also known as surface diagrams or area diagrams.

Types of Two Dimensional Diagrams

The important types of such diagrams are-

- **Square diagram:** They are useful when some items of the series have values much higher than others. In the construction of square diagram, first of all the square roots of the given quantities are taken then with the help of a suitable scale, different squares are made having their side proportion to the square root quantities. The area of the squares would be in the same proportion as the ratio of original figures. All the squares are placed on a common horizontal line at equal distances.

- **Circular diagram:** For making circular diagram, the square roots of the given quantities are taken as the radii of different circles. The centers of all the circles are taken on a common horizontal line. The scale is mentioned both above and below on the vertical line drawn on the left side of this common horizontal line. Circular diagram can be used in all those cases in which square diagram are used. The circles are preferred over the squares because they are easy to drawn and attractive also.

- **Pie diagram or Sector diagram:** When numbers of component quantities at different points of a special series are given and these components seem to be the part of their totals then pie diagram is made. In this diagram by taking totals of component quantities, a simple circle diagram is made. That means, these diagrams are based on the fact that area of each circle is proportional to its radii. Hence, the radii of the circle represent square root of each magnitude of the data. Then each circle is sub-divided into different sectors representing different component parts of the total i.e. area of a sector represents the corresponding component quantities. The area of the sector is proportional to the corresponding angle at the centre of the circle. Therefore, the totals of the different components at each point are first equated with 360^0 and in this way angles for different component quantities are worked out. The total of these angles will be equal to 360^0. These angles on the centre of the circle are taken in the same sequence in each circle. These sectors may also be given different colours. These diagrams are also known as **'Sector diagrams'.**

- **Rectangular diagram:** The area of a rectangle is equal to the width multiplied by the height of the rectangle. Therefore, rectangles can be made either by taking equal widths or by taking equal heights. When two figures are to be shown by the areas of two rectangles, two methods can be adopted, either their width may be kept equal and their lengths in proportion to the two figures or their lengths can be kept equal and their widths in proportion to the size of the two figures. In both the cases the area of the rectangles would be in proportion to the size of the figures.

3) Three dimensional diagrams: In these diagrams the length, width and height have to be taken into account. Such diagrams are used where the range of difference between the smallest and the largest value is very large i.e. **1:1000**. These are also known as volume diagrams. Cylinders, spheres, cubes, etc. are known as **three-dimensional diagrams**. For the construction of **cubes**, cube root of the given quantities is taken. Then with the help of a suitable scale different cubes are made having their sides proportional to the cube root quantities. All the cubes are placed on a common horizontal line of equal distances.

4) Pictograms: In this type pictures of concerned items are made. In other words, it is the pictorial representation of data using pictures, icons, or symbols. These diagrams are most commonly used in exhibitions for propaganda purposes. **For Example**, if the ratio of rabbit strength at three different places is 3:2:4. Then, at first point there will be three pictures of rabbits, at second point two pictures of rabbits, and at third point four pictures of rabbits. For comparison purpose, these pictures may be encircled either by squares or by rectangles with same size or area.

5) Cartograms: These diagrams are made in **geographical situations**. For constructing such diagrams map of the concerned territory is taken. Then on the map the given figures are shown either in the form of "dots" or "crosses" or "colours". The provincial activities are also shown by the cartograms.

Graphical Presentation of Data

The continuous changes in a variable over a period of time cannot be represented by diagrams more suitably. Similarly, frequency distribution cannot be represented by diagrams. Diagrams are generally useful for the purpose of publicity and propaganda. No statistical conclusion can be drowning as these provide only an approximate and rough expression about the phenomenon. Thus, **graphical presentation of data** is more rigorous and improved representation of numerical statements in the form of graphs and charts.

Types of Graphs

Graphs are generally drawn with the help of two perpendicular lines known as x-axis and y-axis. The following are the various types of graphs, which are in common use-

(i) Graphs of time series

(ii) Graphs of frequency distribution

(iii) Zone graph

(iv) Band graph

(i) **Graphs of time series:** The graphs are generally used to represent a data in which the independent variable is time. Here, the time variable is taken on x-axis and the other variable is taken on y-axis. Such a resultant figure is known as Graph of Time Series or Historigram or line graph. There are two types –

(a) **Historigram with one variable:** In this case, the time period is taken on the x-axis and the values of the variable are plotted on y-axis may be taken on the left hand side of the graph.

(b) **Historigram with more variables:** It may further be divided as-

(A) Historigram with homogenous data: In this case, the values of all the variants are plotted on the same y-axis with a common scale against the common time period taken on the x-axis.

(B) Historigram with heterogeneous data: In this case, the time period is taken on the x-axis and the values of one variable are plotted on y_1-axis may be taken on the left hand side of the graph. The value of the second variable will be plotted on y_2-axis may be taken on the right hand side of the graph against the common time period.

(ii) **Graphs of frequency distribution:** In this case, the class-intervals are taken on the x-axis and the rectangles are made on their class-intervals having heights equal to the corresponding frequencies. The resultant figure is known as graph of **frequency distribution** or **Histogram.** Frequency distributions of all types are represented by means of graphs precisely for the same reasons for which graphs are prepared for other types of data. The following types of graphs can be constructed to represent frequency distributions-

(a) **Histogram:** It is a graphical representation of the frequency distribution of a continuous variable. Histogram is a set of vertical bars whose areas are proportional to the frequencies represented. In the construction of histogram, class intervals of continuous data are taken on x-axis and the frequencies depending on it on the y-axis. The y-axis represents the frequencies of each class which constitute the height of its rectangles. In this manner, we get a series of rectangles each having a class interval distance as its width and the frequency distance as its height. There are two types of Histograms:-

(A) **Histogram with continuous data or Histogram with equal class-interval:** In this case, the width of all the rectangles will be equal, *i.e.* equal to the class-interval. The heights of these rectangles will represent the respective frequencies of these class-intervals

(B) **Histogram with discontinuous data or Histogram with unequal class-interval:** In this case, the class intervals are taken again on the x-axis. The heights of the rectangles on these class-interval is instead of equal to the frequencies will be taken as frequencies per unit of class interval i.e. the width of these rectangles will not be equal and heights will represent the frequency per unit of class-interval.

Computation of Mode in Histogram

When the histogram is made then the top corners of the highest rectangle are connected with the corners of the apposite adjoining rectangles from the point of intersection, a perpendicular is dropped on the x-axis. The point where this perpendicular touches the x-axis is the value of **mode**, i.e. the distance between the point of origin and this touching point is the **mode**.

(b) **Frequency polygon:** It is drawn by joining the consecutive points, plotted by taking mid points of the class intervals on x-axis and corresponding frequencies on y-axis. The end points are extended at each end to join the x-axis. The area of the graph between the x-axis and the two extreme ends of the curve is equal to the total frequency of the data. Frequency polygon can also be drawn with the help of histograms by joining the mid points of the upper side of each rectangle of a histogram. The area of a frequency polygon is always equal to the area of the corresponding histogram.

(c) **Frequency curve:** The method of drawing a frequency curve is similar to frequency polygon. The only difference is that instead of joining the successive points by straight lines, a free hand smooth curve is drawn in such a way that it passes through most of the points. A **frequency curve** has no sudden raising and no sudden falls. The area of a frequency curve is always equal to the area of the corresponding frequency polygon and histogram. There are two types of frequency curves:

- **Symmetrical curves:** When a frequency curve is exactly symmetrical on both the sides of the middle ordinate then it is called a **Symmetrical curve** or **Normal curve.**
- **Asymmetrical curves:** When a frequency curve is not symmetrical on the two sides then it is called Asymmetrical curves. There are five different types of asymmetrical curves:

(1) **Positive skewed curve:** When a curve has its tail on the right hand side, it is known as **positive skewed curve**. In this curve, the relationship between mean, median and mode are:

Mode < Mean < Median

(2) **Negative skewed curve:** When a curve has its tail on the left hand side, it is known as **negative skewed curve**. In this curve, the relationship between mean, median and mode are:

Mode > Mean > Median

(3) **L-shaped curve:** If a curve has the shape like the English alphabet 'L', it is known as **L-shaped curve**. In this curve, the relationship between mean, median and mode are:

Mode < Mean < Median

(4) **J – shaped curve:** If a curve has the shape like the English alphabet 'J', it is known as **J-shaped curve**. In this curve, the relationship between mean, median and mode are:

Mode > Mean > Median

(5) **U – shaped curve:** If a curve has the shape like the English alphabet 'U', it is known as **U-shaped curve** or if we joined L-shaped curve and J-shaped curve then it becomes a **U-shaped curve**. That means, it this type of curve the maximum frequency (Two mode value) is at the two ends of the curve and a minimum towards the centre (Mean = Median).

(d) **Cumulative frequency curves or Ogives:** A graph, where the cumulative frequencies (less than or more than) are plotted against the corresponding class limits (upper or lower) and smoothed out in a free hand curve is known as cumulative frequency curve or Ogive. It is known as more than cumulative frequency curve or Ogive if more than cumulative frequencies are plotted on the lower limits of the class intervals and less than cumulative frequency curve or Ogive if less than cumulative frequencies are plotted on the upper limits of the class intervals. The less than and more than cumulative frequency curve or Ogives always intersect at **Median**.

(iii) **Zone graph:** If minimum and maximum values of a variable are given against different intervals may be month, year, etc. Then these intervals are taken on the x-axis and the minimum and maximum values are plotted on the y-axis with a common scale, then minimum and maximum plotted points are joined with vertical straight lines against each point of interval, the resultant graph is known as **Zone graph.**

(iv) **Band graph:** If the components of total at different intervals are given then each line of graph is made over the preceding line of graphs i.e. first we plot the values of first component against different points of intervals and these plotted points are joined with straight line, then this band between the x-axis and line of graph represents the first component. Than the values of second component are taken over the values of first component then these plotted points are joined with a line. Than this band represents the second component. All the bands may be given different colours.

Skewness

The term '**skewness** means 'lack of symmetry', i.e., if the distribution of data is not symmetrical, it is called asymmetrical or skewed. In a skewed distribution these values differ and therefore, the balance of the centre of gravity is shifted to one side or the other, to the left or to the right. The measure of skewness can be classified as follows:

- **Absolute measures of skewness:** It can be measured in absolute term by finding the difference between the mean and the mode or mean and median. Symbolically

  Skewness = Mean – Mode

  or

  Skewness = Mean – Median

  Skewness is **positive** if mean is greater than mode (or median) and is **negative** if mean is less than mode (or median).

- **Relative measures of skewness:** The **relative measure of skewness** known as **coefficient of skewness** is obtained by dividing the absolute measure of skewness by any of the measures of dispersion. It is also known as Karl Pearson's Coefficient of skewness and denoted by "Sk_p". Symbolically,

$$\text{coefficient of skewness } (Sk_p) = \frac{\text{Mean-Mode}}{\text{Standard Deviation (S.D.)}}$$

Moments

In statistics the term "**Moment**" is used to describe the properties of a distribution. They represent a convenient method for summarizing many dispersion and measures of skewness and kurtosis. If we take the first power of the deviations, we get first moment about the mean, denoted by μ_1. Mean of the second power of the deviations gives us second moment about the mean denoted by μ_2 and so on. Symbolically,

$$\mu_1 = \frac{\Sigma(x - \overline{x})}{n} \text{ (It is always zero)}$$

$$\mu_2 = \frac{\Sigma(x - \overline{x})^2}{n} \text{ (It is called variance)}$$

$$\mu_3 = \frac{\Sigma(x - \overline{x})^3}{n} \text{ (It is called } 3^{rd} \text{ moment about mean)}$$

$$\mu_4 = \frac{\Sigma(x-\bar{x})^4}{n} \left(\textit{It is called } 4^{th} \textit{ moment about mean}\right)$$

In practice, we do not go beyond first four moments as the first four moments.

Remark: In a symmetrical distribution, all odd order moments i.e. μ_1, μ_3, μ_5, etc., would always be zero. If they are not equal to zero, the distribution is asymmetrical.

Kurtosis

Kurtosis in Greek means "bulginess". In statistics, kurtosis refers to the degree of **flatness or peakedness** in the region about the mode of a frequency curve. If the curve of a frequency distribution is more peaked than the normal curve, it is called "**leptokurtic**" and if it is more flat-topped than the normal curve, it is called "**platykurtic**". On the other hand, the curve of a frequency distribution is called "**mesokurtic**", if it is neither flat nor sharply peaked.

Karl Pearson defined a coefficient known as β_2 as follows:

$$\beta_2 = \frac{\mu_4}{\mu_2^{\,2}}$$

Where, μ_2 is the second moment and μ_4 is the forth moment about the mean. Kurtosis is measured by β_2 or its derivative γ_2 which is given as:

$\gamma_2 = \beta_2 - 3$

For a symmetrical distribution such as normal distribution, if the value of β_2 is greater than 3 (or $\gamma_2>0$) the curve is more peaked than the normal curve i.e. **leptokurtic**, when β_2 is less than 3 (or $\gamma_2<0$) the curve is less peaked i.e., **platykurtic** and if β_2 is equal to 3 (or $\gamma_2 = 0$) the curve is **mesokurtic**.

MCQ

Q.1. If vertical lines are drawn at every point of straight line in frequency polygon then by this way frequency polygon is transformed into:

(a) Width diagram (b) Length diagram

(c) **Histogram** (d) Dimensional bar charts

Q2. Diagrams such as cubes and cylinders are classified as

(a) One dimension diagrams (b) Two dimension diagram

(c) **Three dimensional diagrams** (d) Dispersion diagrams

Bold means correct answer

Q3. In stem and leaf display diagrams used in exploratory analysis, stems are considered as

(a) Central digits (b) Trailing digits

(c) **Leading digits** (d) Dispersed digits

Q4. Diagrams used to represent grouped and ungrouped data is classified as

(a) Breadth diagrams (b) **Bar diagrams**

(c) Width diagrams (d) Length diagrams

Q5. Three dimensional diagrams are named as so because they considers

(a) Length and breadth (b) Breadth and depth

(c) **Depth, length and breadth** (d) Depth and length

Q6. Histograms and pie charts are classified as two dimensional diagrams because only

(a) **Length is considered** (b) Width is considered

(c) Length plus width is considered (d) Breadth is considered

Q7. Graphs which represent data on maps are considered as:

(a) **Cartograms** (b) Picto-graph

(c) Pictograms (d) Symmetry graph

Q8. Types of histograms includes

(a) Deviation bar charts (b) Paired bar charts

(c) Grouped charts (d) **All of above**

Q9. Area diagrams and surface diagrams are other names of

(a) Single dimension diagrams (b) **Two dimensional diagrams**

(c) Three dimensional diagrams (d) Four dimension diagrams

Q10. Diagrams are for:

(a) Better mental appeal (b) The use of imports

(c) **Better quantitative picture** (d) The use of exports

Q11. Charts and graphs are the presentation of numerical facts by means of:

(a) Symbols (b) Area and other geometrical forms

(c) Points and lines (d) **All of them**

Q12. Trilinear chart is used to portray simultaneously:

(a) One variable
(b) Two variables
(c) **Three variables**
(d) Four variables

Q13. Rectilinear coordinate chart is also referred as:

(a) Rectangular graph
(b) Cartesian coordinate graph
(c) **Both (a) and (b)**
(d) None of them

Q14. The purpose served by diagrams and chart is:

(a) To avoid tabulation
(b) To avoid textual form
(c) **Simple presentation of data**
(d) All of them

Q15. Graphs and charts facilitate:

(a) To know the trend
(b) To know relationship
(c) Comparison of values
(d) **All of them**

Q16. The shape of trilinear charts is that of a:

(a) Cube
(b) Cone
(c) **Equilateral triangle**
(d) Pyramid

Q17. Choice of a particular chart depends on:

(a) Nature of data
(b) Purpose of study
(c) Type of audience
(d) **All of above**

Q18. Non-dimensional diagrams are also known as:

(a) Spheres
(b) Cubes
(c) **Pictograms**
(d) All the above

Q19. Three-dimensional diagrams are known as:

(a) **Cubes**
(b) Pictograms
(c) Cartograms
(d) Both (b) and (c)

Q20. Which of the following is one-dimensional diagram?

(a) Cubes
(b) Sector
(c) Cylinder
(d) **Bar diagram**

Q21. Two-dimensional diagrams are known as:

(a) Cube diagram
(b) Cylinder diagram
(c) Bar diagram
(d) **Square diagram**

Q22. Which of the following is not a one-dimensional diagram?

(a) Multiple bar diagram
(b) **Cylinder diagram**
(c) Sub-divided bar diagram
(d) Simple bar diagram

Q23. Which of the following is not a two-dimensional diagram?

(a) Rectangular diagram
(b) **Multiple bar diagram**
(c) Square diagram
(d) Sector diagram

Q24. Cumulative frequency curve occurs for:

(a) Less than type distribution
(b) More than type distribution
(c) **Both (a) and (b)**
(d) None of the above

Q25. In an ogive curve, the points are plotted for:

(a) The values and frequencies
(b) **The values and cumulative frequencies**
(c) Frequencies and cumulative frequencies
(d) None of the above

Q26. The point of intersection of more than Ogive curve and less than Ogive curve is:

(a) Arithmetic mean
(b) Geometric mean
(c) **Median**
(d) Mode

Q27. In a bar diagram, the base line is:

(a) Horizontal
(b) **Vertical**
(c) False base line
(d) None of them

Q28. When the values are large in magnitude in a chronological series and variation amongst values is small, a graph is better drawn by choosing:

(a) Narrow scale
(b) Wide scale
(c) **False base line**
(d) None of them

Q29. In a column chart, the base line is:

(a) **Horizontal**
(b) Vertical
(c) False base line
(d) None of them

Q30. In a column chart, bars are:

(a) Horizontal
(b) **Vertical**
(c) False base line
(d) None of them

Q31. In a bar diagram, the bars are:

(a) **Horizontal** (b) Vertical

(c) False base line (d) None of them

Q32. In a two dimensional diagram:

(a) Only width is considered

(b) Only height is considered

(c) **Both height and width are considered**

(d) Width, height and thickness are considered

Q33. Diagram and graphs are tools of:

(a) Presentation (b) Summarization

(c) Collection of data (d) **Analysis**

Q34. Which of the following should be avoided as methods of presenting data?

(a) Bars (b) Pie diagram

(c) **Cubs and spheres** (d) Pictograms

Q35. With the help of ogive curve, one can determine:

(a) Median (b) Deciles

(c) Percentiles (d) **All of them**

Q36. Pictograms are:

(a) Mostly used (b) Very accurate

(c) **Least accurate** (d) Scientifically correct

Q37. Pictograms are suitable for the data in:

(a) Intervals (b) **Counts**

(c) Fraction (d) None of these

Q38. Pictograms are shown by:

(a) **Pictures** (b) Dots

(c) Lines (d) Circles

Q39. Cartograms are shown by:

(a) Intervals (b) **Dots**

(c) Fraction (d) Lines

Q40. Pictograms are generally used by:

(a) Scientist
(b) **Dilettante**
(c) Cartographers
(d) All of them

Q41. Lorenz curve is an indicator for the distribution of two factors:

(a) **Being unequal**
(b) Being equal
(c) Both (a) and (b)
(d) Neither (a) nor (b)

Q42. With the help of histogram we can prepare:

(a) Frequency distribution
(b) Frequency Curve
(c) Frequency polygon
(d) **All of them**

Q43. Historigram is suitable for:

(a) Chronological distribution
(b) Time series data
(c) **Both (a) and (b)**
(d) None of them

Q44. In case of frequency distribution with classes of unequal widths, the heights of bars of a histogram are proportional to:

(a) **Class frequency**
(b) Class intervals
(c) Frequency densities
(d) Frequency in percentage

Q45. To show the maximum and minimum values in a time series, the suitable graph is:

(a) **Historigram**
(b) Histogram
(c) Zone graph
(d) Band graph

Q46. In a column chart on circular base, bars are:

(a) Horizontal
(b) Vertical
(c) **Slanting**
(d) Curved

Q47. Pie-diagram represents the components of a factor by:

(a) Circles
(b) Squares
(c) **Sectors**
(d) Percentages

Q48. An alternative chart to pie-chart is:

(a) Sphere
(b) Circles
(c) Cube
(d) **Step bar diagram**

Q49. When there is a pronounced skewness, the desirable scale to plot the frequency distribution is:

(a) **Logarithmic scale** (b) Arithmetic scale

(c) Multiple scale (d) None of these

Q50. If there is an increase in a series at constant rate, the graph will be a:

(a) A straight line from left top to right bottom

(b) **A straight line from left bottom to right top**

(c) Convex curve

(d) Concave curve

Q51. If there is a decrease in a series at constant rate, the graph will be a:

(a) **A straight line from left top to right bottom**

(b) A straight line from left bottom to right top

(c) Convex curve

(d) Concave curve

Q52. The suitable chart to emphasize the difference between two time series, of which one is at a higher level, is:

(a) **Band graph** (b) Zone graph

(c) Paired bar diagram (d) Histogram

Q53. A semi-logarithmic graph of a series increasing by a constant amount will be a:

(a) A convex downward curve (b) **A convex upward curve**

(c) A concave downward curve (d) A concave upward curve

Q54. The graph of the successive points of a distribution joined by straight lines in statistical terminology is known as:

(a) **Frequency curve** (b) Frequency distribution

(c) Cumulative frequency distribution (d) None of these

Q55. An arithmetic chart can:

(a) Have only one amount scale (b) **Have multiple amount scale**

(c) Without any amount scale (d) None of them

Q56. Histogram is suitable for the data presented as:

(a) Individual series

(b) Discrete grouped frequency distribution

(c) **Continuous grouped frequency distribution**

(d) All of them

Q57. If we plot the points of less than type or more than type frequency distribution, the shape of graph is:

(a) Zig-zag curve

(b) Frequency curve

(c) Frequency polygon

(d) **Cumulative frequency curve**

Q58. The most appropriate diagram to represent the data relating to the monthly expenditure on different items by a family is:

(a) Frequency curve
(b) Frequency polygon

(c) Histogram
(d) **Pie-diagram**

Q59. ___________ are used when you want to visually examine the relationship between two quantitative variables.

(a) Bar graphs
(b) Pie graphs

(c) Line graphs
(d) **Scatterplots**

Q60. A graph that uses vertical bars to represent data is called ____.

(a) Line graph
(b) **Bar graph**

(c) Scatterplot
(d) Vertical graph

Q61. Frequency polygon can be drawn with the help of:

(a) Historigram
(b) **Histogram**

(c) Circle
(d) Percentage

Q62. In a cumulative frequency polygon, the cumulative frequency of each class is plotted against:

(a) Mid-point
(b) Lower class boundary

(c) **Upper class boundary**
(d) Upper class limit

Q63. The graph of the cumulative frequency distribution is called:

(a) Histogram (b) Frequency polygon

(c) Pictogram (d) **Ogive**

Q64. When successive mid-points in a histogram are connected by straight lines, the graph is called a:

(a) Historigram (b) Ogive

(c) Frequency curve (d) **Frequency polygon**

Q65. A frequency polygon is a closed figure which is:

(a) One sided (b) Two sided

(c) Three sided (d) **Many sided**

Q66. Ogive curve can be occurred for the distribution of:

(a) Less than type (b) More than type

(c) **Both (a) and (b)** (d) Neither (a) and (b)

Q67. The word ogive is also used for:

(a) Frequency polygon

(b) **Cumulative frequency polygon**

(c) Frequency curve

(d) Histogram

Q68. Cumulative frequency polygon can be used for the calculation of:

(a) Mean (b) **Median**

(c) Mode (d) Geometric mean

Q69. Histogram and frequency polygon are two graphical representations of:

(a) **Frequency distribution** (b) Class boundaries

(c) Class intervals (d) Class marks

Q70. The suitable diagram to represent the data relating to the monthly expenditure on different items by a family is:

(a) Historigram (b) Histogram

(c) Multiple bar diagram (d) **Pie diagram**

Q71. A graph of time series or historical series is called:

(a) Histogram (b) **Historigram**

(c) Frequency curve (d) Frequency polygon

Q72. The historigram is the graphical presentation of data which are classified:

(a) Geographically
(b) Numerically
(c) Qualitatively
(d) **According to time**

Q73. Historigram and histogram are:

(a) Always same
(b) **Not same**
(c) Off and on same
(d) Randomly same

Q74. A distribution in which the observations are concentrated at one end of the distribution is called a:

(a) Symmetric distribution
(b) Normal distribution
(c) **Skewed distribution**
(d) Uniform distribution

Q75. For graphic presentation of a frequency distribution, the paper to be used is:

(a) Carbon paper
(b) Ordinary paper
(c) **Graph paper**
(d) Butter paper

Q76. Histogram can be drawn only for:

(a) Discrete frequency distribution
(b) **Continuous frequency distribution**
(c) Cumulative frequency distribution
(d) Relative frequency distribution

Q77. Histogram is a graph of:

(a) **Frequency distribution**
(b) Time series
(c) Qualitative data
(d) Ogive

Q78. A sector diagram is also called:

(a) Bar diagram
(b) Histogram
(c) Historigram
(d) **Pie diagram**

Q79. Which of the following is not a one-dimensional diagram?

(a) Simple bar diagram
(b) Multiple bar diagram
(c) Component bar diagram
(d) **Pie diagram**

Q80. Which of the following is a two-dimensional diagram?

(a) Sub-divided bar
(b) Percentage component bar chart
(c) **Sub-divided rectangles**
(d) Multiple bar diagram

Q81. Pie diagram represents the components of a factor by:

(a) Circles (b) **Sectors**

(c) Angles (d) Percentages

Q82. A pie diagram is represented by a:

(a) Rectangle (b) **Circle**

(c) Triangle (d) Square

Q83. The budgets of two families can be compared by:

(a) Sub-divided rectangles (b) Pie diagram

(c) **Both (a) and (b)** (d) Histogram

Q84. Total angle of the pie-chart is:

(a) 45 (b) 90

(c) 180 (d) **360**

Q85. Diagrams are another form of:

(a) Classification (b) **Tabulation**

(c) Angle (d) Percentage

Q86. In pie diagram, the angle of a sub-sector is obtained as:

(a) $\frac{Component\ part}{Total} x90^0$ (b) $\frac{Component\ part}{Total} x120^0$

(c) $\frac{Component\ part}{Total} x180^0$ (d) $\boldsymbol{\frac{Component\ part}{Total} x360^0}$

Q87. If you have data on the yearly average temperature at Indira Gandhi International Airport, Delhi from 1990 to 2010, and if you are particularly interested in change over time, what is the most effective graphical display?

(a) Ogive (b) Pie chart

(c) **Line chart** (d) Histogram

Q88. If you suspect that a population is made up of two subpopulations, for example males and females, with different values for a certain continuous variable, which of the following would you use to visually check for such a possible difference?

(a) The median of a histogram (b) **The modes of a histogram**

(c) The skewness of the histogram (d) The modes of a line chart

Q89. If you have data on house prices and the distance of each of those houses from the city centre, and you are curious whether there is an association between distance from the city centre and the price of the house, with which of the following graphical techniques could you most easily see whether there is indeed such a relationship?

(a) Two histograms, one of house prices and one of distance from city centre

(b) **A scatter diagram**

(c) A bar graph with distance on the x-axis and price on the y-axis

(d) A bimodal histogram

Q90. It has been claimed that Vodafone has the highest market share amongst cellphone users in India. A random sample of 250 cellphone users was asked which network they subscribe to. What type of data has been collected and which graphical technique would be the most appropriate to highlight the various market shares, amongst those listed below?

(a) **Qualitative data to be represented in a pie chart**

(b) Quantitative data to be represented in a pie chart

(c) Qualitative data to be represented in a histogram

(d) Quantitative data to be represented in a bar chart

Q91. It has been claimed that B.Com students make up the largest group of students from a single degree programme amongst all students taking the statistics course. You wish to investigate this and ask 1200 statistics students which degree they are currently registered for. What type of data have you collected and how could this be best represented, given the options below?

(a) **Qualitative data to be represented in a pie chart**

(b) Qualitative data to be represented in a histogram

(c) Quantitative data to be represented in a bar chart

(d) Quantitative data to be represented in a pie chart

Q92. The graphical representation of a cumulative relative frequency distribution is called?

(a) Histogram (b) Pie chart

(c) Stem and leaf plot (d) **None of the above**

Q93. All 500 members of a sports club in Mumbai were contacted via email and asked whether they thought that Karate should be added to the list of sports currently offered by the club. Out of 500 members 150 members said yes, 91 said no, 59 said that they were not sure and 200 did not respond. To represent this information graphically, we could use a:

(a) Histogram
(b) Pie chart
(c) **Bar graph**
(d) None of the above

Q94. A histogram is a graphical representation of which of the following:

(a) An ogive
(b) **A frequency distribution**
(c) A cumulative relative frequency distribution
(d) None of the above

Q95. Which of the following is *not* found in a frequency distribution?

(a) Class limits
(b) Class intervals
(c) **Individual observations within each class**
(d) All of the above are found in a frequency distribution

Q96. Which of the following statements is false?

(a) **Ranked/ordinal data is most effectively presented with pie charts**
(b) If a histogram is positively skewed, the mean is larger than the median
(c) An advantage of stem and leaf displays is that the values of the original observations are shown
(d) A histogram can have more than one mode

Q97. A frequency distribution is a(n):

(a) Table of the individual observations collected from a sample
(b) Individual listing of the random values found in a data set
(c) Listing of the individual observations arranged in ascending or descending order
(d) **Table, which classifies the number of data values into classes with counts of the number of data values that fall into each of the classes**

Q98. The difference between a histogram and a bar chart is that:

(a) The histogram reflects qualitative data while the bar chart represents quantitative data
(b) The adjacent rectangles/bars in a histogram have a gap while those for a bar chart do not

(c) **The adjacent rectangles/bars in a bar chart have a gap while those for a histogram do not**

(d) None of the above

Q99. The two graphical techniques that can be used to represent nominal data are:

(a) Bar chart and histogram
(b) Pie chart and ogive
(c) **Bar chart and pie chart**
(d) Histogram and ogive

Q100. The graph of cumulative frequency is called

(a) Histogram
(b) Frequency polygon
(c) **Cumulative frequency polygon**
(d) None of these

Q101.Graph of time series data is called

(a) Polygon
(b) Histogram
(c) Ogive
(d) **Historigram**

Q102.In constructing a histogram, if the class interval size of one class is double than others, then the width of that bar should be:

(a) Quarter
(b) Half
(c) One
(d) **Doubled**

Q103.The graph of a frequency distribution is called:

(a) Historigram
(b) **Histogram**
(c) Ogive
(d) Curve

Q104.Total angles in Pie chart are:

(a) 90
(b) 180
(c) 270
(d) **360**

Q105.A Histogram is a set of adjacent:

(a) Lines
(b) Squares
(c) Triangles
(d) **Rectangles**

Q106. ______________________ use the division of a circle into different sectors:

(a) Frequency polygon
(b) **Sector graph**
(c) Line graph
(d) Conversion graphs

Q107. Cumulative Frequency Curve is also called:

(a) **Ogive** (b) Frequency curve
(c) Frequency polygon (d) Histogram

Q108. If vertical lines are drawn at every point of straight line in frequency polygon then by this way frequency polygon is transformed into:

(a) Width diagram (b) Length diagram
(c) **Histogram** (d) Dimensional bar charts

Q109. Diagrams such as cubes and cylinders are classified as:

(a) One dimension diagrams (b) Two dimension diagram
(c) **Three dimensional diagrams** (d) Dispersion diagrams

Q110. Discrete variables and continuous variables are two types of

(a) Open end classification (b) Time series classification
(c) Qualitative classification (d) **Quantitative classification**

Q111. In stem and leaf display diagrams used in exploratory analysis, stems are considered as

(a) Central digits (b) Trailing digits
(c) **Leading digits** (d) Dispersed digits

Q112. Diagrams used to represent grouped and ungrouped data is classified as

(a) Breadth diagrams (b) **Bar diagrams**
(c) Width diagrams (d) Length diagrams

Q113. Considering types of diagrams, squares, circles and rectangles are classified as

(a) Cumulative diagram (b) Dispersion diagrams
(c) One dimension diagrams (d) **Two dimension diagram**

Q114. Three dimensional diagrams are named as so because they considers both

(a) Length and breadth (b) Breadth and depth
(c) **Depth, length and breadth** (d) Depth and length

Q115. Area diagrams and surface diagrams are other names of

(a) Single dimension diagrams (b) **Two dimensional diagrams**
(c) Three dimensional diagrams (d) Four dimension diagrams

Q116.Types of histograms includes

(a) Deviation bar charts (b) Paired bar charts

(c) Grouped charts (d) **All of above**

Q117.Type of graphical charts that allows user to make direct comparisons between various data sets are called

(a) **Multiple bar charts** (b) Single bar charts

(c) Paired charts (d) Non paired data charts

Q118.In graphical representation of data, ideographs are also called as

(a) Picto-graph (b) **Pictograms**

(c) Symmetry graph (d) Asymmetry graphs

Q119.If midpoints of bars on charts are marked and marked dots are joined by a straight line then this graph is classified as:

(a) Class interval polygon (b) Paired polygon

(c) Marked polygon (d) **Frequency polygon**

Q120.Sub-divided bar charts are considered best to be used if information is presented in

(a) Negative values (b) **Ratios or percentages**

(c) Mean deviations (d) Positive values

Q121.Graphical diagram in which total numbers of observations are represented in percentages rather than absolute values is classified as:

(a) Asymmetrical diagram (b) Ungrouped diagram

(c) Grouped diagram (d) **Pie diagram**

Q122.Type of bar chart that is used to present deficit in loss, excess in exports and deficit in imports is classified as:

(a) Ungrouped bar charts (b) Grouped bar charts

(c) **Deviation bar charts** (d) Dimension bar charts

Q123.If in bar diagram characteristics variable to be measured is written on horizontal axis and frequencies are written on vertical axis then graph is for:

(a) **Grouped data** (b) Ungrouped data

(c) Dimensional data (d) Non dimensional data

Q124,The mode of a frequency distribution can be determined graphically from:

(a) Frequency curve
(b) **Ogive**
(c) Frequency polygon
(d) Histogram

Q125.A many sided closed figure used in frequency distribution is termed as:

(a) **Frequency polygon**
(b) Frequency Hexagon
(c) Frequency pentagon
(d) Frequency decagon

Q126.Frequency polygon is constructed by plotting frequencies against their:

(a) Class boundaries
(b) **Class marks**
(c) Cumulative frequency
(d) None of these

Q127.To draw histogram we take which values on x-axis and on y-axis:

(a) Frequency
(b) Frequency, Attributes
(c) Frequency, class interval
(d) **Attributes, Frequency**

Q128.Diagram and graphs are tools of:

(a) Collection of data
(b) **Analysis**
(c) Presentation
(d) Summarization

Q129.Pictograms are suitable for the data in:

(a) Counts
(b) Intervals
(c) Fraction
(d) All of the above

Q130.In a histogram with equal intervals, heights of bars are proportional to:

(a) Mid-values of the classes
(b) **Frequencies of respective classes**
(c) Either (a) or (b)
(d) Neither (a) nor (b)

Q131.Histogram is used to describe:

(a) **Quantitative data of a group of patients**
(b) Qualitative data of a group of patients
(c) Data collected on nominal scale
(d) Data collected on ordinal scale

Q132.Considering the types of diagrams, squares, circles and rectangles are classified as

(a) Cumulative diagram (b) Dispersion diagrams

(c) One dimension diagrams (d) **Two dimension diagram**

Q133.Specialized processes such as graphical and numerical methods are utilized in which of the following?

(a) Education statistics (b) **Descriptive statistics**

(c) Business statistics (d) Social statistics

Q134.The abscissa of the point of intersection of the less than type and of the more than type cumulative frequency curves of a grouped data gives its

(a) Mean (b) **Median**

(c) Mode (d) All the three above

Q135.Observe the histogram and answer the questions given below :

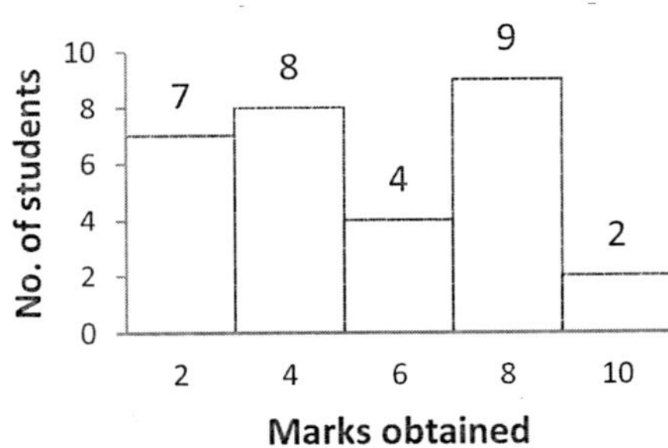

The total number of students is

(a) 10 (b) 20

(c) 25 (d) **30**

Q136.In the above histogram, the number of students getting marks 6 and above is

(a) 9 (b) 10

(c) **11** (d) 12

Q137.Observe the pie chart given below:

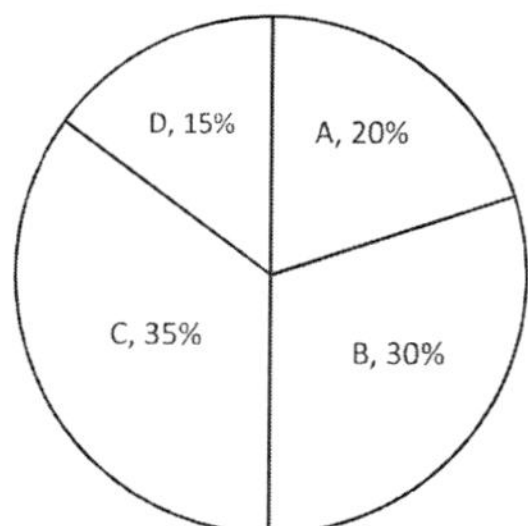

The central angle for sector A is

(a) 72° (b) 108°

(c) 144° (d) 150°.

Hint: Central angle for sector A = (20/100) * 360 = 72^0

Q138.In the above pie chart, which sector has the greatest angle?

(a) A (b) **B**

(c) C (d) D

Q139.In an examination, 20% of candidates got first class. When the data are represented by a pie chart, what is the angle corresponding to first class?

(a) 20^0 (b) 36^0

(c) **72^0** (d) 144^0

Q140.For a histogram based on a frequency distribution with unequal class intervals, the frequency of a class should be proportional to:

(a) The width of the rectangle

(b) The height of the rectangle

(c) **The area of the rectangle**

(d) The parameter of the rectangle

Q141.Ogives can be helpful in locating graphically to:

(a) Mean (b) **Median**

(c) Mode (d) None of these

Q142.For determining of median and mode graphically, one considers:

(a) Histogram and Historigram (b) Historigram and Ogives

(c) **Histogram and Ogives** (d) Zone graph and Band graph

Q143.A diagrammatic representation where a total of the parts make a meaningful whole is:

(a) Simple bar diagram (b) Multiple bar diagram

(c) **Pie diagram** (d) Cube diagram

Q144.The proportions of related items are best shown by a:

(a) Histogram (b) Pictogram

(c) **Pie diagram** (d) Band graph

Q145.The graphical representation of the time series is known as:

(a) **Historigram** (b) Histogram

(c) Ogives (d) Zone graph

Q146.For showing the growth of population for the last 10 years, the best diagram is:

(a) **Simple bar diagram** (b) Multiple bar diagram

(c) Pie diagram (d) Cube diagram

Q147.The measure of the central tendency is given by the X-coordinate of the point of intersection of the more than ogive and less than ogive is:

(a) Mean (b) **Median**

(c) Mode (d) All of them

Q148.A histogram containing a set of:

(a) Adjacent squares (b) Adjacent triangles

(c) **Adjacent rectangles** (d) Adjacent cubes

Q149.For geographically base data, the bars are used:

(a) **Horizontal** (b) Vertical

(c) Diagonal (d) Zigzag

Q150.Frequency curve is:

(a) **Asymptotic to X-axis** (b) Asymptotic to Y-axis

(c) Non-asymptotic to X-axis (d) Non-asymptotic to Y-axis

Q151.In a pie chart, usually, the arrangement of the sector is:

(a) Pair wise (b) Clock wise

(c) **Anti-clock wise** (d) None of these

Q152.A graphical device used for enumerating sample points a multiple-step experiment is:

(a) Histogram (b) Pie chart

(c) Cube chart (d) **None of these**

Q153.The frequency polygon is a closed diagram of:

(a) Two-sides (b) Three-sides

(c) Four-sides (d) **Many-sides**

Q154.Component bar charts are used when data is divided into:

(a) Circles (b) Parts

(c) **Groups** (d) None of these

Q155.The two graphical techniques that can be used to represent nominal data are:

(a) **Bar chart and Pie chart** (b) Bar chart and Histogram

(c) Pie chart and ogives (d) Pie chart and histogram

5

Measure of Central Tendency

Need of an average

Through frequency distribution are the condensed form of raw data, but these are not sufficient for making comparison between two or more data. Therefore further condensation of frequency distribution into a single figure is made. This single figure is always representative of the frequency distribution and such a figure is known as **average.**

Measure of Central Tendency

According to **Croxten and Cowden** "An average is a single value within the range of the data that is used to represent all the values in the series". A statistical measure used for representing the centre of central value of a set of observations is known as "**Measure of Central Tendency**". It is also called as "**Measure of first order**".

Characteristics of an ideal Measure of Central Tendency

An ideal measure of central tendency should possess the following properties:

1) It should be rigidly defined i.e. it has a fixed and finite value
2) It should be easy to understand
3) It should be easy to calculate
4) Its calculation should be based on all the observations
5) It should be capable for further algebraic treatment
6) It should not be affected by extreme values of the observations
7) It should be least affected by fluctuations of sampling

Types of Measure of Central Tendency

Some of the important averages may be classified as:

a) Mathematical averages: It can also be classified into three types:

- Arithmetic mean (A.M.)

- Geometric Mean (G.M.)
- Harmonic Mean (H.M.)
- Weighted Mean (W.M.)

b) Positional averages: It can be classified into three types:

- Median (M_d)
- Mode (M_o)

For simplification and clarity in presentation, following are six different measures of central tendency:

- **Arithmetic Mean (A.M.):** It is the most popular and widely used measure of central tendency and also called '**average**'. The quantity obtained by dividing the sum of values of all the items of a series by their number. Apart from having most of the characteristics of a good average, it also possesses some important properties. So, it is also known as an **ideal measure of central tendency**.

Important Formula of Arithmetic Mean

Following two methods are used for calculating arithmetic mean:

(a) Direct method

(b) Short-cut method

In classification and tabulation of data, we observed that the values of items can be put in the form of any of the following series, namely:

1) Individual series or ungrouped data

(a) **Direct method**

$$A.M.\left(\overline{X}\right)=\frac{x_1+x_2+x_2+.........+x_n}{n}=\frac{\Sigma X}{n}$$

(b) **Short-cut method**

$$A.M.\left(\overline{X}\right)=A+\frac{\Sigma d}{n}$$

Here, A = Assumed mean; Σd = Sum of deviations about assumed mean;

d = (X – A), deviation of each value of the variable from the assumed mean; n = number of observations

2) Discrete frequency distribution

(a) **Direct method**

$$A.M.(\overline{X}) = \frac{\sum fx}{N}$$

Here, Σfx = Sum of the products of observations with their respective frequencies;

N = Σf = sum of the frequencies.

(b) **Short-cut method**

$$A.M.(\overline{X}) = A + \frac{\sum fd}{N}$$

Here, A = Assumed mean; Σfd = Sum of the products of frequencies and their respective deviations;

d = (X – A), deviation of each value of the variable 'x' from the assumed mean 'A';

N = Σf = sum of the frequencies

3) Continuous frequency distribution

(a) **Direct method**

$$A.M.(\overline{X}) = \frac{\sum fm}{N}$$

Here, m = mid value of various classes; N = Σf = sum of the frequencies;

Σfm = Sum of mid values multiplied by their respective frequencies

(b) **Short-cut method**

$$A.M.(\overline{X}) = A + \frac{\sum fd}{N}$$

Here, A = Assumed mean; d = (m – A), deviation of each mid values 'm' from the assumed mean 'A';

N = Σf = sum of the frequencies

- **Geometric Mean:** It is defined to be the nth root of the product of the n items of a series, but when the number of items in a series is large particularly greater than four, then it is computer with the help of logarithm, and then, it can be defined as 'the antilog of the arithmetic mean of logarithm values of all the items of the series.

Important Formula of Geometric Mean

1) **Individual series or ungrouped data**

   $G.M. = (x_1\, x_2\, x_3 \ldots\ldots\ldots\ldots x_n)^{1/n}$

   or

   $$G.M. = Antilog\left(\frac{\sum \log x}{n}\right)$$

2) **Discrete frequency distribution**

   $$G.M. = Antilog\left(\frac{\sum f \log x}{N}\right)$$

   Here, $N = \Sigma f$ = sum of the frequencies

3) **Continuous frequency distribution**

   $$G.M. = Antilog\left(\frac{\sum f \log m}{N}\right)$$

   Here, $N = \Sigma f$ = sum of the frequencies; m = mid value of various classes

- **Harmonic Mean:** It is defined as the reciprocals of the arithmetic mean of reciprocals of all the items of the series.

Important Formula of Harmonic Mean

1) **Individual series or ungrouped data**

   $$H.M. = \frac{n}{\sum \frac{1}{x}}$$

2) **Discrete frequency distribution**

   $$H.M. = \frac{N}{\sum \frac{f}{x}}$$

   Here, $N = \Sigma f$ = sum of the frequencies

3) **Continuous frequency distribution**

   $$H.M. = \frac{N}{\sum \frac{f}{m}}$$

   Here, $N = \Sigma f$ = sum of the frequencies; m = mid value of various classes

Relationship between A.M., G.M. and H.M.

For any series or frequency distribution, the value of arithmetic mean (A.M.), geometric mean (G.M.) and harmonic mean (H.M.) are different and their values always follow the following inequality-

H.M. ≤ G.M. ≤ A.M.

- **Weighted Mean:** When the observations of a series have equal importance (weight) then arithmetic mean is an accurate measure of central tendency for calculating the average. But if items of a series are not equal importance then their arithmetic mean will not be an appropriate measure of central tendency. In such cases, the observations are given due weights according to their relative importance and thus a measure of central tendency will give the correct answer.

Important Formula of Weighted Mean

$$W.M. = \frac{\sum wx}{\sum w}$$

Here, w = relative weights of the variable; x = values of the variable

- **Median:** Median may be defined as the middle most or central value when the items are arranged in ascending or descending order of magnitude or the central value of an array is the **median**. The median divides a series into two equal parts. In one part, all the items are smaller than the median while in other parts are all greater than median.

Important Formula of Median

1) **Individual series or ungrouped data:** Median can be calculated in the following two ways:

   (a) **When number of items is odd:** The determination of median involves the following steps:

   - The data are arranged either in ascending or descending order
   - Locate median by the formula:

$$Median(M_d) = \left(\frac{n+1}{2}\right)^{th} item$$

   (b) **When number of items is even:** The determination of median involves the following steps:

   - The data are arranged either in ascending or descending order
   - Locate median by the formula:

$$Median(M_d) = \frac{\left(\frac{n}{2}\right)^{th} item + \left(\frac{n}{2}+1\right)^{th} item}{2}$$

2) **Discrete frequency distribution:** The determination of median involves the following steps:

- If the order of the discrete frequency distribution is ascending, then we calculate **less than type cumulative frequency**, otherwise, if the order of the discrete frequency distribution is descending, then we calculate **more than type cumulative frequency**.
- Find out the median number. That means, if total frequency is odd, then

  $$Median\, No. = \left(\frac{N+1}{2}\right)^{th} value$$

  And if total frequency is even, then

  $$Median\, No. = \left(\frac{N}{2}\right)^{th} value$$

  Here, N = Σf = sum of the frequencies
- This median number is seen in the cumulative frequency column and the cumulative frequency which is either equal to this median number or first greater than it is encircled.
- The value of item against this encircled cumulative frequency is the **median**.

3) **Continuous frequency distribution:** The determination of median involves the following steps:

- If the order of the class-interval is ascending, then we calculate **less than type cumulative frequency**, otherwise, if the order of the class-interval is descending, then we calculate **more than type cumulative frequency**.
- Find out the median number. That means, if total frequency is odd, then

  $$Median\, No. = \left(\frac{N+1}{2}\right)^{th} value$$

  And if total frequency is even, then

  $$Median\, No. = \left(\frac{N}{2}\right)^{th} value$$

Here, $N = \Sigma f$ = sum of the frequencies

- This median number is seen in the cumulative frequency column and the cumulative frequency which is either equal to this median number or first greater than it is encircled.
- Then, against this encircled cumulative frequency, the class-interval will be the **median class**.
- Now, in order to locate the exact value of median in this median class, the following formula is used:

$$Median = l_1 + \left\{ \frac{(l_2 - l_2)(m - c)}{f} \right\}$$

Here,

l_1 = lower limit of the median class; l_2 = upper limit of the median class;

m = the value of median number;

c = cumulative frequency preceding to the median number;

f = simple frequency of the median class.

- **Mode:** The word 'mode' means **fashion**. In statistical term, in a series, the item which has maximum number of repetition is called **mode**.
- If there is only **one item** which has greatest frequency in a series, than such series is known as **uni-modal series**.
- If in a series, there are **two items** which have maximum as well as equal frequencies, such series is known as **Bi-modal series**.
- If in a series, there are **three items** which have maximum as well as equal frequencies, such series is known as **Tri-modal series.**
- If in a series, there are more than three items having maximum as well as equal frequencies, then such series is known as **Multi-modal series.**

Important Formula of Mode

1) **Individual series or ungrouped data:** The value of mode is obtained simply by inspection, i.e., the value which has greatest frequency in the raw data will be the **mode**.
2) **Discrete frequency distribution:** In this case of a grouped frequency distribution, mode can be located simply by inspection. Here, the variate having maximum frequency will be taken as **mode**. Now, in a uni-modal distribution where the distribution has only one mode and the highest concentration of values is around one value only, we do not face any difficulty in locating the modal value. In such cases attempt is made to

find the value of concentration with the help of **grouping method**. In this method, the values first arranged in ascending order and their frequencies are noted. Normally, the grouping table has the following six columns:

- The maximum simple frequency is encircled.
- The frequencies are grouped in two's and the highest total is encircled.
- Leaving the first frequency, the remaining frequencies are grouped in two's and the highest total is encircled
- The frequencies are grouped in three's and the highest total is encircled.
- Leaving the first frequency, the remaining frequencies are grouped in three's and the highest total is encircled
- Leaving the first two frequencies, the remaining frequencies are grouped in three's and the highest total is encircled

After completing the grouping table, an analysis table is formed for finding the value which is repeated the highest number of times, is the value of mode.

3) **Continuous frequency distribution:** In this case of a continuous frequency distribution, a class having maximum frequency is called **modal class**. In case nearly equal concentration of frequencies is observed in two or more classes, the **grouping method** can be used to determine the modal class. Now, after determining the modal class, the exact value of the mode is obtained by using the following formula:

$$Mode = l_1 + \frac{f_1 - f_0}{2f_1 - (f_0 + f_2)}$$

Here,

l_1 = lower limit of the modal class; l_2 = upper limit of the modal class;

f_0 = the frequency of preceding modal class;

f_1 = the frequency of the modal class;

f_2 = the frequency of succeeding modal class.

Partition values

Median divides the series into two equal parts. Similarly, other measures which divide a series into a number of equal parts are defined as **partition vales**. Suppose, we have to divide a series into four equal parts, we will have three dividing values which are called **quartile**. If we have to divide a series into eight equal parts, we will have seven dividing values which are called **octiles**. Similarly, **deciles** and **percentiles** are other popular measures which divide the series into ten and hundred equal parts, respectively.

Important Formula of partition values

The procedure for calculating quartiles, octiles, deciles and percentiles is the same as that of median. The important difference is that, for calculating partition values, the items must be arranged in an ascending order only. The following formulas will clarify their composition procedure. Now, for computing partition values the data are first arranged in ascending order then:

- The rth quartile Q_r is the value of $r\left(\frac{N+1}{4}\right)^{th}$, in case of N is odd or if N is even than $r\left(\frac{N}{4}\right)^{th}$ item in the series. (r = 1,2,3).
- The rth octile O_r is the value of $r\left(\frac{N+1}{8}\right)^{th}$, in case of N is odd or if N is even than $r\left(\frac{N}{8}\right)^{th}$ item in the series. (r = 1,2,3,4,5,6,7)
- The rth decile D_r is the value of $r\left(\frac{N+1}{10}\right)^{th}$, in case of N is odd or if N is even than $r\left(\frac{N}{10}\right)^{th}$ item in the series. (r = 1,2,3,4,……,9)
- The rth percentile P_r is the value of $r\left(\frac{N+1}{100}\right)^{th}$, in case of N is odd or if N is even than $r\left(\frac{N}{100}\right)^{th}$ item in the series. (r = 1,2,3,……………98,99)

In particular,

$$Q_1 = the\,value\,of\left(\frac{N+1}{4}\right)^{th} \text{ item} \qquad \text{(When N is odd)}$$

Or

$$Q_1 = the\,value\,of\left(\frac{N}{4}\right)^{th} \text{ item} \qquad \text{(When N is odd)}$$

Similarly,

$$Q_3 = the\,value\,of\,3\left(\frac{N+1}{4}\right)^{th} \text{ item} \qquad \text{(When N is odd)}$$

Or

$$Q_3 = the\,value\,of\,3\left(\frac{N}{4}\right)^{th} \text{ item} \qquad \text{(When N is odd)}$$

$$Q_1 = the\,value\,of \left(\frac{N+1}{8}\right)^{th} \text{ item} \qquad \text{(When N is odd)}$$

Or

$$Q_1 = the\,value\,of \left(\frac{N}{8}\right)^{th} \text{ item} \qquad \text{(When N is odd)}$$

$$D_3 = the\,value\,of\; 3\left(\frac{N+1}{10}\right)^{th} \text{ item} \qquad \text{(When N is odd)}$$

Or

$$D_3 = the\,value\,of\; 3\left(\frac{N}{10}\right)^{th} \text{ item} \qquad \text{(When N is odd)}$$

$$P_{40} = the\,value\,of\; 40\left(\frac{N+1}{100}\right)^{th} \text{ item} \qquad \text{(When N is odd)}$$

Or

$$P_{40} = the\,value\,of\; 40\left(\frac{N}{100}\right)^{th} \text{ item} \qquad \text{(When N is odd)}$$

Here, $N = \Sigma f$ = sum of the frequencies

MCQ

Q.1. In quartiles, central tendency median to be measured must lie in

(a) First quartile
(b) **Second quartile**
(c) Third quartile
(d) Four quartile

Q2. Arithmetic mean is 12 and number of observations are 20 then sum of all values is

(a) 8
(b) 32
(c) **240**
(d) 1.667

Q3. Method used to compute average or central value of collected data is considered as

(a) Measures of positive variation
(b) **Measures of central tendency**
(c) Measures of negative skewness
(d) Measures of negative variation

Bold means correct answer

Q4. Mean or average used to measure central tendency is called

(a) Sample mean (b) **Arithmetic mean**

(c) Negative mean (d) Population mean

Q5. If mean of percentages, rates and ratios is to be calculated then central tendency measure which must be used in this situation is:

(a) **Weighted arithmetic mean** (b) Paired arithmetic mean

(c) Non-paired arithmetic mean (d) Square of arithmetic mean

Q6. When data is arranged, middle value in set of observations is classified as

(a) **Median** (b) Mean

(c) Variance (d) Standard deviation

Q7. If frequencies widely varies between different classes, then measure of central tendency must be used is

(a) Non-paired arithmetic mean (b) Square of arithmetic mean

(c) **Weighted arithmetic mean** (d) Paired arithmetic mean

Q8. Measure of central tendency which represents over time multiplicative effects for inflation and compound interest is considered as

(a) Deviation square mean (b) Paired mean

(c) **Geometric mean** (d) Harmonic mean

Q9. Around central value of observations, extent to which values depart from normal distribution is classified as:

(a) Negative variation (b) Positive variation

(c) **Skewness** (d) Positive trailing

Q10. In measure of central tendency, population parameter is denoted by

(a) **Greek letter μ** (b) Greek letter α

(c) Greek letter σ (d) Roman letter $\overline{x}$

Q11. Product W has per unit contribution of 8 with sold quantity of 124 units, product X has per unit contribution of 5 with sold quantity of 105 units, product Y has per unit contribution of 9 with sold quantity of 135 units, product Z has per unit contribution of 12 with sold quantity of 140 units then weighted average mean is

(a) 11.75 (b) 10.75

(c) 9.75 (d) **8.75**

Q12. Per day wage of 15 employees of different departments is as 620, 640, 750, 850, 650, 720, 730, 785, 630, 740, 900, 880, 780, 690, 850 then value of $\overline{x}$ is

(a) 647.67 (b) 947.67

(c) 847.67 (d) **747.67**

Q13. Number of observations are 24 and value of $\overline{x}$ is 28 then sum of all values is

(a) 24 (b) 28

(c) 52 (d) **672**

Q14. Service time (in minutes) at airport ticket counter is as 4.5, 5.5, 6, 7, 8, 8.5, 4, 3, 3.5, 2.5, 3.8 then median of data is

(a) 3.8 (b) 4.5

(c) 4 (d) **4.75**

Q15. In a hospital in Delhi, number of heart patients who visited cardiologist in 17 days are 10, 16, 16, 18, 17, 19, 14, 18, 22, 20 then arithmetic mean of data is

(a) **10** (b) 14

(c) 16 (d) 15.5

Q16. In statistics out of 100, marks of 21 students in final exams are as 90, 95, 95, 94, 90, 85, 84, 83, 85, 81, 92, 93, 82, 78, 79, 81, 80, 82, 85, 76, 85 then mode of data is

(a) **85** (b) 81

(c) 90 (d) 95

Q17. Type of central tendency measures which divides data set into 100 equal parts is classified as

(a) Quartiles (b) Deciles

(c) **Percentiles** (d) Multiple pile of data

Q18. Sample stability and ability to be easily understandable are requirements to measure

(a) positive tendency (b) population tendency

(c) **central tendency** (d) sample tendency

Q19. Distribution in which values of median, mean and mode are not equal is considered as

(a) Experimental distribution (b) **Asymmetrical distribution**

(c) Symmetrical distribution (d) Exploratory distribution

Q20. If value of three measures of central tendencies median, mean and mode then distribution is considered as

(a) Multimodal (b) Tri-modal

(c) **Uni-modal** (d) Bi-modal

Q21. If central tendency is found by using sample data from population then this is classified as

(a) Tendency statistic (b) Average statistic

(c) **Sample statistic** (d) Population statistic

Q22. Value of Σfd is 165, A= 25, and width of class interval is 10, arithmetic mean is 145 then number of observations are:

(a) 35 (b) **36**

(c) 34 (d) 32

Q23. Σxw is divided by Σw is used to calculate

(a) **Weighted mean** (b) Geometric mean

(c) Harmonic mean (d) Median

Q24. Type of central tendency measures which divides data set into four equal parts is

(a) **Quartiles** (b) Deciles

(c) Percentiles (d) Median

Q25. If value of mode is 14 and value of arithmetic mean is 5 then value of median is

(a) 12 (b) 18

(c) **8** (d) 14

Q26. Properties such as variation, central tendency and shape of distribution of frequency are basis to extract information from:

(a) Extended measures (b) Skewed measure

(c) Ordinal measures (d) **Descriptive measures**

Q27. Numerical value which shows tendency around central value of cluster is classified as

(a) **Central tendency** (b) Cluster tendency

(c) Group tendency (d) Numerical tendency

Q28. Concept used in calculation of index numbers and where smaller observations must be taken into consideration is called:

(a) Arithmetic mean (b) Weighted mean

(c) **Geometric mean** (d) Harmonic mean

Q29. Distribution which has outliers with relatively lower values is considered as

(a) Experimentally skewed (b) Exploratory skewed

(c) Positively skewed (d) **Negatively skewed**

Q30. Calculation of average which is calculated by pooling data together from different data sets is classified as

(a) **Geometric mean** (b) Harmonic mean

(c) Deviation square mean (d) Paired mean

Q31. In measure of central tendency, sample statistic is denoted by

(a) Greek letter μ (b) Greek letter α

(c) Greek letter σ (d) **Roman letter $\overline{x}$**

Q32. For individual observations, reciprocal of arithmetic mean is called

(a) Geometric mean (b) **Harmonic mean**

(c) Arithmetic mean (d) Median

Q33. Median, mode, deciles and percentiles are all considered as measures of

(a) Mathematical averages (b) Population averages

(c) Sample averages (d) **Averages of position**

Q34. In two units of company, employees in unit one are 650 and monthly salary is Rs.2750/-, employees in unit two are 700 and monthly salary is Rs. 2500/- then combined arithmetic mean is

(a) **Rs. 2,620** (b) Rs. 2,520

(c) Rs. 2,420 (d) Rs. 2,320

Q35. If most repeated observations recorded are outliers of data then mode is considered as

(a) Intended measure (b) Percentage measure

(c) Best measure (d) **Poor measure**

Q36. Number of observations are 30 and value of arithmetic mean is 15 then sum of all values is

(a) 15 (b) **450**

(c) 200 (d) 45

Q37. Value of Σfx is 180, A= 22, and width of class interval is 5, arithmetic mean is 120 then number of observations are:

(a) 59 (b) **30**

(c) 39.5 (d) 49.5

Q38. Value of Σfx is 300, A= 35, number of observations are 15 and width of class interval is 5 then arithmetic mean is:

(a) **135** (b) 145

(c) 150 (d) 235

Q39. Considering probability distribution, if mode is greater than median then distribution is classified as

(a) Variable model (b) **Right skewed**

(c) Left skewed (d) Constant model

Q40. Types of descriptive measures includes

(a) Measures of skewness (b) Measures of dispersion

(c) Measures of central tendency (d) **All of above**

Q41. According to percentiles, median to be measured must lie in

(a) 80th (b) 40th

(c) **50th** (d) 100th

Q42. Frequency distribution whose most values are dispersed to left or right of mode is classified as

(a) **Skewed** (b) Explored

(c) Bimodal (d) Unimodal

Q43. Harmonic mean, arithmetic mean and geometric mean are all considered as

(a) **Mathematical averages** (b) Population averages

(c) Sample averages (d) Extended measures

Q44. Sum of values of data is divided by total number of values is used to calculate

(a) **Arithmetic mean** (b) Weighted mean

(c) Geometric mean (d) Harmonic mean

Q45. Mode is equal to

(a) 3(mean - median) (b) **3 median – 2 mean**

(c) 3 mean – 2 median (d) 3(median - mean)

Q46. If all the observations in the series are same, than the relationship of arithmetic mean, geometric mean and harmonic mean will have:

(a) AM = GM > HM (b) AM > GM = HM

(c) **AM = GM = HM** (d) AM < GM < HM

Q47. If mean is 11 and median is 13 then value of mode is

(a) 15 (b) 13

(c) 11 (d) **17**

Q48. If central tendency is found by using whole population as input data then this is classified as

(a) Sample statistic (b) Population statistic

(c) Population tendency (d) **Population parameters**

Q49. Criteria of inferential statistics which considers sum of squared deviations is classified as

(a) Central squares criterion (b) Outliers square criterion

(c) Multiple squares criterion (d) **Least squares criterion**

Q50. In a negative skewed distribution, order of mean, median and mode is as

(a) mean<median>mode (b) mean>median>mode

(c) **mean<median<mode** (d) mean>median<mode

Q51. In a positive skewed distribution, order of mean, median and mode is as

(a) mean<median>mode (b) **mean>median>mode**

(c) mean<median<mode (d) mean>median<mode

Q52. Measure which describes detailed characteristic of whole data set is classified as

(a) **Average or central value** (b) Positive skewed value
(c) Negative skewed value (d) Positive extended value

Q53. Extent to which values are dispersed around central observation is considered as

(a) Trailing (b) **Variation**
(c) Extension (d) Centralized valuation

Q54. In a moderately asymmetrical distribution:

(a) **A.M>G.M>H.M** (b) A.M>G.M<H.M
(c) A.M<G.M<H.M (d) A.M<G.M>H.M

Q55. Mode is best measure of tendency if analysis is

(a) **Descriptive** (b) Exploratory
(c) Experimental (d) Set of deciles

Q56. In deciles, central tendency median to be measured must lie in

(a) Fourth deciles (b) Seventh deciles
(c) Sixth deciles (d) **Fifth deciles**

Q57. Type of central tendency measures which divides data set into ten equal parts is classified as

(a) Percentiles (b) Multiple pile of data
(c) Quartiles (d) **Deciles**

Q58. In statistics out of 100, marks of 21 students in final exams are as 90, 95, 95, 94, 90, 85, 84, 83, 85, 82, 92, 93, 82, 78, 79, 81, 80, 82, 85, 82, 85 then mode of data is

(a) 81 and 90 (b) 95 and 90
(c) 90 and 94 (d) **82 and 85**

Q59. Measure of central tendency which is calculated by considering most frequent occurring value as central value is classified as

(a) Central mode (b) **Mode**
(c) Frequent value (d) Percent value

Q60. Number of observations are 10 and value of arithmetic mean is 19 then sum of all values is

(a) **190** (b) 30

(c) 29 (d) 10

Q61. In arithmetic mean, sum of deviations of all recorded observations must always be

(a) Two (b) Minus one

(c) One (d) **Zero**

Q62. Distribution whose outliers are lower values is considered as

(a) Variable model (b) Right skewed

(c) **Left skewed** (d) Constant model

Q63. Arithmetic mean is 25 and sum of all observations is 350 then number of observations is:

(a) 25 (b) 70

(c) **14** (d) 75

Q64. Considering all observations of arithmetic mean, sum of squares of deviations must be less than

(a) Sum of multiples of other quantity

(b) Sum of deviations from other quantity

(c) **Sum of squares from other quantity**

(d) Cumulative frequency from other quantity

Q65. In a set of observations, unusual lower and higher values are called

(a) **Outliers** (b) Free liners

(c) Central liners (d) Median liner

Q66. The algebraic sum of the deviations from mean is:

(a) Least (b) **Zero**

(c) Maximum (d) One

Q67. In a batch of 25 students, 6 have failed. The marks of the successful candidates were 57, 52, 65, 63, 75, 83, 56, 88, 91, 72, 50, 65, 77, 82, 73, 64, 82, 57, 80. Find the median marks:

(a) 56 (b) **64**

(c) 74 (d) 77

Q68. In a symmetric distribution:

(a) **(Median – Q_1) = (Q_3 – Median)**

(b) (Median – Q_1) < (Q_3 – Median)

(c) (Median – Q_1) > (Q_3 – Median)

(d) None of these

Q69. If an observation in a series is zero, then its G.M. will be:

(a) Positive (b) Negative

(c) **Zero** (d) Infinite

Q70. If an observation in a series is negative, then its G.M. will be:

(a) Positive (b) Negative

(c) Zero (d) **Infinite**

Q71. 80th percentile is equal to:

(a) Q_1 (b) Q_3

(c) **8th decile** (d) Q_2

Q72. The following information- Mean = 45, Median = 48 and Mode = 63 describe a:

(a) Positive skewed distribution

(b) **Negative skewed distribution**

(c) Symmetrical distribution

(d) None of these

Q73. The following information- Mean = 63, Median = 48 and Mode = 45 describe a:

(a) **Positive skewed distribution** (b) Negative skewed distribution

(c) Symmetrical distribution (d) None of these

Q74. The following information- Mean = 63, Median = 63 and Mode = 63 describe a:

(a) Positive skewed distribution (b) Negative skewed distribution

(c) **Symmetrical distribution** (d) None of these

Q75. In a moderately asymmetrical distribution, the mean and median are 20 and 22, respectively. Then, the value of mode will be:

(a) 08 (b) 18

(c) **28** (d) None of these

Q76. The most stable measure of central tendency is:

(a) **A.M.** (b) H.M.

(c) Median (d) Mode

Q77. For a symmetrical distribution, Q_1 and Q_3 are 20 and 40, respectively. The value of 50th percentile will be:

(a) 20 (b) **30**

(c) 40 (d) 50

Q78. For a symmetrical distribution, P_{25} and P_{75} are 30 and 50, respectively. The value of median will be:

(a) 20 (b) 30

(c) **40** (d) 50

Q79. The median of a series is 10. Two additional observations, 8 and 20 are added to the series. The median of the new series was:

(a) 7 (b) 9

(c) **10** (d) 20

Q80. The sum of the deviations of individual observations is zero from:

(a) Geometric mean (b) Median

(c) Mode (d) **None of these**

Q81. Which average is affected most by extreme observations?

(a) **Geometric mean** (b) Median

(c) Mode (d) Harmonic mean

Q82. Which average is not affected by extreme observations?

(a) Geometric mean (b) **Median**

(c) Arithmetic mean (d) Harmonic mean

Q83. Which of the following is the most unstable average?

(a) Arithmetic mean (b) Geometric mean

(c) Median (d) **Mode**

Q84. The positional measure of central tendency is:

(a) Arithmetic mean (b) Geometric mean
(c) **Median** (d) Mode

Q85. The geometric mean of two numbers 4 and 16 shall be:

(a) 6 (b) **8**
(c) 10 (d) 12

Q86. Mode is a:

(a) **Measure of Central tendency** (b) Measure of Dispersion
(c) Test of significance (d) None of these

Q87. The median of 10, 6, 8, 11 and 12 shall be:

(a) 8 (b) 9
(c) **10** (d) 11

Q88. The median of 10, 6, 8 and 12 shall be:

(a) 8 (b) **9**
(c) 10 (d) 11

Q89. As a general rule, the _______ is the best measure of central tendency because it is more precise.

(a) **Mean** (b) Median
(c) Mode (d) Range

Q90. The median is ______.

(a) **The middle point** (b) The highest number
(c) The average (d) Affected by extreme scores

Q91. If the grouped data has open end classes, one can be calculate:

(a) Arithmetic mean (b) Geometric mean
(c) Harmonic mean (d) **Median**

Q92. If the grouped data has open end classes, one cannot calculate:

(a) Median (b) Percentile
(c) **Mean** (d) Quartile

Q93. The _______ is often the preferred measure of central tendency if the data are severely skewed.

(a) Mean (b) **Median**

(c) Mode (d) None of these

Q94. The ______ is the value you calculate when you want the arithmetic average.

(a) **Mean** (b) Median

(c) Mode (d) All of the above

Q95. The most frequently occurring number in a set of values is called the ____.

(a) Mean (b) Median

(c) **Mode** (d) None of these

Q96. The sum of the deviations of individual observations is zero from:

(a) Geometric mean (b) Median

(c) Mode (d) **None of these**

Q97. If an observation in a series is zero, then its G.M. will be:

(a) **zero** (b) one

(c) positive (d) negative

Q98. The arithmetic mean of 9 observations is 100 and that of 6 is 80, the combined mean of all the 15 observations will be:

(a) 80 (b) 90

(c) **92** (d) 100

Q99. In case of weighted mean, the accuracy or utility of the mean:

(a) **Increases** (b) Decreases

(c) Unaffected (d) None of these

Q100.In certain situations, weighted mean and un-weighted mean can:

(a) Never be equal (b) **Be equal**

(c) Both (a) and (b) (d) Neither (a) nor (b)

Q101.The number of partition values in case of quartiles is:

(a) 1 (b) 2

(c) **3** (d) 4

Q102.The number of partition values in case of decile is:

(a) 5 (b) 7

(c) **9** (d) 10

Q103.The number of partition values in case of percentile is:

(a) 10 (b) 50

(c) **99** (d) 100

Q104.The first quartile divides a frequency distribution in the ratio:

(a) 1 : 4 (b) 4 : 1

(c) **1 : 3** (d) 3 : 1

Q105.The middle value of an ordered series is called:

(a) 50 percentile (b) 5th decile

(c) 2nd quartile (d) **All of the above**

Q106.If modal value is not clear in a distribution, it can be ascertained by the method of:

(a) **Grouping** (b) Trial and error

(c) Guessing (d) Summarizing

Q107.If the two observations are -4 and 4, then their harmonic mean is:

(a) 0 (b) 4

(c) -4 (d) **∞ (Infinite)**

Q108.If the two observations are -10 and 10, then their geometric mean is:

(a) 0 (b) 10

(c) -10 (d) **None of the above**

Q109.If the two observations are -10 and 10, then their arithmetic mean is:

(a) **0** (b) 10

(c) -10 (d) None of the above

Q110.If the sum of N observations is 330 and their mean is 22, then the value of N is:

(a) 5 (b) 11

(c) **15** (d) 20

Q111.The mean of six observations is 7. A new observation 14 is added. The mean of seven observations is:

(a) 7 (b) **8**

(c) 9 (d) 10

Q112.The mean of six observations is 7. Two new observations 12 and 18 are added. The mean of eight observations is:

(a) 7 (b) 8

(c) **9** (d) 10

Q113.The arithmetic mean of two observations is 6.5 and their geometric mean is 6. The two numbers are:

(a) 6, 9 (b) 5, 9

(c) **4, 9** (d) 6, 7

Q114.The arithmetic mean of two observations is 6.5 and their geometric mean is 6. Then the harmonic mean is:

(a) **4** (b) 5

(c) 6 (d) 6.5

Q115.The average of 6 numbers 7, 9, 12, x, 5, 10 is 9. The missing number x is:

(a) 9 (b) 10

(c) **11** (d) 12

Q116.Extreme values have no effect on:

(a) A.M. (b) G.M.

(c) H.M. (d) **Median**

Q117.Harmonic mean is better than other means if

(a) **Speed or rates** (b) Heights and lengths

(c) Ratio or proportions (d) None of these

Q118.If a constant 5 is added to each observation of a set, the mean is:

(a) **Increased by 5** (b) Decreased by 5

(c) 5 times the original mean (d) Not affected

Q119.If a constant 5 is subtracted from each observation of a set, the mean of the set is:

(a) Increased by 5 (b) **Decreased by 5**

(c) 5 times the original mean (d) Not affected

Q120.If each observation of a set is multiplied by 5, the mean of the new set of observations:

(a) Increased by 5 (b) Decreased by 5

(c) **5 times the original mean** (d) Not affected

Q121.If we plot the more than type and less than type frequency distributions of the same set of data, their graphs intersect at the point which is known as:

(a) A.M. (b) G.M.

(c) H.M. (d) **Median**

Q122.Which mean is most affected by extreme values?

(a) **A.M.** (b) G.M.

(c) H.M. (d) None of these

Q123.The measure of central value which cannot be calculated with open end classes in case of grouped data is:

(a) **A.M.** (b) Median

(c) Third quartile (d) 75th percentile

Q124.Harmonic mean gives more weightage to:

(a) Large values (b) **Small values**

(c) Positive values (d) Negative values

Q125.Geometric mean gives more weightage to:

(a) Large values (b) **Small values**

(c) Positive values (d) Negative values

Q126.For further algebraic treatment geometric mean is:

(a) Suitable (b) **Not suitable**

(c) Sometimes suitable (d) None of these

Q127. For further algebraic treatment harmonic mean is:

(a) Suitable (b) **Not suitable**

(c) Sometimes suitable (d) None of these

Q128.For further algebraic treatment arithmetic mean is:

(a) **Suitable** (b) Not suitable

(c) Sometimes suitable (d) None of these

Q129.The partition value, which divides a series into two equal parts is known as:

(a) First quartile (b) **Second quartile**

(c) Third quartile (d) None of these

Q130.Sum of the absolute deviations about median is:

(a) Zero (b) One

(c) **Minimum** (d) Maximum

Q131.The suitable measure of central tendency for qualitative data is:

(a) A.M. (b) G.M.

(c) H.M. (d) **Median**

Q132.The value of term is formula of:

(a) **First quartile** (b) Second quartile

(c) Third quartile (d) Mode

Q133.Middle observation in an arranged set of data is called:

(a) A.M. (b) G.M.

(c) **Median** (d) Mode

Q134.The nth root of the product of values $x_1, x_2, x_3, \ldots\ldots, x_n$ is called:

(a) A.M. (b) **G.M.**

(c) Median (d) Mode

Q135.The value of $\left\{\frac{3(n+1)}{4}\right\}^{th}$ term is formula of:

(a) First quartile (b) Second quartile

(c) **Third quartile** (d) Mode

Q136.Third quartile is also known as:

(a) Lower quartile (b) **Upper quartile**

(c) Geometric mean (d) Median

Q137.By adding all numbers in set together and then total is divided by number of scores in that set, we obtained:

(a) **A.M.** (b) G.M.

(c) Median (d) Mode

Q138.Most frequent observation in a data set is called:

(a) A.M. (b) G.M.

(c) Median (d) **Mode**

Q139.If >10, 10-20, 20-30, 30-40, < 40 are the class intervals of the data, one can be calculate:

(a) Arithmetic mean (b) Geometric mean

(c) Harmonic mean (d) **Median**

Q140.Which of the measure of central tendency is not affected by extreme values?

(a) Mode (b) Median

(c) Sixth decile (d) **All of them**

Q141.If an observation in a series is negative, then its Geometrical mean will be:

(a) Negative (b) Positive

(c) Finite (d) **Infinite**

Q142.The following scores were obtained by Twelve Hockey players in a goal-shoot competition: 5, 3, 6, 8, 7, 8, 3, 11, 6, 3, 2, 4. The modal score was:

(a) 2 (b) **3**

(c) 8 (d) 11

Q143.The following scores were obtained by Twelve Hockey players in a goal-shoot competition: 5, 3, 6, 8, 7, 8, 3, 11, 6, 3, 2, 4. The median score was:

(a) 2 (b) 3

(c) **5.5** (d) 11

Q144.The mean of ten numbers is 58. If one of the numbers is 40, what is the mean of the other nine?

(a) 18 (b) **60**

(c) 162 (d) 540

Q145.The mean of 11 numbers is 7. One of the numbers, 13, is deleted. What is the mean of the remaining 10 numbers?

(a) 7.7 **(b) 6.4**

(c) 6.0 (d) 5.8

Q146.Which measure of central tendency is derived from the most common value?

(a) Mean (b) Median

(c) **Mode** (d) None of these

Q147.Which measure of central tendency is obtained using the middle score when all scores are organized in numerical order?

(a) Mean (b) **Median**

(c) Mode (d) None of these

Q148.Which measure of central tendency is obtained by calculating the sum of values and dividing this figure by the number of values there are in the data set?

(a) **Mean** (b) Median

(c) Mode (d) None of these

Q149.Which measure is the most unreliable indicator of central tendency if data are skewed?

(a) **Mean** (b) Median

(c) Mode (d) None of these

Q150.The arithmetic mean is highly affected by:

(a) Moderate values (b) Extremely small values

(c) Odd values (d) **Extremely large values**

Q151.Change of origin and scale is used for calculation of the:

(a) **Arithmetic mean** (b) Geometric mean

(c) Weighted mean (d) Lower and upper quartiles

Q152.Any measure indicating the centre of a set of data, arranged in an increasing or decreasing order of magnitude, is called a measure of:

(a) Skewness (b) Symmetry

(c) **Central tendency** (d) Dispersion

Q153.The sample mean $\bar{X}$ is calculated by the formula:

(a) $\frac{\sum fx}{\sum f}$

(b) $a+\frac{\sum fd}{\sum f}$

(c) $a+\frac{\sum fu}{\sum f}\times h$

(d) **All of the above**

Q154.Which of the following statements is always true?

(a) The mean has an effect on extreme scores

(b) The median has an effect on extreme scores

(c) **Extreme scores have an effect on the mean**

(d) Extreme scores have an effect on the median

Q155.The elimination of extreme scores at the bottom of the set has the effect of:

(a) Lowering the mean

(b) **Raising the mean**

(c) No effect

(d) None of the above

Q156.The elimination of extreme scores at the top of the set has the effect of:

(a) **Lowering the mean**

(b) Raising the mean

(c) No effect

(d) Difficult to tell

Q157.The sum of the squares of the deviations about mean is:

(a) Zero

(b) Maximum

(c) **Minimum**

(d) All of the above

Q158.If $\sum_{i=1}^{10}(x_i-50)=100$, then sample mean $\bar{X}$ will be:

(a) 10

(b) 50

(c) 60

(d) 100

Q159.If $x_1, x_2, ..., x_n$, be n observations having arithmetic mean and if $Y = 4X \pm 2$, then $\bar{Y}$ is equal to

(a) 4X

(b) 4

(c) 4 ± 2

(d) **$4\bar{X} \pm 2$**

Q160.If $\bar{X}$ =100 and Y=2X – 200, then mean of Y values will be:

(a) 0 (b) 2

(c) 100 (d) 200

Q161.Ten families have an average of 2 boys. How many boys do they have together?

(a) 2 (b) 10

(c) 12 **(d) 20**

Q162.If the arithmetic mean of the two numbers X_1 and X_2 is 5 if X_1=3, then X_2 is:

(a) 3 (b) 5

(c) 7 (d) 10

Q163.The mean of 10 observations is 10. All the observations are increased by 10%. The mean of increased observations will be:

(a) 10 (b) 1.1

(c) 10.1 **(d) 11**

Q164.The sample mean of first n natural numbers is:

(a) n(n+ 1) / 2 (b) (n+ 1) / 2

(c) n/2 (d) **(n+ 1) / 2**

Q165.The mean of first 2n natural numbers is:

(a) $\frac{n(n+1)}{2}$ (b) $\frac{n(2n+1)}{2}$

(c) $\frac{2\boldsymbol{n}+1}{2}$ (d) $\frac{(n+1)}{2}$

Q166.If $\bar{X}_1, \bar{X}_2, \ldots, \bar{X}_k$ be the arithmetic means of k distributions with respective frequencies n_1, n_2, ..., n_k, then the mean of the whole distribution $\bar{X}_c$ is given by:

(a) $\sum n\bar{X} / \sum n$ (b) $\sum n / \sum \bar{X}$

(c) $\boldsymbol{\sum n\bar{X} / \sum n}$ (d) $\sum(n+\bar{X}) / \sum n$

Q167.The first step in calculating the median of a discrete variable is to determine the:

(a) Cumulative frequencies (b) Relative weights

(c) Relative frequencies **(d) Array**

Q168. The suitable average for qualitative data is:

(a) Mean | **(b) Median**
(c) Mode | (d) Geometric mean

Q169. Extreme scores will have the following effect on the median of an examination:

(a) They may have no effect on it | (b) They may tend to raise it
(c) They may tend to lower it | (d) None of the above

Q170. If the smallest observation in a data is decreased, the average which is not affected is:

(a) Mode | **(b) Median**
(c) Mean | (d) Harmonic mean

Q171. The values of the variate that divide a set of data into four equal parts after arranging the observations in ascending order of magnitude are called:

(a) Quartiles | (b) Deciles
(c) Percentiles | (d) Difficult to tell

Q172. If in a discrete series 75% values are less than 30, then:

(a) $Q_3 < 75$ | (b) $Q_3 < 30$
(c) $Q_3 = 30$ | (d) $Q_3 > 30$

Q173. If in a discrete series 25% values are greater than 75, then:

(a) $Q_1 > 75$ | (b) $Q_1 = 75$
(c) $Q_3 = 75$ | (d) $Q_3 > 75$

Q174. If in a discrete series 40% values are less than 40, then :

(a) $D_4 \neq 40$ | (b) $D_4 < 40$
(c) $D_4 > 40$ | **(d) $D_4 = 40$**

Q175. If in a discrete series 15% values are greater than 40, then:

(a) $P_{15} = 70$ | (b) $P_{85} = 15$
(c) $P_{85} = 70$ | (d) $P_{70} = 70$

Q176. If in a discrete series 50% values are less than 50, then:

(a) $Q_2 = 50$ | (b) $D_5 = 50$
(c) $P_{50} = 50$ | **(d) All of the above**

Q177.The mode or model value of the distribution is that value of the variate for which frequency is:

(a) Minimum
(b) **Maximum**
(c) Odd number
(d) Even number

Q178.Suitable average for averaging the shoe sizes for children is:

(a) Mean
(b) Mode
(c) Median
(d) Geometric mean

Q179.Extreme scores on an examination have the following effect on the mode:

(a) They tend to raise it
(b) They tend to lower it
(c) They have no effect on it
(d) Difficult to tell

Q180.Which of the following average cannot be calculated for the observations 2, 2, 4, 4, 6, 6, 8, 8, 10, 10 ?

(a) Mean
(b) Median
(c) Mode
(d) All of the above

Q181.A distribution with one mode is called:

(a) **Unimodal**
(b) Bimodal
(c) Trimodal
(d) Multimodal

Q182.A distribution with two modes is called:

(a) Unimodal
(b) Bimodal
(c) Trimodal
(d) Multimodal

Q183.A distribution with more than three modes is called:

(a) Unimodel
(b) Bimodal
(c) Trimodal
(d) Multimodal

Q184.The model letter of the word "STATISTICS" is:

(a) S
(b) T
(c) Both S and I
(d) Both S and T

Q185.If the values of mean, median and mode coincide in a unimodel distribution, then the distribution will be:

(a) Skewed to the left
(b) Skewed to the right
(c) Multimodal
(d) Symmetrical

Q186.Taking the relevant root of the product of all non-zero and positive values are called:

(a) Arithmetic mean **(b) Geometric mean**

(c) Harmonic mean (d) Combined mean

Q187.The best average in percentage rates and ratios is:

(a) Arithmetic mean (b) Lower and upper quartiles

(c) **Geometric mean** (d) Harmonic mean

Q188.The suitable average for computing average percentage increase in population is:

(a) **Geometric mean** (b) Harmonic mean

(c) Arithmetic mean (d) Mode

Q189.If 10% is added to each value of variable, the geometric mean of new variable is added by:

(a) 10 (b) 1/100

(c) 10% (d) 1.1

Q190.If each observation of a variable X is increased by 20%, then geometric mean is also increased by:

(a) 20 (b) 1/20

(c) 20% (d) 100%

Q191.If any value in a series is negative, then we cannot calculate the:

(a) Mean (d) Median

(c) **Geometric mean** (d) Mode

Q192.If any value in a series is zero, then we cannot calculate the:

(a) Mean (d) Median

(c) **Geometric mean** (d) Mode

Q193.Geometric mean is suitable when the values are given as:

(a) Proportions (b) Ratios

(c) Percentage rates **(d) All of the above**

Q194.If the two observations are a = 2 and b = -2, then their geometric mean will be:

(a) Zero (b) Infinity

(c) Impossible (d) Negative

Q195.If the two observations are a = 0 and b = 4, then their geometric mean will be:

(a) Zero (b) Infinity

(c) Impossible (d) Negative

Q196.Geometric mean of -4, -2 and 8 is:

(a) 4 (b) 0

(c) -2 **(d) Impossible**

Q197.The ratio among the number of items and the sum of reciprocals of items is called:

(a) Arithmetic mean (b) Geometric mean

(c) **Harmonic mean** (d) Mode

Q198.The appropriate average for calculating the average speed of a journey is:

(a) Median (b) Arithmetic mean

(c) Mode (d) **Harmonic mean**

Q199.Harmonic mean gives less weightage to:

(a) Small values (b) **Large values**

(c) Positive values (d) Negative values

Q200.The harmonic mean of the values 5, 9, 11, 0, 17, 13 is:

(a) 9.5 (b) 6.2

(c) 0 **(d) Impossible**

Q201.If the harmonic mean of the two numbers X_1 and X_2 is 6.4 if X_2=16, then X_1 is:

(a) 4 (b) 10

(c) 16 (d) 20

Q202. If a = 5 and b = -5, then their harmonic mean is:

(a) -5 (b) 5

(c) 0 **(d) ∞**

Q203. For an open-end frequency distribution, it is not possible to find:

(a) Arithmetic mean (b) Geometric mean

(c) Harmonic mean **(d) All of the above**

Q204. The geometric mean of a set of positive numbers $X_1, X_2, X_3, \ldots, X_n$ is less than or equal to their arithmetic mean but is greater than or equal to their:

(a) Harmonic mean (b) Median

(c) Mode (d) Lower and upper quartiles

Q205. Which pair of averages cannot be calculated when one of numbers in the series is zero?

(a) Geometric mean and Median

(b) Harmonic mean and Mode

(c) Arithmetic mean and Weighted mean

(d) **Geometric mean and Harmonic mean**

Q206. In a given data the average which has the least value is:

(a) Arithmetic mean (b) Median

(c) Harmonic mean (d) Geometric mean

Q207. If all the values in a series are same, then:

(a) A.M = G.M = H.M (b) A.M ≠ G.M ≠ H.M

(c) A.M > G.M > H.M (d) A.M < G.M < H.M

Q208. The averages are affected by change of:

(a) Origin (b) Scale

(c) Both (a) and (b) (d) None of the above

Q209. Geometric mean and harmonic mean for the values 3, -11, 0, 63, -14, 100 are:

(a) 0 and 3 (b) 3 and -3

(c) 0 and 0 **(d) Impossible**

Q210.Since the mode is the most frequently occurring data value, it:

(a) Can never be larger than the mean

(b) Is always larger than the median

(c) Is always larger than the mean

(d) **None of these**

Q211.In a moderately skewed distribution, mean is equal to:

(a) **(3Median - Mode) / 2** (b) (2Mean + Mode) / 3

(c) 3Median – 2Mean (d) 3Median – Mode

Q212.In a moderately asymmetrical distribution, the value of median is given by:

(a) 3Median + 2Mean (b) 2Mean + Mode

(c) **(2Mean + Mode) / 3** (d) (3Median - Mode) / 2

Q213.If the arithmetic mean and harmonic mean of two positive numbers are 4 and 16, then their geometric mean will be:

(a) 4 **(b) 8**

(c) 16 (d) 64

Q214.The arithmetic mean and geometric mean of two observations are 4 and 8 respectively, then harmonic mean of these two observations is:

(a) 4 (b) 8

(c) **16** (d) 32

Q215.The geometric mean and harmonic mean of two values are. 8 and 16 respectively, then arithmetic mean of values is:

(a) 4 (b) 16

(c) 24 (d) 128

Q216.If the values in a series are 2, 2, 2, 2, then we cannot calculate:

(a) Geometric mean (b) Harmonic mean

(c) Arithmetic mean **(d) Mode**

Q217.Which of the following measures of central tendency will always change if a single value in the data changes?

(a) **Mean** (b) Median

(c) Mode (d) All of them

Q218. If a positively skewed distribution has a median of 50, which of the following statement is true?

(a) **Mean is greater than 50** (b) Mode is greater than 50

(c) Mean is less than 50 (d) None of these

Q219. In a frequency distribution the last cumulative frequency is 300, Median shall lie in:

(a) 125th item (b) **150th item**

(c) 175th item (d) 300th item

Q220. Suppose a researcher is concerned with a nominal scale that identifies users versus nonusers of bank credit cards. The measure of central tendency appropriate to this scale is the:

(a) A.M. (b) Median

(c) **Mode** (d) G.M.

Q221. In a grouped frequency distribution, the mid-values of the classes are used to measure which of the following central tendency?

(a) **Mean** (b) Median

(c) Mode (d) All of them

Q222. In the formula, $\overline{x} = a + \frac{\sum f_i d_i}{\sum f_i}$ for finding of the grouped data, d_i are deviations from a assumed mean of:

(a) Frequencies of the classes (b) **Mid-point of the classes**

(c) Lower limit of the classes (d) Upper limit of the classes

Q223. Construction of a cumulative frequency distribution table is useful in determining the:

(a) Mean (b) **Median**

(c) Mode (d) All of them

Q224. Consider the following frequency distribution:

Classes	6 - 11	12 - 17	18 - 23	24 - 29	30 - 35
Frequency	7	10	15	8	5

The upper limit of the median class is:

(a) **23.5** (b) 23

(c) 18 (d) 18.5

Q225.For the following frequency distribution:

Classes	0 - 5	5 - 10	10 - 15	15 - 20	20 - 25	25 - 30
Frequency	7	10	12	18	5	2

The sum of lower limit of the median class and modal class is:

(a) 10 (b) 15

(c) 20 (d) **25**

Q226.If mean and mode of the data are 6 and 12 respectively, its median will be:

(a) 6 (b) **8**

(c) 10 (d) 12

Q227.If the mode of the following data is 7, than the value of X in the data set 7, 8, 6, 3, 7, 6, 10, 1, 6, 5X + 2, 7, 14, 8 is:

(a) **1** (b) 2

(c) 3 (d) 4

Q228.If average of three numbers be 15. If two of the numbers are 10 and 26, what is the remaining number?

(a) – 6 (b) **9**

(c) 10 (d) 26

Q229.The mean of 5 numbers is 41, If one number is excluded, their mean become 32. The excluded number is:

(a) 32 (b) 41

(c) **77** (d) 87

Q230.If k(3Median – Mode) = Mean, then k is:

(a) **1/2** (b) 1/3

(c) 2 (d) 3

Q231.Find the value ‘k’ if the mean of the set of the numbers 8, 13, 17, k, 5 is given as 10.

(a) 5 (b) **7**

(c) 8 (d) 10

Q232.Find the median of the series of all the even terms from 2 to 298.

(a) 120 (b) 130

(c) 140 (d) **150**

Hint: Here, the sequence is 2, 4, 6, ……, 298. Therefore, the common difference is 2 (Which is constant)

The given sequence is AP. Therefore, Median = (First term + Last term)/2 = (2+298)/2 = 150

Q233. The two methods of finding mode in a discrete series are:

(a) Table method and mid-point method

(b) Grouping method and ascending method

(c) **Grouping method and inspecting method**

(d) None of these

Q234. When the values in a series do not have equal importance, we calculate the:

(a) Arithmetic mean
(b) Geometric mean
(c) Harmonic mean
(d) **Weighted mean**

Q235. To calculate the median, all the items of a series have to be arranged in a/an:

(a) Ascending order
(b) Descending order
(c) **Both (a) & (b)**
(d) None of them

Q236. The number of observations smaller than …….is the same as the number of observations larger than it:

(a) Mean
(b) **Median**
(c) Mode
(d) None of these

Q237. Which of the following is a characteristic of a mean?

(a) It is affected by extreme observations

(b) The sum of deviations about mean is zero

(c) It minimize the sum of square deviation

(d) **All of them**

Q238. Percentiles divide a series into:

(a) Two equal parts
(b) Four equal parts
(c) Ten equal parts
(d) **Hundred equal parts**

Q239. …………… divides the data into four equal parts

(a) Median
(b) **Quartile**
(c) Decile
(d) Percentile

Q240................ divides the data into ten equal parts

(a) Median (b) Quartile

(c) **Decile** (d) Percentile

Q241.In the given data:

Class Interval	5 - 15	15 - 25	25 - 35	35 - 45	45 - 55
Frequency	5	12	18	9	6

The difference of the upper limit of the median class and the lower limit of modal class is:

(a) 5 (b) **10**

(c) 25 (d) 35

Q242.For the following distribution

Marks	Less than 20	Less than 40	Less than 60	Less than 80	Less than 100
No. of students	5	17	35	44	50

The modal class is

(a) 0 – 20 (b) 20 – 40

(c) **40 – 60** (d) 60 -80

Q243.For the following distribution

Marks	0 - 10	10 - 20	20 - 30	30 - 40	40 - 50
No. of students	20	30	24	40	18

the sum of lower limits of the modal class and the median class is

(a) 20 (b) 30

(c) 40 (d) **50**

Q244.Mode is the

(a) Middle most frequent value (b) Least frequent value

(c) **Maximum frequent value** (d) None of these

6

Measure of Dispersion

Dispersion

Measure of central tendency tells about the central position of the series. They do not throw like on the formation of series because in some cases, two series may have a same measure of central tendency but differ in their composition. In order to study the formation, composition, structure and scatteredness of the items of the series, measure of second order are necessary. The measure of second order will give idea about the formation of the series. As the representation of a series on the basis of its measure of central tendency is not sufficient but when a series is represented on the basis of its measure of central tendency as well as its measure of dispersion, then it will give the complete adequate representation of the series. Therefore, it is defined as the extent or degree of which data tend to spread around an average is called the **dispersion**. It is also known as **variability** or **measure of second order**.

Characteristics of an ideal Measure of Dispersion

An ideal measure of dispersion should possess the following properties:

1) It should be rigidly defined i.e. it has a fixed and finite value
2) It should be easy to understand
3) It should be easy to calculate
4) Its calculation should be based on all the observations
5) It should be capable for further algebraic treatment
6) It should not be affected by extreme values of the observations
7) It should be least affected by fluctuations of sampling

Types of Measure of Dispersion

There are two types of measure of dispersion;

(i) Absolute measure of dispersion

(ii) Relative measure of dispersion

(i) **Absolute measure of dispersion:** It is useful for comparing variation in two or more distributions where units of measurement are the same. Such types of measures are not suitable for comparing the variability of the distributions expressed in different units of measurement. There are five measure of dispersions, namely:

- **Range**
- **Mean Deviation (M.D.)**
- **Standard Deviation (S.D.)**
- **Variance**
- **Standard Error (S.E.)**

(ii) **Relative measure of dispersion:** They are expressed as ratio or percentage or the coefficient of the absolute measure of dispersion. In this way, relative measures are pure unit less number and are generally called coefficient of dispersion. They are useful for comparing variability in two or more distributions where units of measurement may be different. There are one measure of dispersion:

- **Coefficient of variation (C.V.)**

  So, the scatteredness of a series can be measured by six different kinds of measures of dispersion:

- **Range**
- **Mean Deviation (M.D.)**
- **Standard Deviation (S.D.)**
- **Variance**
- **Standard Error (S.E.)**
- **Coefficient of variation (C.V.)**
- **Range:** The difference between the two extreme values of the series is known as **range**. Thus,

$$\text{Range} = X_{max} - X_{min}$$

Here, X_{max} = the largest observation; X_{min} = the smallest observation

In case of a **continuous frequency distribution**, the range is the difference between the lower limit of lowest class interval and upper limit of highest class interval.

Coefficient of Range: Range is an absolute measure of dispersion which is unsuitable for comparing variation in two or more distributions expressed in different units. To overcome this difficulty, a relative measure of dispersion called the **coefficient of range** is calculated by the following formula:

$$Coefficient\ of\ Range = \frac{X_{max} - X_{min}}{X_{max} + X_{min}}$$

Inter-quartile Range: Range is based on two extreme observations and it does not take into account the variation within the range. **Inter-quartile range** is the difference between the third quartile (Q_3) and the first quartile (Q_1). That means;

Inter-quartile range = $Q_3 - Q_1$

Semi- Inter quartile Range: It is defined as the average amount by which the two quartiles differ from the median. It is also known as **quartile deviation (Q.D.)**. Therefore,

$$\text{Semi} - \text{Inter quartile Range} = \frac{Q_3 - Q_1}{2}$$

Now, for comparing two or more distributions in respect of variation, the coefficient of quartile deviation is defined as:

$$\text{Coefficient of Quartile Deviation}\left(\text{Q.D.}\right) = \frac{Q_3 - Q_1}{Q_3 + Q_1}$$

• **Mean Deviation (M.D.):** It is defined as the arithmetic mean of all the deviations measures for all the items of the series from their arithmetic average by ignoring the signs of the deviation. In other words, **mean deviation** can be defined as the average distance of all the items from their arithmetic mean of all the items, or, **mean deviation** can be defined as arithmetic mean of absolute deviation from all the items from their arithmetic average, median or mode.

Important Formula of Arithmetic Mean

The mean deviation about mean, median and mode can be expressed as:

a) **Mean deviation about mean**

(i) **Individual series or ungrouped data**

$$Mean\ deviation\ about\ mean = \frac{\Sigma\left|x - \bar{x}\right|}{n}$$

Here; n = number of observations

(ii) **Discrete frequency distribution**

$$Mean\ deviation\ about\ mean = \frac{\Sigma f\left|x - \bar{x}\right|}{N}$$

Here; N = Σf = sum of the frequencies

(iii) **Continuous frequency distribution**

$$Mean\,deviation\,about\,mean = \frac{\sum f|m-\bar{x}|}{N}$$

Here; m = mid value of various classes; N = Σf = sum of the frequencies

b) **Mean deviation about median (M_d):**

(i) **Individual series or ungrouped data**

$$Mean\,deviation\,about\,median = \frac{\sum|x-M_d|}{n}$$

Here; n = number of observations

(ii) **Discrete frequency distribution**

$$Mean\,deviation\,about\,median = \frac{\sum f|x-M_d|}{N}$$

Here; N = Σf = sum of the frequencies

(iii) **Continuous frequency distribution**

$$Mean\,deviation\,about\,mode = \frac{\sum|x-M_o|}{n}$$

Here; m = mid value of various classes; N = Σf = sum of the frequencies

c) **Mean deviation about mode(M_o):**

(i) **Individual series or ungrouped data**

$$Mean\,deviation\,about\,mode = \frac{\sum|x-M_o|}{n}$$

Here; n = number of observations

(ii) **Discrete frequency distribution**

$$Mean\,deviation\,about\,mode = \frac{\sum f|x-M_o|}{N}$$

Here; N = Σf = sum of the frequencies

(iii) **Continuous frequency distribution**

$$Mean\,deviation\,about\,mode = \frac{\sum f|m-M_o|}{N}$$

Here; m = mid value of various classes; N = Σf = sum of the frequencies

• **Standard Deviation (S.D.):** It is defined as the square root of arithmetic mean of square deviations of all the items of the series measure from the arithmetic average. This concept was first introduced by **Karl Pearson**. It is denoted by σ.

Important Formula of Standard Deviation (S.D.)

1) **Individual series or ungrouped data**

(a) **Direct method**

$$Standard\ deviation(\sigma) = \sqrt{\frac{\Sigma(x-\overline{x})^2}{n}}$$

(b) **Short-cut method**

$$Standard\ deviation(\sigma) = \sqrt{\frac{\Sigma(x-a)^2 - n(a-\overline{x})^2}{n}}$$

Here, a = assumed mean; n = number of observations; $\overline{x} = \frac{\Sigma x}{n}$

2) **Discrete frequency distribution**

(a) **Direct method**

$$Standard\ deviation(\sigma) = \sqrt{\frac{\Sigma f(x-\overline{x})^2}{N}}$$

Here, N = Σf = sum of the frequencies; $\overline{x} = \frac{\Sigma fx}{N}$

(b) **Short-cut method**

$$Standard\ deviation(\sigma) = \sqrt{\frac{\Sigma f(x-a)^2 - N(a-\overline{x})^2}{N}}$$

Here, a = assumed mean; N = Σf = sum of the frequencies; $\overline{x} = \frac{\Sigma fx}{N}$

3) **Continuous frequency distribution**

(a) **Direct method**

$$Standard\ deviation(\sigma) = \sqrt{\frac{\Sigma f(x-a)^2 - N(a-\overline{x})^2}{N}}$$

Here, m = mid value of various classes; N = Σf = sum of the frequencies; $\overline{x} = \frac{\Sigma fm}{N}$

(b) **Short-cut method**

$$Standard\ deviation(\sigma) = \sqrt{\frac{\Sigma f(x-a)^2 - N(a-\overline{x})^2}{N}}$$

Here, m = mid value of various classes; N = Σf = sum of the frequencies; $\overline{x} = \frac{\sum fm}{N}$

Remark: When we derive the standard deviation, we divide the sum of square deviation by "n" but when we calculate the value of standard deviation, and then we divide the sum of square deviation by (n-1).

• **Variance:** It is the quantity obtained by squaring the standard deviation. It is denoted by σ^2.

Important Formula of Variance

1) **Individual series or ungrouped data**

a) Direct method

$$Variance(\sigma^2) = \frac{\sum(x-\overline{x})^2}{n}$$

b) **Short-cut method**

$$Variance(\sigma^2) = \frac{\sum(x-a)^2 - n(a-\overline{x})^2}{n}$$

Here, a = assumed mean; n = number of observations; $\overline{x} = \frac{\sum x}{n}$

2) **Discrete frequency distribution**

a) **Direct method**

$$Variance(\sigma^2) = \frac{\sum f(x-\overline{x})^2}{N}$$

Here, N = Σf = sum of the frequencies; $\overline{x} = \frac{\sum fx}{N}$

b) **Short-cut method**

$$Variance(\sigma^2) = \frac{\sum f(x-a)^2 - N(a-\overline{x})^2}{N}$$

Here, a = assumed mean; N = Σf = sum of the frequencies; $\overline{x} = \frac{\sum fx}{N}$

3) **Continuous frequency distribution**

a) **Direct method**

$$Variance(\sigma^2) = \frac{\sum f(m-\overline{x})^2}{N}$$

Here, m = mid value of various classes; N = Σf = sum of the frequencies;

$$\overline{x} = \frac{\sum fm}{N}$$

b) **Short-cut method**

Here, m = mid value of various classes; N = Σf = sum of the frequencies;

$$\overline{x} = \frac{\sum fm}{N}$$

$$Variance\left(\sigma^2\right) = \frac{\sum f\left(m-a\right)^2 - N\left(a-\overline{x}\right)^2}{N}$$

Remark: When we derive the standard deviation, we divide the sum of square deviation by "n" but when we calculate the value of standard deviation, and then we divide the sum of square deviation by (n-1).

- **Standard Error (S.E.):** It is defined as the standard deviation of sampling distribution of any statistics calculated from different samples of equal sizes taken from a parent population.
- **Standard error of mean (SEM):** It is the standard deviation of sampling distribution of arithmetic mean of different samples of equal size taken from one population.

The bigger the standard error means there will be less accurate statistics. In any experiment, less standard error is desirable for an optimum accuracy. Now, for less standard error (S.E.):-

(a) Number of observations should be more

(b) The observations are homogeneous, then standard deviation is less.

Important Formula of Standard Error (S.E.)

1) **Individual series or ungrouped data**

$$Standard\ Error\left(S.E.\right) = \frac{Standard\ Deviation\left(S.D.\right)}{\sqrt{n}}$$

Here, n = number of observations

2) **Discrete frequency distribution**

$$Standard\ Error\left(S.E.\right) = \frac{Standard\ Deviation\left(S.D.\right)}{\sqrt{N}}$$

Here, N = Σf = sum of the frequencies

3) **Continuous frequency distribution**

$$Standard\ Error\left(S.E.\right)=\frac{Standard\ Deviation\left(S.D.\right)}{\sqrt{N}}$$

Here, N = Σf = sum of the frequencies

• **Coefficient of Variation (C.V.):** It is defined as the ratio of standard deviation to the arithmetic mean expressed in percentage. In other words, **coefficient of variation (C.V.)** is also defined as the percentage variation in the mean when standard deviation is taken as the total variation in mean.

Important Formula of Coefficient of Variation (C.V.)

$$\text{Coefficient of Variation}\left(\text{C.V.}\right)=\frac{\text{Standard Deviation}\left(\text{S.D.}\right)}{\text{Arithmatic Mean}\left(\text{A.M.}\right)}\text{x}100$$

Remark

(i) It have special used in the case when the two sets of data belong to **different of units of measurement**, that means, it is useful for making comparison between two or more series in respect of their variability, uniformity or consistency.

(ii) It is a pure number independent of the units of measurements.

(iii) The distribution having greater **coefficient of variation (C.V.)** is considered more variable than the other, and the distribution with lesser **coefficient of variation (C.V.)** shows greater consistency, uniformity and homogeneity.

MCQ

Q.1. Statistically, the scatterness of observations in a data is called:

(a) Range
(b) Central tendency
(c) **Dispersion**
(d) Standard deviation

Q2. Positive square root of the mean of the squared deviations of some observations from their arithmetic mean is called:

(a) Mode
(b) Standard error
(c) Mean deviation
(d) **Standard deviation**

Q3. The standard deviation

(a) Is the most valuable measure of spread or dispersion

(b) Does not have the limitations of the average deviation

Bold means correct answer

(c) Does not have the drawbacks of the variance because its unit of measure is not squared, but is in the same unit of measure as the mean

(d) **All of the above**

Q4. The variance:

(a) Is a poor index of the degree of dispersion

(b) **Has a major drawback because it reflects a unit of measurement that has been squared**

(c) Is the squared root of the standard deviation

(d) Is the average deviation squared

Q5. Dispersion measures:

(a) **The scatterness of a set of observations**

(b) The concentration of a set of observations

(c) Question does not provide sufficient data

(d) None of these

Q6. A shift of origin has no impact on:

(a) S.D.

(b) M.D.

(c) Q.D.

(d) **Range**

Q7. Sum of all squared deviations is divided by total number of observations to calculate:

(a) Population deviation

(b) **Population variance**

(c) Sample deviation

(d) Sample variance

Q8. For recorded observation, ratios measured by absolute variation are considered as

(a) **Relative measures**

(b) Non-relative measures

(c) High uniform measures

(d) Low uniform measures

Q9. If arithmetic mean is multiplied to coefficient of variation then resulting value is classified as

(a) Coefficient of deviation

(b) Coefficient of mean

(c) Variance

(d) **Standard deviation**

Q10. Standard deviation is always measured from:

(a) **Arithmetic Mean** (b) Median

(c) Mode (d) Geometric mean

Q11. In terms of mean absolute deviation, quartile deviation is equal to:

(a) 2⁄3 M.D. (b) 3⁄2 M.D.

(c) 6⁄5 M.D. (d) **5⁄6 M.D.**

Q12. Considering two projects for investment X and Y, standard deviation of project A is Rs.8,450/- and standard deviation of project B is Rs. 15, 750/- then project that must be selected for investment is:

(a) **Project A** (b) Project B

(c) Both projects (d) None of above

Q13. If dispersion in distribution is less on each side then this indicates:

(a) Low uniformity of data (b) **High uniformity of data**

(c) Dispersion of data (d) Outliers of data

Q14. Arithmetic mean is multiplied to coefficient of mean absolute deviation to calculate the:

(a) **Absolute mean deviation** (b) Absolute median deviation

(c) Relative mean deviation (d) Relative median deviation

Q15. Which one easier to compute:

(a) **Range** (b) M.D.

(c) S.D. (d) C.V.

Q16. If arithmetic mean is considered as average of deviations then resultant measure is considered as

(a) Close end deviation (b) **Mean absolute deviation**

(c) Mean deviation (d) Variance deviation

Q17. If positive square root is taken of population variance then calculated measure is transformed into

(a) Standard root (b) **Standard deviation**

(c) Standard variance (d) Sample variance

Q18. Examples of applications of range in real world includes

(a) Weather forecasts
(b) Quality control
(c) Fluctuation in share prices
(d) **All of above**

Q19. Formula of coefficient of range is

(a) L+L/H+H
(b) H+H/L+L
(c) **H-L/H+L**
(d) H+L/H-L

Q20. In a set of observations, amount of variation can be shown in form of figures with help of

(a) **Absolute measures**
(b) Uniform measures
(c) Non-uniform measures
(d) Exploratory measures

Q21. If value of first quartile is 49 and value of third quartile is 60 then value of inter quartile range is

(a) 21
(b) 31
(c) **11**
(d) 41

Q22. If set of observations is 11, 13, 15, 12, 16, 18, 19, 14, 20, 17 and absolute mean deviation is 12 then percentage of coefficient of mean absolute deviation is

(a) 47.41%
(b) 57.41%
(c) 67.41%
(d) **77.41%**

Q23. Variability which is defined as difference between third and first quartile is considered as

(a) Quartile range
(b) Deciles range
(c) Percentile range
(d) **Inter quartile range**

Q24. Mean of squared deviations which is calculated from arithmetic mean is called

(a) **Mean square average**
(b) Standard square average
(c) Population average
(d) Sample square average

Q25. Measure of distance which is greatly influenced by extreme values in data is considered as

(a) **Range**
(b) Average
(c) Positive uniformity
(d) Negative uniformity

Q26. Mean absolute deviation is 5 and arithmetic mean is 110 then coefficient of mean absolute deviation is

(a) 1.054 (b) **0.045**

(c) 0.054 (d) 0.064

Q27. Number of patients who visited cardiologists are as 63, 57, 51, 65 in four days then absolute mean deviation (approximately) is

(a) 8 patents (b) **4 patients**

(c) 10 patients (d) 15 patients

Q28. In terms of dispersion difference, measurement of dispersion for available data is classified as

(a) Average measures (b) **Distance measures**

(c) Average deviation measures (d) Availability measures

Q29. The range includes:

(a) 25% of items (b) 50% of items

(c) 75% of items (d) **100% of items**

Q30. Considering standard deviation, mean absolute deviation is equal to

(a) $5/4\sigma$ (b) $5/8\sigma$

(c) **$4/5\sigma$** (d) $7/8\sigma$

Q31. Considering sales, coefficient of variation for product X is 9.3% and coefficient of variation for product Y is 8.9% then sales fluctuation of

(a) **Product X is higher** (b) Product Y is higher

(c) Product X is lower (d) Product X and Y is lower

Q32. If Q.D. is 16, the most likely value of S.D. will be:

(a) 16 (b) **24**

(c) 32 (d) 42

Q33. Difference between highest and lowest observation is 20 and coefficient of range is 0.077 then sum of highest and lowest value is:

(a) 210 (b) 220

(c) **260** (d) 240

Q34. Which of the following is a unit free number?

(a) M.D. (b) S.D.

(c) Variance (d) **C.V.**

Q35. If quartile deviation of given set of data of 20 observations is 12 then value of standard deviation is

(a) 1.7 (b) **18**

(c) 8 (d) 32

Q36. High uniformity of 50% observations around median value is indicated with help of

(a) Larger value of quartile deviation

(b) **Smaller value of quartile deviation**

(c) Larger value of range deviation

(d) Smaller value of range deviation

Q37. Relative measures in measures of dispersion are also considered as

(a) Coefficient of deviation (b) Coefficient of average

(c) **Coefficient of variation** (d) Coefficient of uniformity

Q38. Standard deviation is divided by coefficient of variation to calculate

(a) **Arithmetic mean** (b) Coefficient of arithmetic

(c) Coefficient of variance (d) Multiplier of deviation

Q39. If standard deviation is 7 then mean absolute deviation is

(a) 9.75 (b) **5.6**

(c) 7 (d) 8.75

Q40. According to empirical rule, standard deviation and mean interval that covers approximately 99.75% of data from a frequency distribution is

(a) $4\mu\pm4\sigma$ (b) $3\mu\pm3\sigma$

(c) $\boldsymbol{\mu\pm3\sigma}$ (d) $2\mu\pm2\sigma$

Q41. Categories of measures of dispersion are classified as

(a) Uniform measures (b) Relative measures

(c) Absolute measures (d) **Both (b) and (c)**

Q42. Population variance is also called

(a) **Sigma squared** (b) Negative sigma

(c) Square root (d) Cubic root

Q43. Lesser uniformity of 50% observations around median value is indicated with help of

(a) Larger value of range deviation

(b) Smaller value of range deviation

(c) **Larger value of quartile deviation**

(d) Smaller value of quartile deviation

Q44. Formula which considers relationship between set of observations, standard deviation and mean is classified as

(a) Empirical value (b) Three way rule

(c) Normal rule (d) **Both (a) and (c)**

Q45. If calculated value of total sum of squares in sample variance is larger than variation in data set is considered as:

(a) Smaller (b) **Greater**

(c) Zero (d) Negative

Q46. Standard deviation of data is 12 and mean is 72 then coefficient of variation is

(a) 14.67% (b) **16.67%**

(c) 12.67% (d) 13.67%

Q47. Value of third quartile is 61 and inter quartile range of set of observation is 48 then value of first quartile is

(a) 24 (b) 34

(c) 64 (d) **13**

Q48. If in a formula, mean absolute deviation is numerator and arithmetic mean is denominator then resultant value is classified as:

(a) Coefficient of mean deviation

(b) Coefficient of absolute quartile deviation

(c) Coefficient of quartile range deviation

(d) **Coefficient of mean absolute deviation**

Q49. Technique which implies in statistical process to measure variation in data is called

(a) Measures of dispersion (b) Measures of statistics

(c) Measures of process (d) **None of above**

Q50. If standard deviation is 5 then quartile deviation is

(a) 0.123 (b) 0.334

(c) 0.234 (d) **0.134**

Q51. Value of third quartile is 72, second quartile is 52 and first quartile is 45 then quartile deviation is

(a) **13.5** (b) 14

(c) 16.5 (d) 18.5

Q52. Undesirable consequences which cause estimated population variance to appear less as compared to real results are classified as:

(a) Undesired error (b) **Bias**

(c) Non-calculate error (d) Non-zero error

Q53. Measuring theorem which helps in determining proportion of observations for specific interval of mean and standard deviation is classified as:

(a) Pearson Theorem (b) **Chebyshev's Theorem**

(c) Sampling theorem (d) Population theorem

Q54. Considering set of observations, percentage of values that lies within population mean plus two standard deviations is

(a) 60% (b) 55%

(c) **75%** (d) 85%

Q55. Measure of variation which is useful for highly skewed distribution is

(a) Inter quartile deviation (b) **Quartile deviation**

(c) Inter quartile range (d) Quartile range

Q56. Sum of highest and lowest value is 80 and coefficient of range is 0.625 then difference between highest and lowest value is:

(a) 70 (b) 100

(c) 150 (d) **50**

Q57. Standard deviation of first 50 natural numbers is

(a) 45.43 (b) **14.43**

(c) 20.43 (d) 16.43

Q58. Standard deviation of population is denoted by

(a) Ω (b) ω

(c) $\boldsymbol{\sigma}$ (d) Σ

Q59. Value of first quartile is 23 and inter quartile range is 20 then value of third quartile is

(a) 63 (b) 53

(c) **43** (d) 73

Q60. If arithmetic mean is 78 and coefficient of variation is 12.3% then standard deviation is

(a) 10.594 (b) **9.594**

(c) 8.59 (d) 11

Q61. Output of 20 workers in hand made pot painting store is as 55, 65, 62, 60, 74, 75, 65, 70, 70, 72, 67, 78, 79, 80, 68, 54, 56, 63, 69, 71 then coefficient of range is:

(a) 0.29 (b) **0.19**

(c) 0.49 (d) 0.39

Q62. Sum of squared deviation of sample mean is 48 and total number of observation is 13 then population variance is:

(a) 61 (b) 48

(c) 13 (d) **4**

Q63. Sum of observations is 12 and coefficient of absolute mean deviation is 18 then value of mean absolute deviation is

(a) 516 (b) 716

(c) **216** (d) 616

Q64. According to empirical rule, mean and standard deviation interval that covers approximately 95.45% of data from a frequency distribution is:

(a) $\mu \pm \sigma$ (b) $2\mu \pm 2\sigma$

(c) $3\mu \pm 2\sigma$ (d) $\mu \pm \mathbf{2\sigma}$

Q65. If scatter or dispersion in the distribution is less on each side then this indicates

(a) Outliers of data
(b) Low uniformity of data
(c) **High uniformity of data**
(d) Dispersion of data

Q66. If scatter or dispersion in distribution is high on each side then this indicates

(e) Outliers of data
(f) **Low uniformity of data**
(g) High uniformity of data
(h) Dispersion of data

Q67. Technique used in measures of variations to show direction of variation in set of observations is classified as measure of:

(a) Dispersion
(b) Statistics
(c) **Skewness**
(d) Process

Q68. Total revenue (in crores) of five rubber goods companies are as two companies have revenues between 10-20, one company has revenue between 20-30 and one company has revenue between 30-40 then standard deviation is:

(a) 7.9
(b) **4.9**
(c) 5.9
(d) 6.9

Q69. In classes of grouped data such as 10-15, 15-20, 20-25, 25-30 with respective frequencies of each class as 3, 5, 4, 3, then range is

(a) 5
(b) **10**
(c) 15
(d) 20

Q70. Measures which considers mean or median to calculate average deviation does not include:

(a) Mean absolute deviation
(b) Standard deviation
(c) Variance
(d) **Median deviation**

Q71. Variability measuring tool in which standard deviation is divided by arithmetic mean and multiplied by 100 is classified as coefficient of:

(a) **Variation**
(b) Standard deviation
(c) Deviation
(d) Mean

Q72. Mean $\pm$ 3σ covers:

(a) **99.73%**
(b) 95.45%
(c) 93.45%
(d) 68.54%

Q73. Difference between smallest observation in data set and largest observation in data set is classified as

(a) Positive uniformity
(b) negative uniformity
(c) **Range**
(d) Average

Q74. Considering standard deviation, quartile deviation is equal to:

(a) 3⁄2 σ
(b) **2⁄3 σ**
(c) 2μ⁄3 σ
(d) 4μ⁄3 σ

Q75. For a given set of number of customers who visit a shoes shop in 5 consecutive days, mean absolute deviation is 10 then standard deviation of data set is:

(a) 2
(b) 10
(c) **12.5**
(d) 50

Q76. Mean absolute deviation which is used as relative measure is called coefficient of:

(a) Quartile range deviation
(b) **Mean absolute deviation**
(c) Mean deviation
(d) Absolute quartile deviation

Q77. In statistics, distance or dispersion from central value is classified as

(a) Standard variance
(b) Sample variance
(c) Standard root
(d) **Standard deviation**

Q78. Average deviation measures and distance measures are classified as measures of dispersion on basis of

(a) Relative processing
(b) Information compiled
(c) Data available
(d) **Method employed**

Q79. Numbers of emergency cases in hospital for five days are as 12, 15, 18, 16, 14 then variance of sample is

(a) **5**
(b) 20
(c) 4
(d) 15

Q80. According to empirical rule, standard deviation and mean interval that covers approximately 68.27% of data from a frequency distribution is:

(a) **μ±σ**
(b) 2μ±2σ
(c) 3μ±2σ
(d) 4μ±4σ

Q81. If M.D. is 16, the value of S.D. will be:

(a) 12 (b) 15

(c) **20** (d) 24

Q82. Considering sample rather than population, standard deviation is thus denoted by

(a) **Small s** (b) Capital s

(c) Ω (d) σ

Q83. Inter quartile range and coefficients of range are two categories of

(a) Average deviation measures (b) Availability measures

(c) Average measures (d) **Distance measures**

Q84. Measure of variation which is useful in large deviations occurrences on occasional basis is considered as

(a) **Mean absolute deviation** (b) Standard deviation

(c) Variance (d) Median deviation

Q85. Quartile deviation for number of X-ray patients who visits orthopedic in 7 consecutive days is 14 then mean absolute deviation is

(a) 18.8 (b) 15.8

(c) 10.8 (d) **16.8**

Q86. Numbers or values that are considered independent of measurement units are called

(a) **Coefficient** (b) Uniformity

(c) Variation (d) Exploration

Q87. If first quartile is subtracted from median then answer must be equal to

(a) **Third quartile minus median**

(b) Third quartile plus median

(c) First quartile plus median

(d) Median multiply third quartile

Q88. Which measure is based on only the central fifty percent of the observations?

(a) S.D. (b) M.D.

(c) **Q.D.** (d) None of these

Q89. If coefficient of variation is 13.4% and standard deviation is 12 then arithmetic mean of that set of observations is:

(a) 69.5 (b) 99.5

(c) 59.5 (d) **89.55**

Q90. Coefficient of quartile deviation is 10.34% and difference of first and third quartile is 15 then sum of third and first quartile is:

(a) 155 (b) 165

(c) 175 (d) **145**

Q91. Letter used to denote variance of population is

(a) $2\sigma^2$ (b) $2\Sigma xy$

(c) $2\Sigma^2$ (d) $\boldsymbol{\sigma^2}$

Q92. One of advantages of using quartile deviation is to use it for

(a) **Open end class interval** (b) Close end class interval

(c) Open end class width (d) Close end class width

Q93. Mean absolute deviation is 23 and coefficient of mean absolute deviation is 18 then arithmetic mean is

(a) 2.28 (b) **1.28**

(c) 3.28 (d) 4.28

Q94. In formula of sample variance $\sum\left(x-\overline{x}\right)^2$ is called

(a) **Total sum of squares** (b) Total differences of squares

(c) Multiplier of deviation (d) Divisor of mean deviation

Q95. The scatter in a series of values about the average is called:

(a) Central tendency (b) **Dispersion**

(c) Skewness (d) Symmetry

Q96. The measurements of spread or scatter of the individual values around the central point is called:

(a) **Measures of dispersion** (b) Measures of central tendency

(c) Measures of skewness (d) Measures of kurtosis

Q97. The measures used to calculate the variation present among the observations in the unit of the variable is called:

(a) Relative measures of dispersion

(b) Coefficient of skewness

(c) **Absolute measures of dispersion**

(d) Coefficient of variation

Q98. The measures used to calculate the variation present among the observations relative to their average is called:

(a) Coefficient of kurtosis

(b) Absolute measures of dispersion

(c) Quartile deviation

(d) **Relative measures of dispersion**

Q99. The degree to which numerical data tend to spread about an average value called:

(a) Constant
(b) Flatness
(c) **Variation**
(d) Skewness

Q100.The measures of dispersion can never be:

(a) Positive
(b) Zero
(c) **Negative**
(d) None of these

Q101.If all the scores on examination cluster around the mean, the dispersion is said to be:

(a) **Small**
(b) Large
(c) Normal
(d) Symmetrical

Q102.If there are many extreme scores on all examination, the dispersion is:

(a) **Large**
(b) Small
(c) Normal
(d) Symmetric

Q103.Given below the four sets of observations, which set has the minimum variation?

(a) 22, 24, 26, 28, 30
(b) 20, 30, 40, 50, 60
(c) 10, 15, 20, 25, 30
(d) **40, 41, 42, 43, 44**

Q104.Which of the following is an absolute measure of dispersion?

(a) Coefficient of variation
(b) Coefficient of dispersion
(c) **Standard deviation**
(d) Coefficient of skewness

Q105.The measure of dispersion which uses only two observations is called:

(a) Mean
(b) Median
(c) **Range**
(d) Coefficient of variation

Q106.A measure of dispersion is an indicator of the reliability of:

(a) Variability
(b) **An average**
(c) Mean deviation
(d) Standard deviation

Q107.In quality control of manufactured items, the most common measure of dispersion is:

(a) **Range**
(b) Average deviation
(c) Standard deviation
(d) Quartile deviation

Q108.The range of the scores 29, 3, 143, 27, 99 is:

(a) **140**
(b) 143
(c) 146
(d) 70

Q109.If the observations of a variable X are, -4, -20, -30, -44 and -36, then the value of the range will be:

(a) -48
(b) **40**
(c) -40
(d) 48

Q110.The range of the values -5, -8, -10, 0, 6, 10 is:

(a) 0
(b) 10
(c) -10
(d) **20**

Q111.If $Y = aX \pm b$, where a and b are any two numbers and $a \neq 0$, then the range of Y values will be:

(a) Range(X)
(b) a range(X) + b
(c) a range(X) – b
(d) **|a| range(X)**

Q112.If the maximum value in a series is 25 and its range is 15, the minimum value of the series is:

(a) **10**
(b) 15
(c) 25
(d) 35

Q113. Half of the difference between upper and lower quartiles is called:

(a) Interquartile range
(b) **Quartile deviation**
(c) Mean deviation
(d) Standard deviation

Q114. If Q_3=20 and Q_1=10, the coefficient of quartile deviation is:

(a) 3
(b) **1/3**
(c) 2/3
(d) 1

Q115. Which measure of dispersion can be computed in case of open-end classes?

(a) Standard deviation
(b) Range
(c) **Quartile deviation**
(d) Coefficient of variation

Q116. If Y = aX ± b, where a and b are any two constants and a ≠ 0, then the quartile deviation of Y values is equal to:

(a) a Q.D(X) + b
(b) **|a| Q.D(X)**
(c) Q.D(X) – b
(d) |b| Q.D(X)

Q117. The sum of absolute deviations is minimum if these deviations are taken from the:

(a) Mean
(b) Mode
(c) **Median**
(d) Upper quartile

Q118. If mean () = 15, S.D.(σ) = 6, all observations multiplied by 2 would give the result:

(a) Mean = 15, S.D. = 12
(b) **Mean = 30, S.D. = 12**
(c) Mean = 15, S.D. = 6
(d) Mean = 7.5, S.D. = 6

Q119. If Y = aX ± b, where a and b are any two numbers but a ≠ 0, then M.D(Y) is equal to:

(a) M.D(X)
(b) M.D(X) ± b
(c) **|a| M.D(X)**
(d) M.D(Y) + M.D(X)

Q120. The mean deviation of the scores 12, 15, 18 is:

(a) 6
(b) 0
(c) 3
(d) 2

Q121.Mean deviation computed from a set of data is always:

(a) Negative

(b) Equal to standard deviation

(c) More than standard deviation

(d) **Less than standard deviation**

Q122.The average of squared deviations from mean is called:

(a) Mean deviation

(b) **Variance**

(c) Standard deviation

(d) Coefficient of variation

Q123.The sum of squares of the deviations is minimum, when deviations are taken from:

(a) **Mean**

(b) Mode

(c) Median

(d) Zero

Q124.Which of the following measures of dispersion is expressed in the same units as the units of observation?

(a) Variance

(b) **Standard deviation**

(c) Coefficient of variation

(d) Coefficient of standard deviation

Q125.Which measure of dispersion has a different unit other than the unit of measurement of values?

(a) Range

(b) Standard deviation

(c) **Variance**

(d) Mean deviation

Q126.Which of the following is a unit free quantity?

(a) Range

(b) Standard deviation

(c) **Coefficient of variation**

(d) Arithmetic mean

Q127.If the dispersion is small, the standard deviation is:

(a) Large

(b) Zero

(c) **Small**

(d) Negative

Q128.The value of standard deviation changes by a change of:

(a) Origin
(b) **Scale**
(c) Algebraic signs
(d) None of these

Q129.The standard deviation one distribution dividedly the mean of the distribution and expressing in percentage is called:

(a) Coefficient of Standard deviation
(b) Coefficient of skewness
(c) Coefficient of quartile deviation
(d) **Coefficient of variation**

Q130.The positive square root of the mean of the squares of the eluviations of observations from their mean is called:

(a) Variance
(b) Range
(c) **Standard deviation**
(d) Coefficient of variation

Q131.The variance is zero only if all observations are the:

(a) Different
(b) Square
(c) Square root
(d) **Same**

Q132.The standard deviation is independent of:

(a) **Change of origin**
(b) Change of scale of measurement
(c) Change of origin and scale of measurement
(d) Difficult to tell

Q133.If there are ten values each equal to 10, then standard deviation of these values is:

(a) 100
(b) 20
(c) 10
(d) 0

Q134.If X and Y are independent random variables, then S.D(X ± Y) is equal to:

(a) S.D(X) ± S.D(Y)
(b) Var(X) ± Var(Y)
(c) $\sqrt{Var(X)-Var(Y)}$
(d) $\sqrt{\boldsymbol{Var(X)+Var(Y)}}$

Q135.S.D(X) = 6 and S.D(Y) = 8. If X and Yare independent random variables, then S.D(X-Y) is:

(a) 2 (b) **10**

(c) 14 (d) 100

Q136.For two independent variables X and Y if S.D(X) = 1 and S.D(Y) = 3, then Var(3X - Y) is equal to:

(a) 0 (b) 6

(c) **18** (d) 12

Q137. If Y = aX ± b, where a and b are any two constants and a ≠ 0, then Vat (Y) is equal to:

(a) a Var(X) (b) a Var(X) + b

(c) **a^2 Var(X) – b** (d) a^2 Var(X)

Q138.If Y = aX + b, where a and b are any two numbers but a ≠ 0, then S.D(Y) is equal to:

(a) S.D(X) (b) a S.D(X)

(c) **|a| S.D(X)** (d) a S.D(X) + b

Q139.The ratio of the standard deviation to the arithmetic mean expressed as a percentage is called:

(a) Coefficient of standard deviation (b) Coefficient of skewness

(c) Coefficient of kurtosis (d) **Coefficient of variation**

Q140.Which of the following statements is correct?

(a) The standard deviation of a constant is equal to unity

(b) The sum of absolute deviations is minimum, if these deviations are taken from the mean.

(c) The second moment about origin equals variance

(d) **The variance is positive quantity and is expressed in square of the units of the observations**

Q141.Which of the following statements is false?

(a) The standard deviation is independent of change of origin

(b) If the moment coefficient of kurtosis kurtosis $\beta_2 = 3$, the distribution is mesokurtic or normal.

(c) If the frequency curve has the same shape on both sides of the centre line which divides the curve into two equal parts, is called a symmetrical distribution.

(d) **Variance of the sum or difference of any two variables is equal to the sum of their respective variances**

Q142.If Var(X) = 25, then $S.D.\left(\frac{2x+5}{2}\right)$ is equal to:

(a) 15/2 (b) 50

(c) 25 **(d) 5**

Q143.To compare the variation of two or more than two series, we use:

(a) Combined standard deviation (b) Corrected standard deviation

(c) **Coefficient of variation** (d) Coefficient of skewness

Q144.The standard deviation of -5, -5, -5, -5, and -5 is:

(a) -5 (b) +5

(c) 0 (d) -25

Q145.Standard deviation is always calculated from:

(a) **Mean** (b) Median

(c) Mode (d) Lower quartile

Q146.The mean of an examination is 69, the median is 68, the mode is 67, and the standard deviation is 3, the measures of variation for this examination is:

(a) 67 (b) 68

(c) 69 (d) **3**

Q147.The variance of 19, 21, 23, 25 and 27 is 8. The variance of 14, 16, 18, 20 and 22 is:

(a) Greater than 8 **(b) 8**

(c) Less than 8 (d) 3

Q148.The moments about origin are called:

(a) Moments about zero (b) Raw moments

(c) **Both (a) and (b)** (d) Neither (a) nor (b)

Q149. All odd order moments about mean in a symmetrical distribution are:

(a) Positive (b) Negative

(c) **Zero** (d) Three

Q150.An automobile manufacturer obtains data concerning the sales of six of its deals in the last week of 1996. The results indicate the standard deviation of their sales equals 6 autos. If this is so, the variance of their sales equals:

(a) (b) 6

(c) (d) **36**

Q151.If standard deviation of the values 2, 4, 6, 8 is 2.236, then standard deviation of the values 4, 8, 12, 16 is:

(a) 0 (b) **4.472**

(c) 4.236 (d) 2.236

Q152.Var(X) = 4 and Var(Y) =9. If X and Y are independent random variable then Var(2X + Y) is:

(a) 13 (b) 17

(c) **25** (d) -1

Q153.If = Rs.20, S.D.= Rs.10, then coefficient of variation is:

(a) 45% (b) **50%**

(c) 60% (d) 65%

Q154.Which of the following measures of dispersion is independent of the units employed?

(a) **Coefficient of variation** (b) Quartile deviation

(c) Standard deviation (d) Range

Q155.In sheppard's correction μ_2 is equal to:

(a) $\mu_2 - \frac{\mathbf{h^2}}{\mathbf{12}}$ (b) $\mu_2 + \frac{h^2}{12}$

(c) $\mu_2 - \frac{h^2}{2}$ (d) $\mu_2 - \frac{h}{12}$

Q156.The moments about mean are called:

(a) Raw moments (b) **Central moments**

(c) Moments about origin (d) All of the above

Q157. In a set of observations the variance is 50. All the observations are increased by 100%. The variance of the increased observations will become:

(a) 50 (b) **200**

(c) 100 (d) No change

Q158.Three factories A, B, C have 100, 200 and 300 workers, respectively. The mean of the wages is the same in the three factories. Which of the following statements is true?

(a) There is greater variation in factory C.

(b) Standard deviation in. factory A is the smallest.

(c) Standard deviation in all the three factories is equal

(d) **None of the above**

Q159.The second moment about arithmetic mean is 16, the standard deviation will be:

(a) 16 **(b) 4**

(c) 2 (d) 0

Q160.The first and second moments about arbitrary constant are -2 and 13 respectively, the standard deviation will be:

(a) -2 **(b) 3**

(c) 9 (d) 13

Q161.Moment ratios β_1 and β_2 are:

(a) Independent of origin and scale of measurement

(b) Expressed in original unit of the data

(c) **Unit less quantities**

(d) Both (a) and (c)

Q162.The first moment about X = 0 of a distribution is 12.08. The mean is:

(a) 10.80 (b) 10.08

(c) **12.08** (d) 12.88

Q163.First two moments about the value 2 of a variable are 1 and 16. The variance will be:

(a) 13 (b) **15**

(c) 16 (d) None of these

Q164.The first three moments of a distribution about the mean are 1, 4 and 0. The distribution is:

(a) **Symmetrical** (b) Positive skewed

(c) Negative skewed (d) None of these

Q165.If the third central is negative, the distribution will be:

(a) Symmetrical (b) Positively skewed

(c) **Negatively skewed** (d) None of these

Q166.If the third moment about mean is zero, then the distribution is:

(a) Positively skewed (b) Negatively skewed

(c) **Symmetrical** (d) Mesokurtic

Q167.Departure from symmetry is called:

(a) Second moment (b) Kurtosis

(c) **Skewness** (d) Variation

Q168.In a symmetrical distribution, the coefficient of skewness will be:

(a) 0 (b) Q_1

(c) Q_3 (d) 1

Q169.The lack of uniformity or symmetry is called:

(a) Skewness (b) Dispersion

(c) Kurtosis (d) Standard deviation

Q170.For a positively skewed distribution, mean is always:

(a) Less than the median (b) Less than the mode

(c) **Greater than the mode** (d) None of these

Q171.For a symmetrical distribution:

(a) $\beta_1 > 0$ (b) $\beta_1 < 0$

(c) $\boldsymbol{\beta_1 = 0}$ (d) $\beta_1 = 3$

Q172.If mean=50, mode=40 and standard deviation=5, the distribution is:

(a) **Positively skewed** (b) Negatively skewed

(c) Symmetrical (d) None of these

Q173.If 25% of the items are less than 20 and 25% are more than 40, the quartile deviation is:

(a) **10** (b) 20

(c) 30 (d) 40

Q174.Which of the following is a unit free number?

(e) M.D. (f) S.D.

(g) Variance (h) **C.V.**

Q175.If mean ($\overline{x}$) = 15, S.D.(σ) = 6, all observations multiplied by 2 would give the result:

(e) Mean = 15, S.D. = 12 (f) **Mean = 30, S.D. = 12**

(g) Mean = 15, S.D. = 6 (h) Mean = 7.5, S.D. = 6

Q176.The mean of absolute deviations from an average is called:

(a) Range (b) **M.D.**

(c) S.D. (d) C.V.

Q177.The mean of squared deviations about the mean is called:

(a) M.D. (b) S.D.

(c) **Variance** (d) C.V.

Q178.The dispersion measure least affected by extreme values is:

(a) Range (b) **M.D.**

(c) S.D. (d) C.V.

Q179.If each and every value of a set of observations is multiplied by a constant, the variance of the resultant variable is:

(a) Decreases (b) **Increases**

(c) Unaltered (d) None of these

Q180.Which one of the given measure of dispersion is considered an ideal?

(a) Range (b) M.D.

(c) **S.D.** (d) C.V.

Q181.Formula for coefficient of variation (C.V.) is equal to:

(a) $\frac{A.M.}{S.D.}x100$ (b) $\mathbf{\frac{S.D.}{A.M.}x100}$

(c) $\frac{A.M. x S.D.}{100}$ (d) $\frac{100}{A.M.\ x\ S.D.}$

Q182.Suppose $x_1, x_2, x_3, \ldots\ldots, x_n$ are n observations of a series, then range (R) is:

(a) $R = X_{min} - X_{max}$ (b) $\mathbf{R = X_{max} - X_{min}}$

(c) $R = |X_{min} - X_{max}|$ (d) $R = X_n - X_1$

Q183.Quartile deviation is given by the formula:

(a) Q.D. = $(Q_1 + Q_3)/2$ (b) Q.D. = $(Q_1 + Q_3)$

(c) Q.D. = $(Q_3 - Q_1)$ (d) **Q.D. = $(Q_3 - Q_1)/2$**

Q184.If each observation of a set is divided by 2, the S.D. of the new observations is:

(a) 2 times of S.D. of original observations

(b) **(½)th of S.D. of original observations**

(c) Unchanged

(d) 4 times of S.D. of original observations

Q185.The relation between quartile deviation (Q.D.) and mean deviation (M.D.) is:

(a) 3 Q.D. = 5 M.D. (b) 5 Q.D. = 3 M.D.

(c) **6 Q.D. = 5 M.D.** (d) 5 Q.D. = 6 M.D.

Q186.The relation between quartile deviation (Q.D.), mean deviation (M.D.) and standard deviation (S.D.) is:

(a) **6 Q.D. = 5 M.D. = 4 S.D.** (b) 4 Q.D. = 5 M.D. = 6 S.D.

(c) 6 Q.D. = 4 M.D. = 5 S.D. (d) 5 Q.D. = 6 M.D. = 4 S.D.

Q187.Which measure of dispersion ensures lowest degree of reliability?

(a) M.D. (b) S.D.

(c) C.V. (d) **Q.D.**

Q188.Which measure of dispersion ensures highest degree of reliability?

(a) M.D. (b) **S.D.**

(c) C.V. (d) Q.D.

Q189.Average wages of workers of the poultry industry are Rs. 8000/- per month and the standard deviation of wages is 400, then the coefficient of variation is:

(a) **5%** (b) 10%

(c) 15% (d) 20%

Q190.If the mean and standard deviation of X and Y are, , and , , which of the two series is more consistent?

(a) Series X (b) **Series Y**

(c) Both (a) and (b) (d) None of these

Q191.Suppose $x_1, x_2, \ldots\ldots, x_{n1}$ are the observations in a series of size n1 and $y_1, y_2, \ldots\ldots, y_{n2}$ are the another series of size n_2, then pooled variance of two series is:

(a) Pooled mean (b) Variance of individual series

(c) Means of individual series (d) **All of them**

Q192.The standard deviation of a set of values will be:

(a) **Always positive**

(b) Positive only when the values are positive

(c) Positive only when the values are negative

(d) Negative only when the values are positive

Q193.The relationship between the standard deviation of median and standard deviation of mean of a normal population is:

(a) S.D.(Median) = S.D. (Mean)

(b) S.D.(Median) < S.D. (Mean)

(c) **S.D.(Median) > S.D. (Mean)**

(d) 2 S.D.(Median) = S.D. (Mean)

Q194.Kurtosis is adjudged around:

(a) Median (b) **Mode**

(c) Mean (d) None of these

Q195. If for a distribution, coefficient of Kurtosis $\beta_2 = 3$, the frequency curve is:

(a) Platykurtic (b) **Mesokurtic**

(c) Leptokurtic (d) None of these

Q196.If for a distribution, coefficient of Kurtosis $\beta_2 < 3$, the frequency curve is:

(a) **Platykurtic** (b) Mesokurtic

(c) Leptokurtic (d) None of these

Q197.If for a distribution, coefficient of Kurtosis $\beta_2 > 3$, the frequency curve is:

(a) Platykurtic (b) Mesokurtic

(c) **Leptokurtic** (d) None of these

Q198.For a set of values:

(a) M.D. is always greater than S.D. (b) M.D. is always equal to S.D.

(c) **M.D. is always less than S.D.** (d) All of them

Q199.The value of coefficient of kurtosis λ_2 can be:

(a) $\lambda_2 = 0$ (b) $\lambda_2 < 0$

(c) $\lambda_2 > 0$ (d) **All of them**

Q200.If $Q_1 = 20$ and $Q_3 = 40$, the coefficient of quartile deviation is:

(a) **1/3** (b) 3

(c) 10 (d) 20

Q201.If $Q_1 = 20$ and $Q_3 = 40$, the quartile deviation is:

(a) 1/3 (b) 3

(c) **10** (d) 20

Q202.The relationship between variance and standard deviation is:

(a) Variance = Standard deviation

(b) **Variance = (Standard deviation)^2**

(c) Variance = (Standard deviation)$^{1/2}$

(d) (Variance)^2 = Standard deviation

Q203.In case of negative skewed distribution, the extreme values lie in the:

(a) **Left tail** (b) Right tail

(c) Middle (d) Anywhere

Q204.In case of positive skewed distribution, the extreme values lie in the:

(a) Left tail (b) **Right tail**

(c) Middle (d) Anywhere

Q205.Which measure of dispersion has a different unit other than the unit of measurement of values?

(a) M.D. (b) S.D.

(c) **Variance** (d) Range

Q206.If each value of a series is divided by 10, its coefficient of variation will be increased by:

(a) 20% (b) 10%

(c) 5% (d) **0%**

Q207.If each value of a series is multiplied by 4, its coefficient of variation will be reduced by:

(a) 20% (b) 10%

(c) 5% (d) **0%**

Q208.If each value of a series is multiplied by a constant 'a', its coefficient of variation will be increased by:

(a) Less than original value (b) More than original value

(c) **Same as original value** (d) None of these

Q209.In poultry feed industry, the average wages of workers are Rs 8000/- per month and the standard deviation of wages is 400, then the standard error (S.E.) is:

(a) Rs. 20/- (b) **Rs. 40/-**

(c) Rs. 80/- (d) Rs. 800/-

Q210.If the values of a set are measured in inches, the unit of coefficient of variation will be:

(a) Inches (b) Inches^2

(c) Inches^3 (d) **Percent**

Q211.If the values of a set are measured in inches, the unit of variance will be:

(a) Inches (b) **Inches^2**

(c) Inches^3 (d) Percent

Q212.If the values of a set are measured in inches, the unit of standard error will be:

(a) **Inches** (b) Inches^2

(c) Inches^3 (d) Percent

Q213.The range of the following frequency distribution,

Classes	Frequency
5 – 10	3
10 – 15	6
15 – 20	4
20 – 25	2

is equal to:

(a) 2 (b) 5

(c) 15 **(d) 20**

Q214.The mean deviation about median of the following frequency distribution,

Classes	Frequency
2 – 4	4
4 – 6	8
6 – 8	5
8 – 10	3

is equal to:

(a) 1 **(b) 1.6**

(c) 2.4 (d) 4

Q215.Coefficient of variation is calculated by the formula:

(a) $\frac{\overline{X}}{\sigma}$ (b) $\frac{\overline{X}}{\sigma} x100$

(c) $\frac{\overline{\sigma}}{\overline{X}}$ (d) $\frac{\sigma}{\overline{X}} x100$

Q216.The sum of squares of 40 observations was worked out as 2500. But while calculating it, an observation 32 was wrongly considered as 23, and then the corrected sum of squares is:

(a) 2800 **(b) 2995**

(c) 3050 (d) 3750

Q217.If 25% of the items are less than 20 and 25% are more than 40, the quartile deviation is:

(e) 10 (f) 20

(g) 30 (h) 40

Q218.The mean deviation is minimum when deviations are taken from:

(a) Mean (b) Mode

(c) **Median** (d) Zero

Q219.Mean $\pm$ 3σ covers:

(a) 90% (b) 94.67%

(c) 99.73% (d) **All of them**

Q220.The standard deviation of the following frequency distribution,

Classes	Frequency
2 – 4	2
4 – 6	6
6 – 8	8
8 – 10	4

is approximately equal to:

(a) 1.4 **(b) 1.9**

(c) 2.3 (d) 5.4

Q221.The standard error of mean for the following frequency distribution,

Classes	Frequency
2 – 4	2
4 – 6	6
6 – 8	8
8 – 10	4

is approximately equal to:

(a) 0.10 **(b) 0.43**

(c) 0.63 (d) 0.87

Q222.The coefficient of quartile deviation for the following frequency distribution,

Classes	Frequency
2 – 4	2
4 – 6	6
6 – 8	8
8 – 10	4

is approximately equal to:

(a) 0.11 **(b) 0.22**

(c) 0.33 (d) 0.44

Q223. The quartile deviations for number of X-ray patient visits orthopedic in 7 consecutive days is 14 then mean absolute deviation is:

(a) 10.8 (b) 15.8

(c) **16.8** (d) 18.8

Q224. The quartile deviation for the following frequency distribution,

Classes	Frequency
2 – 4	2
4 – 6	6
6 – 8	8
8 – 10	4

is approximately equal to:

(a) 0.22 (b) 0.43

(c) 1.58 (d) 1.90

Q225. Which of the following is not a measure of variability in a data set?

(a) Range (b) Variance

(c) Inter-quartile range (d) **Median**

Q226. Which measure of dispersion is based on the absolute deviations only?

(e) Range (f) Variance

(g) Inter-quartile range (a) **Mean deviation**

Q227. Which measure of dispersion is based on all the observations?

(a) S.D. (b) M.D.

(c) Q.D. (d) **Both (a) and (b)**

Q228. The appropriate measure of dispersion for open end classes is:

(a) S.D. (b) M.D.

(c) **Q.D.** (d) Both (a) and (b)

Q229.If the profit of a company remains same for last six months, then the standard deviation of profit for tee six months would be:

(a) **Zero** (b) Positive

(c) Negative (d) Both (b) and (c)

Q230.The standard deviation of 10, 12, 10, 12, 10, 12, 10, 12 is:

(a) 0 (b) **1**

(c) 2 (d) 3

Q231.Which measure of dispersion is considered for finding a pooled measure of dispersion after combining several groups?

(a) **S.D.** (b) M.D.

(c) Q.D. (d) Both (a) and (b)

Q232.If quartile range is 24 then quartile deviation is

(a) 48 (b) **12**

(c) 24 (d) 72

Q233.For any two numbers standard deviation is always:

(a) Twice the range (b) Square of the range

(c) **Half of the range** (d) None of these

Q234.Difference between first and third quartile is 20 and sum of first and third quartile is 60 then coefficient of quartile deviation is

(a) 18% (b) 15%

(c) **3%** (d) 13%

Q235.If mean absolute deviation of set of observations is 8.5 then value of quartile deviation is

(a) **7.08** (b) 9.08

(c) 10.2 (d) 11.2

Q236.In a hospital in Delhi, number of heart patients who visited cardiologist in 5 days is 15, 12, 16, 18, 14 then variance of data is:

(a) 2 (b) 3

(c) 4 (d) **5**

Q237.Output of 15 workers in hand made leather Shoes Company is as 50, 65, 70, 55, 62, 74, 75, 65, 70, 78, 79, 80, 68, 72, 67 then range is

(a) **30** (b) 80

(c) 75 (d) 79

Q238.The A.Ms and S.Ds. of runs scored by four cricket batsmen I, II, III, IV are 45, 12, 36, 24 and 14, 4, 9, 7, respectively, then which player is most consistent?

(a) I (b) II

(c) **III** (d) IV

Q239.Numbers of emergency cases in a hospital for three days are as 20, 21, 22, and then value of standard deviation with assumed mean method is:

(a) **1** (b) 2

(c) 3 (d) 10

Q240.Coefficient of Variation (CV) of four separate sets of data is given, i.e. CV of Set A is 4.8, CV of Set B is 4.3, CV of Set C is 3.9 and CV of Set D is 2.9, which Set is more consistent?

(a) Set A (b) Set B

(c) Set C (d) **Set D**

Q241.____ is used to compare the consistency of two or more sets of data.

a) **Coefficient of Variation** b) Coefficient of Kurtosis

c) Coefficient of Skewness d) None of these

Q242.______is used when only a rough measure of dispersion is required.

(a) Mean Deviation (b) Standard Deviation

(c) Quartile Deviation (d) **Range**

Q243.If all the observations within a series are multiplied by five, then:

(a) The new standard deviation would be decreased by five

(b) The new standard deviation would be increased by five

(c) The new standard deviation would be half of the previous standard deviation

(d) **The new standard deviation would be multiplied by five**

Q244.The numerical value of a standard deviation can never be:

(a) Zero
(b) One
(c) **Negative**
(d) Larger than the variance

Q245 If the scatter or dispersion in the distribution is less on each side then this indicates:

(a) Dispersion of data
(b) Low uniformity of data
(c) **High uniformity of data**
(d) Outliers of data

Q246.The quartile deviation for number of X-ray patients who visited orthopedic in 7 consecutive days is 14 then mean absolute deviation is:

(a) 10.8
(b) 15.8
(c) **16.8**
(d) 18.8

Q247.If the first quartile is subtracted from median then the answer must be equal to:

(a) First quartile plus median
(b) Third quartile plus median
(c) First quartile minus median
(d) **Third quartile minus median**

Q248.The Arithmetic mean is multiplied to the coefficient of mean absolute deviation to calculate:

(a) Relative mean deviation
(b) **Absolute mean deviation**
(c) Absolute median deviation
(d) Relative median deviation

7

Correlation and Regression

Correlation

In a statistical analysis, when two variables are studied simultaneously, then it is called "**Bivariate study**'. Now, in a bivariate study, if it is found that change in the values of one variable then in sympathy with the change in the values of other variable, then these two variables are said to be associated and the association is termed as **correlation**. In other words, the term **correlation** indicates the relationship between two such variables in which with changes in the values of one variable, the values of the other variable also change. In short, the tendency of simultaneous variation between two variables is called **correlation**.

Types of Correlation

Correlation can be classified by-

- **Positive and Negative correlation:** The correlation may be classified according to the direction of change in the two variables. In this regard, correlation may be either positive or negative. **Positive correlation** refers to the movement of variables in the same direction. **Negative correlation** efers to the movement of variables in the opposite direction.
- **Simple, Partial and Multiple correlations:** When only two variables are studied it is a problem of **simple correlation**. When three or more variables are studied simultaneously, it is a problem of **multiple correlations**. On the other hand, in **partial correlation** we recognize more than two variables, but consider only two variables to be influencing each other the effect of other influencing variables being kept constant.
- **Linear and non-linear correlations:** In **linear correlation** for every unit change in the values of one variable, there is constant change in the value of other variable. The perfect positive and negative correlations are also linear correlations. In **non-linear correlation** the change in the values of one variable does not have a constant ratio to the change in the other variable.

Methods of studying correlation

Methods of studying correlation between two variables in the case of ungrouped data may be studies as follows-

- **Graphical method:** Under this method, the period values of the two variables are plotted on a graph paper by taking independent variable on X-axis and dependent variable on Y-axis. If the trend of the plotted points, if found **upward rising from left bottom to right top** then there is a **positive correlation** between the two variables. On the other hand, If the trend of plotted points is found **downward dealing from left top to right bottom** then there is **negative correlation** between two variables. Now, if there is no such trend then it is concluded that the **two variables are independent** of each other means there is **no correlation** between the variables. The graphical representation of bivariate ungrouped data is known as **scattered diagram** or **dot diagram.**

- **Numerical method:** By the graphical method we can only know the direction of correlation and not the amount (magnitude) of correlation. For determining the amount of correlation between the two variables, the numerical method is used. **Karl Pearson** (1867-1936) gave a coefficient known as '**Coefficient of correlation**' which is denoted by 'r', gives an exact idea about the degree of linear relationship between the two variables x and y. Therefore, Coefficient of correlation is the ratio of co-variance between two variables and geometric mean (G.M.) of the variances of the two variables and is defined as-

$$Coefficient\ of\ correlation(r) = \frac{Cov(x,y)}{\sqrt{Var(x).Var(y)}}$$

On simplification,

$$Coefficient\ of\ correlation(r) = \frac{\Sigma(x-\bar{x}).(y-\bar{y})}{\sqrt{\Sigma(x-\bar{x})^2.\Sigma(y-\bar{y})^2}}$$

In this way, **correlation coefficient** can also be defined as the value obtained by dividing the sum of product of deviations of paired values from their respective mean with the square root of product of sum of square deviations of two variables from their respective means.

Remarks

1) Cov(x,y) means covariance of x and y, is the product of deviations of x and y from their respective means.
2) The coefficient of correlation has no unit.
3) The value of the coefficient of correlation 'r' ranges in between -1 to +1.
4) When r = -1, then there is **perfect negative correlation** between the two variables.

5) When r = +1, then there is **perfect positive correlation** between the two variables.

6) When r lies between -1 to 0, then there is **limited negative correlation** between the two variables.

7) When r = 0, then there is **no correlation** between the two variables. That means, the two variables are independent.

8) When r lies between 0 to +1, then there is **limited positive correlation** between the two variables.

9) The value of coefficient of correlation is independent of the change in origin and scale.

- **Spearman's coefficient of rank correlation:** It is useful in situations when precise quantified information on the two variables is not available. After assigning ranks to various items in the two series in order of size or importance, we measure the extent or degree or correlation between two sets of rank by using **Spearman's coefficient of rank correlation.** which is denoted by Greek letter 'ρ' and is given by-

$$\text{Rank correlation}(\rho) = 1 - \frac{6\sum D^2}{n(n^2-1)}$$

Here, ρ = coefficient of rank correlation; D = difference of ranks; n = number of observations

$$\text{Rank correlation}(\rho) = 1 - \frac{6\left[\sum D^2 + \sum \frac{m(m^2-1)}{12}\right]}{n(n^2-1)}$$

Rank correlation for tied ranks

In some cases, we may have two or more equal observations in either of the two series or in both the series. The Spearman's coefficient of rank correlation is then corrected for these tied ranks and then the formula being used-

$$\text{Rank correlation}(\rho) = 1 - \frac{6\sum D^2}{n(n^2-1)}$$

Here, m = the number of equal observations with common ranks;

D = difference between paired ranks; n = number of items ranked

Remarks

1) It is nothing but Karl Pearson's coefficient of correlation calculated for two sets of ranks and therefore, its value also lies between -1 to +1.

2) When ρ = -1, implies a perfect negative agreement or relationship between two sets of ranks.

3) When ρ = +1, then there is perfect positive agreement or relationship between two sets of ranks.

4) When ρ = 0, then there is no relationship or agreement between two sets of ranks.

5) Where ranks are treated as scores, coefficient of rank correlation (ρ) will be equal to coefficient of correlation 'r'.

6) This method is applicable even when precise quantitative information on the two variables is available.

- **Coefficient of concurrent deviation:** In this method, only the direction of change (+ or -) and not the magnitude of variations, is taken into consideration. The result obtained by this method gives a general idea of the direction of covariation between the two variables. The **coefficient of concurrent deviation** denoted by 'r_c' and is obtained by the formula-

$$r_c = \pm\sqrt{\left[\pm\left(\frac{2C-n}{n}\right)\right]}$$

Here, C = the number of concurrent deviations, i.e., the number of positive products of D_x and D_y. D_x, D_y = Stands the deviations of x and y-series. Here each value in the series is compared with the preceding one. If the value is greater than the preceding, deviation is taken positive (+) otherwise (-). If the value is equal to the preceding, deviation will be zero (0); n = the number of pairs of deviations compared. This number is one less than N (The number of paired observations in the series), i.e., n = (N-1).

Note: The significance of ± signs, both inside the under-root and outside the under-root, is that we cannot take the under-root of minus sign. Therefore, if (2C-n)/n is negative, this negative value multiplied with the minus sign inside would make it positive and we can take the under-root. But the ultimate result would be negative. If (2C-n)/n is positive then, we get a positive value of the coefficient of correlation.

Standard Error of correlation coefficient [S.E.(r)] and Probable Error of correlation coefficient [P.E.(r)]: The Standard Error of correlation coefficient [S.E.(r)] and Probable Error of correlation coefficient [P.E.(**r**)] are used for interpreting about the significance of the coefficient of correlation. These are also useful for finding limits of the population coefficient of correlation on the basis of a sample drawn out of it. The standard Error of correlation coefficient [S.E.(r)] is obtained by the formula-

$$S.E.(r) = \frac{1-r^2}{\sqrt{n}}$$

The **probable Error of correlation coefficient [P.E.(r)]** is obtained by the formula-

$$P.E.(r) = 0.6745 \frac{1 - r^2}{\sqrt{n}}$$

Here, r = coefficient of correlation compute from a sample of size n drawn from a bivariate normal population.

Relationship between S.E.(r) and P.E.(r)

The S.E.(r) and P.E.(r) have the following relationship:

P.E.(r) = 0.6745 × S.E.(r)

Coefficient of determination: It is defined as the square of the coefficient of correlation (r) and indicates the percentage variation in the dependent variable which is accounted for by the independent variable. It should be analysed for comparing the values of coefficient of correlation (r). Symbolically,

$$r^2 = \frac{Explained\ variation}{Total\ variation}$$

Remark

- If r = ± 1, that means, $r^2 = 1$ and then we say that 100% variation in the dependent variable has been expected by the independent variable.
- As the value of r decreases, the value of r^2 also decreases at a faster rate.
- The value of r^2 is always a positive number as such it is unable to explain the direction of relationship between the two variables.

Coefficient of non-determination: The ration of the unexplained variation to the total variation is called as **coefficient of non-determination.** It is usually denoted by K^2. Thus,

$$K^2 = \frac{Unexplained\ variation}{Total\ variation} = 1 - \frac{Explained\ variation}{Total\ variation}$$

i.e. $K^2 = 1 - r^2$

The square root of the coefficient of non-determination is called the coefficient of alienation. That means,

$K = \pm \sqrt{(1 - r^2)}$

Regression

The term 'regression' was coined by **Sir Francis Galton (1885)** to study between the height of offspring and their parents for a few generations for the conclusion that the height of offspring tend to approach median position. So, the average relationship between two or more variables, which can be used for estimating the value of one variable from the given values of one or more variables, is called **regression**. When only two variables are considered it is termed as **simple regression**.

Lines of Regression

In a bivariate study, when the two variables are **interdependent** then different pair values are plotted on a graph paper by taking **independent variable on the X-axis** and the **dependent variable on the Y-axis**. If the plotted points are found concentrated around a straight line then the two variables have linear relationship and the line is known as **line of regression or regression line**. This line is also known as '**Line of best fit**'. In a bivariate study, we have two lines of regression, namely-

- **Regression line of y on x**
- **Regression line of x on y**

Regression line of Y on X: When a line is drawn by minimizing the sum of square deviations in y-values then the line is known as regression line of y on x, which may be written as:

$y = a + bx$

Where, a = Constant; b= regression coefficient of y on x; and x= independent variable; y = dependent variable.

The values of a (Constant) and b (Regression coefficient) can be obtained from the following normal equations:

$\sum y = na+b\sum x$ (i)

$\sum xy = a\sum x+b\sum x^2$ (ii)

Regression line of x on y: When a line is drawn by minimizing the sum of square deviations parallel to x-values then the line is known as regression line of x on y, which may be written as:

$x = a + by$

Where, x and y denotes the two variables, a = Constant; b= regression coefficient of x on y

The values of a (Constant) and b (Regression coefficient) can be obtained from the following normal equations:

$\sum x = na + b\sum y$ (i)

$\sum xy = a\sum y + b\sum y^2$ (ii)

Regression Coefficients

The amount of change in the dependent variable with a unit change in the independent variable is known as '**regression coefficient of y on x**', denoted by b_{yx}. Thus-

$$Regression\,coefficient\,of\;y\,on\,x\left(b_{yx}\right)=\frac{Cov\left(x,y\right)}{Var\left(x\right)}=\frac{\Sigma\left(x-\bar{x}\right)\left(y-\bar{y}\right)}{\Sigma\left(x-\bar{x}\right)^2}=r\frac{\sigma_y}{\sigma_x}$$

If we reverse, the amount of change in the independent variable with a unit change in the dependent variable is known as '**regression coefficient of x on y**', denoted by b_{xy}. Thus-

$$Regression\,coefficient\,of\;x\,on\,y\left(b_{xy}\right)=\frac{Cov\left(x,y\right)}{Var\left(y\right)}=\frac{\Sigma\left(x-\bar{x}\right)\left(y-\bar{y}\right)}{\Sigma\left(y-\bar{y}\right)^2}=r\frac{\sigma_x}{\sigma_y}$$

Remarks

1) Cov(x,y) means covariance of x and y, is the product of deviations of x and y from their respective means.
2) The regression coefficient has always unit.
3) The value of regression coefficient ranges in between -∞ to +∞.
4) The signs of both the regression coefficients are the same. That is, either both will be negative or both will be positive.
5) It is always calculated when the two variables in a bivariate study are interdependent.
6) It can be calculated in both ways i.e., by changing the two variables.
7) The values of both the regression coefficient can never exceed unity i.e., If one regression coefficient is greater than unity then other will be definitely smaller than unity. So that their product cannot exceed unity. Both the regression coefficients can be smaller than unity.

Relationship between coefficient of correlation and regression coefficients

The coefficient of correlation is the geometric mean of the two regression coefficients. Thus-

$$r=\sqrt{b_{yx}.b_{xy}}$$

Note: The sign of the coefficient of correlation is the same as that of the two regression coefficients. Thus, r will be positive, if b_{yx} and b_{xy} are positive. Similarly, r will be negative, if b_{yx} and b_{xy} are negative. In this case, this relationship well be written as-

$$r = -\sqrt{b_{yx}.b_{xy}}$$

Regression equations

The best average value of one variable associated with the given value of another variable may also be estimated or predicted by means of an equation known as regression equation. On the basis of theory of least square there may be two such regression lines-

- Regression equation of y on x
- Regression equation of x on y

• **Regression equation of y on x:** The line of regression of y on x is used to predict or estimate the value of y for the given value of the variable x. Thus, y is the dependent variable and x is an independent variable. It can be written in the following form-

$$(y-\bar{y}) = b_{yx}(x-\bar{x})$$

Where, $\bar{x}$ and $\bar{y}$ are the means of independent variable x and dependent variable y, respectively.

From this equation, values of dependent variable y are estimated for different values of independent variable x and therefore, this equation is also known as '**Prediction equation for y when values of x are given**'.

• **Regression equation of y on x:** The line of regression of x on y is used to predict or estimate the value of x for the given value of the variable y. In this case, x is the dependent variable and y is an independent variable. It can be written in the following form-

$$(x-\bar{x}) = b_{xy}(y-\bar{y})$$

Where, $\bar{x}$ and $\bar{y}$ are the means of independent variable x and dependent variable y, respectively.

From this equation, values of independent variable x are estimated for different values of dependent variable y and therefore, this equation is also known as 'Prediction equation for x when values of y are given'.

MCQ

Q1. Correlation coefficient (r) will be positive when:

(a) X increases, Y decreases (b) X decreases, Y increases

(c) **Both X and Y increases** (d) None of these

Bold means correct answer

Q2. Correlation coefficient (r) will be negative when:

(a) Both X and Y decreases
(b) **X decreases, Y increases**
(c) Both X and Y increases
(d) None of these

Q3. If r = 0, the P.E.(r) will be:

(a) 0
(b) $\frac{0.6745}{\sqrt{n}}$
(c) 1
(d) $\frac{\sqrt{}}{0.6745}$

Q4. If r = 0, the S.E.(r) will be:

(a) 0
(b) $\frac{1}{\sqrt{n}}$
(c) 1
(d) $\sqrt{n}$

Q5. If r = 1, the P.E.(r) will be:

(a) **0**
(b) $\frac{0.6745}{\sqrt{n}}$
(c) 1
(d) $\frac{\sqrt{n}}{0.6745}$

Q6. If r = 1, the S.E.(r) will be:

(a) **0**
(b) $\frac{1}{\sqrt{n}}$
(c) 1
(d) $\sqrt{n}$

Q7. If S.E.(r) = 0, the P.E.(r) will be:

(a) **0**
(b) $\frac{0.6745}{\sqrt{n}}$
(c) 1
(d) $\frac{\sqrt{n}}{0.6745}$

Q8. If S.E.(r) = 1, the P.E.(r) will be:

(a) 0
(b) 1
(c) **0.6745**
(d) 1/0.6745

Q9. If r = 0.4, the coefficient of determination is:

(a) 0.4 (b) 0.02

(c) **0.16** (d) 0.84

Q10. If r = 0.4, the coefficient of non-determination is:

(a) 0.4 (b) 0.02

(c) 0.16 (d) **0.84**

Q11. 50% limits for population correlation coefficient are:

(a) r ± S.E.(r) (b) **r ± P.E.(r)**

(c) r ± 3 S.E.(r) (d) r ± 3 P.E.(r)

Q12. 68.26% limits for population correlation coefficient are:

(a) **r ± S.E.(r)** (b) r ± P.E.(r)

(c) r ± 3 S.E.(r) (d) r ± 3 P.E.(r)

Q13. 99.73% limits for population correlation coefficient are:

(a) r ± S.E.(r) (b) r ± P.E.(r)

(c) **r ± 3 S.E.(r)** (d) r ± 3 P.E.(r)

Q14. If r = 1, then:

(a) **Explained variation = Total variation**

(b) Un-explained variation = Total variation

(c) Explained variation < Total variation

(d) Explained variation > Total variation

Q15. If r = 0, then explained variation will be:

(a) **0** (b) 1

(c) - 1 (d) ∞

Q16. If r = 1, then unexplained variation will be:

(a) **0** (b) 1

(c) - 1 (d) ∞

Q17. If r = 0.6, the coefficient of determination is:

(a) 0.6 (b) 0.12

(c) **0.36** (d) 0.18

Q18. If coefficient of determination is 0.64, then r is equal to:

(a) 0.6 (b) **0.8**

(c) 0.32 (d) 0.16

Q19. If r = 0.8, the P.E.(r) will be:

(a) 0.4 (b) 0.8

(c) 0.64 (d) **Cannot be determined**

Q20. The coefficient of correlation:

(a) 0 to 1 (b) $-\infty$ to 0

(c) 0 to ∞ (d) **-1 to 1**

Q21. If Cov(x,y) = 0, the value of r will be:

(a) -1 (b) **0**

(c) 1 (d) ∞

Q22. If two variables have the linear relationship x + y = 100, the correlation will be:

(a) **-1** (b) 1

(c) 0.20 (d) 0.80

Q23. If x – y = 50, the correlation between x and y will be:

(a) -1 (b) **1**

(c) 0.40 (d) 0.50

Q24. If x' = 2x and y' = y+3 and r_{xy} = +0.80, then $r_{x'y'}$ will be:

(a) – 0.80 (b) **+ 0.80**

(c) + 0.40 (d) – 0.40

Q25. If the sum of the product of deviations of x and y series from their means is zero, the correlation coefficient will be:

(a) -1 (b) 1

(c) **0** (d) + 0.5

Q26. The coefficient of concurrent deviation gives correlation coefficient which varies between:

(a) – 1 and 0 (b) **– 1 and +1**

(c) 0 and 1 (d) None of these

Q27. If $\sum D^2 = 0$, the coefficient of rank correlation will be:

(a) - 1 **(b) +1**

(c) 0 (d) ∞

Q28. If $\sum D^2 = 1$, n = 4, the coefficient of rank correlation will be:

(a) 0 (b) 0.10

(c) **0.90** (d) 1

Q29. If $\rho = 0.2$ and n = 10, then $\sum D^2$ is equal to:

(a) 100 (b) **132**

(c) 394 (d) 990

Q30. If $\rho = 2/3$ and $\sum D^2 = 55$, then the number of pairs of observations (n) is equal to:

(a) 5 (b) **10**

(c) 15 (d) 20

Q31. If $\sum D^2 = 57.50$, n = 6, $m_1 = 3$, and $m_2 = 2$, then the coefficient of rank correlation will be:

(a) 0.50 (b) 0.60

(c) **0.71** (d) 0.81

Q32. If n = 8, C = 5, then the coefficient of concurrent deviation (r_c) will be:

(a) -1 (b) 0

(c) **+0.5** (d) +1

Q33. If r = 1, then

(a) **Cov (x, y) =** $\boldsymbol{\sigma_x \sigma_y}$ (b) Cov (x, y) < $\sigma_x \sigma_y$

(c) Cov (x, y) > $\sigma_x \sigma_y$ (d) None of these

Q34. Coefficient of correlation is significant if:

(a) r < P.E.(r) (b) r > P.E.(r)

(c) r < 6 P.E.(r) (d) **r > 6 P.E.(r)**

Q35. If r = 0.6, then coefficient of non-determination is:

(a) 0 (b) 0.32

(c) 0.64 (d) **0.8**

Q36. Given that explained variation = 20 and unexplained variation = 40, the coefficient of determination is:

(a) 1/2 (b) **1/3**

(c) 1/4 (d) 1/8

Q37. Given that unexplained variation = 60 and total variation = 100, the coefficient of determination is:

(a) **0.40** (b) 0.60

(c) 40 (d) 60

Q38. Let the coefficient of determination computed to be 0.40 in a problem involving one independent variable and one dependent variable. This result means that:

(a) Relationship between the two variables is positive

(b) Correlation coefficient is also equal to 0.40

(c) **40% of the total variation is explained by the independent variable**

(d) 40% of the total variation is explained by the dependent variable

Q39. The ratio of explained variation to the total variation is called:

(a) Coefficient of correlation

(b) **Coefficient of determination**

(c) Coefficient of rank correlation

(d) Coefficient of concurrent correlation

Q40. If the scatter diagram is drawn the scatter points lie on a straight line then it indicates:

(a) No correlation (b) Skewness

(c) **Perfect correlation** (d) None of these

Q41. Spearman's rank correlation is used for calculating correlation of coefficient of:

(a) Discrete variables (b) Continuous variables

(c) Qualitative data (d) **Quantitative data**

Q42. No correlation is found when:

(a) Two variables are dependent

(b) **Two variables are independent**

(c) Scattered points move in same direction

(d) Scattered points move in opposite direction

Q43. Scatter diagram shows that:

(a) Positive correlation (b) Negative correlation

(c) No correlation (d) **All of them**

Q44. Concurrent deviation method is associated with:

(a) Qualitative measurement

(b) Quantitative measurement

(c) **Correlation coefficient of variables of two qualitative characters**

(d) None of these

Q45. The term correlation was introduced by:

(a) R. A Fisher

(b) **Karl Pearson**

(c) Sir Francis Galton

(d) W. S. Gossett

Q46. If r is the simple correlation coefficient, the quantity r^2 is known as:

(a) Coefficient of alienation

(b) **Coefficient of determination**

(c) Coefficient of non-determination

(d) None of these

Q47. If r is the simple correlation coefficient, the quantity $\sqrt{(1-r^2)}$ is known as:

(a) **Coefficient of alienation** (b) Coefficient of determination

(c) Coefficient of non-determination (d) None of these

Q48. If r is the simple correlation coefficient, the quantity $(1 - r^2)$ is known as:

(a) Coefficient of alienation (b) Coefficient of determination

(c) **Coefficient of non-determination** (d) None of these

Q49. If $r_{xy} = 0$, then the variables X and Y is:

(a) Dependent (b) Linearly related

(c) **Not linearly related** (d) None of these

Q50. If $r_{xy} = 1$, then the relation between two variables X and Y is:

(a) Y is equal to X

(b) **When Y increases, X proportionately increases**

(c) When Y increases, X proportionately decreases

(d) None of these

Q51. If $r_{xy} = -1$, then the relation between two variables X and Y is:

(a) Y is equal to X

(b) When Y increases, X proportionately increases

(c) **When Y increases, X proportionately decreases**

(d) None of these

Q52. S.E.(r) based on n paired values is:

(a) $(1 - r^2)$

(b) $\sqrt{(1 - r^2)}$

(c) $\mathbf{\frac{1 - r^2}{\sqrt{n}}}$

(d) $\frac{1 + r^2}{\sqrt{n}}$

Q53. The limit of correlation ratio is:

(a) -1 to 0

(b) **0 to 1**

(c) – 1 to 1

(d) 0 to ∞

Q54. The relation between correlation coefficient r and correlation ratio η^2 is:

(a) $r^2 > \eta^2$

(b) $\mathbf{r^2 < \eta^2}$

(c) $r^2 = \eta^2$

(d) None of these

Q55. Which of the following relation is correct?

(a) $r_{13} = r_{23}$

(b) $r_{12.34} = r_{13.24}$

(c) $\mathbf{r_{12.3} = r_{21.3}}$

(d) $r_{12.3} = r_{13.2}$

Q56. If the covariance between two variables is positive, it means that:

(a) The variables would change in the opposite direction

(b) **The variables would change in the same direction**

(c) The variables would not change

(d) None of these

Q57. If the covariance between two variables is negative, it means that:

(a) **The variables would change in the opposite direction**

(b) The variables would change in the same direction

(c) The variables would not change

(d) None of these

Q58. Correlation between the directions of deviations is calculated by:

(a) Simple correlation coefficient

(b) Rank correlation

(c) **Concurrent correlation**

(d) Partial correlation

Q59. If all the observations are in one group, then correlation ratio E^2 is:

(a) -1 to 0 (b) **0**

(c) 1 (d) -1 to 1

Q60. If each group consists of only one observation, then correlation ratio E^2 is:

(a) -1 to 0 (b) 0

(c) **1** (d) -1 to 1

Q61. If r is 0.8, then coefficient of alienation shall be:

(a) 0.4 (b) **0.6**

(c) 0.8 (d) 0.64

Q62. If r is 0.8, then coefficient of determination shall be:

(a) 0.4 (b) 0.6

(c) 0.8 (d) **0.64**

Q63. If r is 0.8, then coefficient of non-determination shall be:

(a) 0.64 (b) **0.36**

(c) 0.18 (d) 0.6

Q64. In a scatter diagram, if the plotted points come to a straight line falling from the lower left hand corner to the upper right hand corner, then the correlation is said to be:

(a) **Perfect positive** (b) Perfect negative

(c) No correlation (d) Curvi-linear

Q65. If sum of the product of deviations of X and Y series from their means is zero, then r shall be:

(a) -1
(b) **0**
(c) 1
(d) ∞

Q66. In a scatter diagram, if the plotted points come to a straight line rising from the upper left hand corner to the lower right hand corner, then the correlation is said to be:

(a) Perfect positive
(b) **Perfect negative**
(c) No correlation
(d) Curvi-linear

Q67. In a scatter diagram, if the plotted points lie on a straight line parallel to the X-axis or in a haphazard manner, then the relationship between variables is said to be:

(a) Perfect positive
(b) Perfect negative
(c) No correlation
(d) Curvi-linear

Q68. Rank correlation coefficient was given by:

(a) R. A. Fisher
(b) Karl Pearson
(c) W.S. Gossett
(d) **Spearman**

Q69. If two variates X and Y are independent, then the angle between the two lines of regression is:

(a) π
(b) 2π
(c) **$\pi/2$**
(d) $\pi/3$

Q70. If two variates X and Y are perfect correlated, then the angle between the two lines of regression is:

(a) **0**
(b) π
(c) $\pi/2$
(d) $\pi/3$

Q71. A relationship where the flow of data points is best represented by a curve is called:

(a) **Linear relationship**
(b) Non-linear relationship
(c) Residual
(d) None of these

Q72. All data points falling along a straight line is called:

(a) **Linear relationship**
(b) Non-linear relationship
(c) Residual
(d) None of these

Q73. The value we would predict for the dependent variable when the independent variables are all equal to zero is called:

(a) **Intercept** (b) Slope

(c) Sum of residual (d) None of these

Q74. The predicted rate of response of the dependent variable to changes in the independent variable is called:

(a) Intercept (b) **Slope**

(c) Sum of residual (d) None of these

Q75. In simple regression equation the number of variables involved is:

(a) 1 (b) **2**

(c) 3 (d) 4

Q76. If the value of any regression coefficient is zero, the two variables are:

(a) **Independent** (b) Dependent

(c) Qualitative (d) None of these

Q77. In simple linear regression, the numbers of unknown constants are:

(a) 0 (b) 1

(c) **2** (d) 3

Q78. The slope of the regression line of Y on X is also called:

(a) Correlation coefficient of Y on X

(b) Regression coefficient of Y on X

(c) Correlation coefficient of X on Y

(d) Regression coefficient of X on Y

Q79. A process by which we estimate the value of independent variable on the basis of one or more independent variables is called:

(a) Residual (b) Slope

(c) Regression (d) Correlation

Q80. The coefficient of correlation is:

(a) Square of the coefficient of determination

(b) **Square root of the coefficient of determination**

(c) Same as coefficient of determination

(d) Square root of the coefficient of non-determination

Q81. If the coefficient of determination is equal to 1, then the correlation coefficient:

(a) Must be equal to 1

(b) Must be equal to -1

(c) **Either -1 or +1**

(d) Any value in between -1 to +1

Q82. According to least squares method, the residual sum of the square is:

(a) Maximum

(b) **Minimum**

(c) Increased

(d) Decreased

Q83. The straight line graph of the linear equation Y = a+ bX, slope will be upward if:

(a) $b = 0$

(b) $b < 0$

(c) $\mathbf{b > 0}$

(d) $b \neq 0$

Q84. The straight line graph of the linear equation Y = a+ bX, slope will be downward if:

(a) $b = 0$

(b) $\mathbf{b < 0}$

(c) $b > 0$

(d) $b \neq 0$

Q85. The straight line graph of the linear equation Y = a+ bX, slope will be downward if:

(a) $a = b$

(b) $\mathbf{b = 0}$

(c) $b = 1$

(d) $b \neq 0$

Q86. If Y = 2 - 0.2X, then the value of Y intercept is equal to:

(a) -0.2

(b) **2**

(c) 0.2X

(d) All of the above

Q87. If regression line of = 5, then value of regression coefficient of Y on X is:

(a) **0**

(b) 0.5

(c) 1

(d) 5

Q88. If one regression coefficient is greater than one, then other will be:

(a) **Less than one**

(b) More than one

(c) Equal to one

(d) Equal to zero

Q89. In the regression equation X = a + bY, the X is called:

(a) Independent variable
(b) **Dependent variable**
(c) Qualitative variable
(d) None of the above

Q90. In the regression equation Y = a + bX, the Y is called:

(a) Independent variable
(b) **Dependent variable**
(c) Qualitative variable
(d) None of the above

Q91. In the regression equation X = a + bY, the Y is called:

(a) **Independent variable**
(b) Dependent variable
(c) Qualitative variable
(d) None of the above

Q92. In the regression equation Y = a + bX, the X is called:

(a) **Independent variable**
(b) Dependent variable
(c) Qualitative variable
(d) None of the above

Q93. In the regression equation Y = a +bX, here a is called:

(a) Independent variable
(b) Dependent variable
(c) **Y-intercept**
(d) X-intercept

Q94. In the regression equation X = a +bY, here a is called:

(a) Independent variable
(b) Dependent variable
(c) Y-intercept
(d) **X-intercept**

Q95. In the regression equation X = a + bY, here b is called:

(a) **Regression coefficient of X on Y**
(b) Regression coefficient of Y on X
(c) Correlation coefficient of X and Y
(d) Correlation coefficient of Y and X

Q96. In the regression equation Y = a + bX, here b is called:

(a) Regression coefficient of X on Y
(b) **Regression coefficient of Y on X**
(c) Correlation coefficient of X and Y
(d) Correlation coefficient of Y and X

Q97. The regression equation always passes through:

(a) (X, Y) (b) (a, b)

(c) **$(\bar{X},\bar{Y})$** (d) $(\bar{X},Y)$

Q98. The graph showing the paired points of (X_i , Y_i) is called:

(a) **Scatter diagram** (b) Histogram

(c) Bar diagram (d) Pie diagram

Q99. The independent variable in a regression line is:

(a) Random variable (b) **Non-random variable**

(c) Qualitative variable (d) None of the above

Q100.When b_{XY} is positive, then byx will be:

(a) Zero (b) One

(c) Negative (d) **Positive**

Q101.When regression line passes through the origin, then:

(a) Association is zero (b) **Intercept is zero**

(c) Regression coefficient is zero (d) Correlation is zero

Q102 The correlation coefficient is the............... of two regression coefficients:

(a) Arithmetic mean (b) **Geometric mean**

(c) Harmonic mean (d) Median

Q103.It is possible that two regression coefficients have:

(a) Opposite signs (b) **Same signs**

(c) No sign (d) None of these

Q104.Regression coefficient is independent of:

(a) Scale (b) Origin

(c) **Both (a) and (b)** (d) Nether (a) nor (b)

Q105.In the regression line Y = a+ bX:

(a) $\sum X = \sum Y$ (b) $\sum X = \sum \hat{X}$

(c) **$\sum Y = \sum \hat{Y}$** (d) $X = Y$

Q106.In the regression line Y = a + bX, the following is always true:

(a) $\sum(X-\hat{X}) = \sum(Y-\hat{Y})$ (b) $\sum(X-\hat{X}) = 0$

(c) **$\sum(Y-\hat{Y}) = 0$** (d) $\sum(Y - \hat{Y})^2 = 0$

Q107.In the regression line X = a + bY, the follo wing is always true:

(a) $\sum(X-\hat{X}) = \sum(Y-Y^{\wedge})$
(b) **$\sum(X-\hat{X}) = 0$**
(c) $\sum(Y-\hat{Y}) = 0$
(d) $\sum(Y-\hat{Y})^2 = 0$

Q108.When the ratio of variations in the related variables is constant, it is called:

(a) **Linear correlation**
(b) Nonlinear correlation
(c) Positive correlation
(d) Negative correlation

Q109.The sum of the difference between the actual values of Y and its values obtained from the fitted regression line is always:

(a) **Zero**
(b) Positive
(c) Negative
(d) Minimum

Q110.If all the actual and estimated values of Y are same on the regression line, the sum of squares of error will be:

(a) **Zero**
(b) Minimum
(c) Maximum
(d) Unknown

Q111.A measure of the strength of the linear relationship that exists between two variables is called:

(a) Slope
(b) Intercept
(c) **Correlation coefficient**
(d) Regression equation

Q112.If both variables X and Y increase or decrease simultaneously, then the coefficient of correlation will be:

(a) **Positive**
(b) Negative
(c) Zero
(d) One

Q113.If one variable X increase or decrease and Y decrease or increase, then the coefficient of correlation will be:

(a) Positive
(b) **Negative**
(c) Zero
(d) One

Q114.If one item is fixed and unchangeable and the other item varies, the correlation coefficient will be:

(a) Positive
(b) Negative
(c) **Zero**
(d) Undecided

Q115.In scatter diagram, if most of the points lie in the first and third quadrants, then coefficient of correlation is:

(a) Negative (b) **Positive**

(c) Zero (d) All of the above

Q116.If the two series move in reverse directions and the variations in their values are always proportionate, it is said to be:

(a) Limited negative correlation (b) Limited positive correlation

(c) **Perfect negative correlation** (d) Perfect positive correlation

Q117.If both the series move in the same direction and the variations are in a fixed proportion, correlation between them is said to be:

(a) Perfect negative correlation (b) **Perfect positive correlation**

(c) Linear correlation (d) Nonlinear correlation

Q118.If X is measured in kg and Y is measured in inches, then correlation coefficient has the unit:

(a) kg (b) Inches

(c) Both (a) and (b) (d) **No unit**

Q119.The signs of regression coefficients and correlation coefficient are always:

(a) Different (b) **Same**

(c) Positive (d) Negative

Q120.The range of regression coefficient is:

(a) -1 to +1 (b) 0 to 1

(c) **-∞ to +∞** (d) 0 to ∞

Q121.The arithmetic mean of the two regression coefficients is greater than or equal to:

(a) -1 (b) +1

(c) 0 (d) **r**

Q122.If the value of regression coefficient is zero, then the two variables are called:

(a) Independent (b) Dependent

(c) **Both (a) and (b)** (d) Neither (a) nor (b)

Q123.The term regression was used by:

(a) Fisher (b) Pearson

(c) Spearman (d) **Galton**

Q124.When the two regression lines are parallel to each other, then their slopes are:

(a) Zero (b) Different

(c) **Same** (d) Positive

Q125.In correlation problem both variables are:

(a) Unknown (b) Fixed

(c) **Random** (d) Equal

Q126.In the regression equation Y = a + bX, where a and b is called:

(a) Constants (b) Estimates

(c) **Both (a) and (b)** (d) Neither (a) nor (b)

Q127.The correlation coefficient between X and -X is:

(a) **- 1** (b) 0

(c) 1 (d) ∞

Q128.If $b_{yx} = b_{xy} = r_{xy}$, then:

(a) $\sigma_x \neq \sigma_y$ (b) $\boldsymbol{\sigma_x = \sigma_y}$

(c) $\sigma_x > \sigma_y$ (d) $\sigma_x < \sigma_y$

Q129.If $r_{xy} = 0.2$, then $r_{(2x,\, 2y)}$ is equal to:

(a) **0.2** (b) 0.4

(c) 0 (d) 1

Q130.If $b_{yx} = -4$ and $r_{xy} = -1$, then b_{xy} is equal to:

(a) **-1/4** (b) -4

(c) -1 (d) 1/4

Q131.If $b_{yx} = -0.8$ and $b_{xy} = -0.2$, then r_{yx} is equal to:

(a) **-0.4** (b) 0.4

(c) -0.8 (d) 0.8

Q132.If $\hat{Y} = 4 - X$, then r will be:

(a) 0 (b) 1

(c) **-1** (d) 1/4

Q133.If $\hat{Y} = X + 5$, then r equal to:

(a) **1** (b) -1

(c) 1/2 (d) -1/2

Q134.If Y = 10X and X = 0.1Y, then r is equal to:

(a) 0.1 (b) **1**

(c) -1 (d) 10

Q135.If $r_{xy} = 1$, then:

(a) $b_{yx} = b_{xy}$ (b) $b_{yx} > b_{xy}$

(c) $b_{yx} < b_{xy}$ (d) $\mathbf{b_{yx} \cdot b_{xy} = 1}$

Q136.The relation between the regression coefficient b_{yx} and correlation coefficient(r) is:

(a) σ_y/σ_x (b) σ_x/σ_y

(c) $r.\sigma_x/\sigma_y$ (d) $\mathbf{r.\sigma_y/\sigma_x}$

Q137.The relation between the regression coefficient b_{xy} and correlation coefficient(r) is:

(a) σ_y/σ_x (b) σ_x/σ_y

(c) $\mathbf{r.\sigma_x/\sigma_y}$ (d) $r.\sigma_y/\sigma_x$

Q138.If the sum of the product of the deviation of X and Y from their means is zero, the correlation coefficient between X and Y is:

(a) **Zero** (b) Maximum

(c) Minimum (d) Undecided

Q139.If the coefficient of correlation between the variables X and Y is r, the coefficient of correlation between X^2 and Y^2 is:

(a) r (b) $\mathbf{r^2}$

(c) 0 (d) 1

Q140.If $r_{xy} = 0.4$, then r_{yx} will be:

(a) **0.4** (b) - 0.4

(c) 1/0.4 (d) -1/0.4

Q141.If $\hat{Y} = a$, then b_{yx} is equal to:

(a) **0** (b) 1

(c) a (d) None of these

Q142.If $\hat{Y} = bX$, then a is equal to:

(a) **0** (b) 1

(c) b (d) None of these

Q143.If $\hat{X} = bY$, then a is equal to:

(a) **0** (b) 1

(c) b (d) None of these

Q144.When $r_{xy} < 0$, then b_{yx} and b_{xy} will be:

(a) Zero (b) Not equal to zero

(c) **< 0** (d) > 0

Q145.When $r_{xy} > 0$, then b_{yx} and b_{xy} are both:

(a) Zero (b) Not equal to zero

(c) < 0 (d) **> 0**

Q146.If $r_{xy} = 0$, then:

(a) $b_{yx} = 0$ (b) $b_{xy} = 0$

(c) $b_{yx} \neq b_{xy}$ (d) **Both (a) and (b)**

Q147.A regression model may be:

(a) Linear (b) Non-linear

(c) **Both (a) and (b)** (d) Neither (a) and (b):

Q148.In regression equation y=α+βx+e, both x and y variables are:

(a) Y is fixed and X is random (b) **X is fixed and Y is random**

(c) Fixed (d) Random

Q149.If the regression equation is equal to Y=23.0+43.0 X, then 23.0 is the _____ while 43.0 is the ____ of the regression line:

(a) Slope, Intercept (b) **Intercept, slope**

(c) Slope, regression coefficient (d) Regression coefficient, slope

Q150.If the equation of regression line is y=3, then what result will you take out from it?

(a) **The line is parallel to X-axis**

(b) The line is parallel to Y-axis

(c) The line is perpendicular to X-axis

(d) The line is perpendicular to Y-axis

Q151.The product of two regression coefficients is:

(a) 1 (b) Greater than 1

(c) **Less than 1** (d) None of these

Q152.For minimizing the error sum of squares, we obtain:

(a) Standard error of estimate (b) Correlation coefficient

(c) **Normal equations** (d) None of these

Q153.The standard form of the line of regression of Y on X is:

(a) $\mathbf{Y = a + bX}$ (b) $X = a + bX$

(c) $aX + bY + c = 0$ (d) $X + Y + c = 0$

Q154.The standard form of the line of regression of X on Y is:

(a) $Y = a + bX$ (b) $\mathbf{X = a + bX}$

(c) $aX + bY + c = 0$ (d) $X + Y + c = 0$

Q155.If $3X + 2Y = 6$ is the regression of Y on X, then b_{yx} is,

(a) 2/3 (b) 3/2

(c) **– 3/2** (d) -2/3

Q156.If $3X + 2Y = 6$ is the regression of X on Y, then b_{xy} is,

(a) 2/3 (b) 3/2

(c) – 3/2 (d) **-2/3**

Q157. In a bivariate study, the line of regression of Y on X is 4X – 3Y = 7, the correlation between X and Y will be:

(a) **Positive** (b) Negative

(c) Zero (d) One

Q158. In a bivariate study, the line of regression of X on Y is 4X – 3Y = 7, the correlation between X and Y will be:

(a) **Positive** (b) Negative

(c) Zero (d) One

Q159. In a bivariate study, the line of regression of X on Y is 3Y + 4X = 5, the correlation between Y and X will be:

(a) Positive (b) **Negative**

(c) Zero (d) One

Q160. If b_{yx} = - 3/2 and b_{xy} = - 1/6, the value of correlation coefficient (r) will be:

(a) 1/4 (b) -1/4

(c) 1/2 (d) **-1/2**

Q161. If b_{yx} = 1/3 and b_{xy} = 3/4, the value correlation coefficient (r) will be:

(a) **1/2** (b) -1/2

(c) 1/4 (d) -1/4

Q162. If b_{xy} is positive, then b_{yx} is:

(a) Negative (b) **Positive**

(c) Undetermined (d) None of these

Q163. The change in the variable X corresponding to a unit change in the variable Y is measured by:

(a) b_{yx} (b) $\mathbf{b_{xy}}$

(c) r_{yx} (d) r_{xy}

Q164. The line of best fit is obtained by:

(a) **Method of least squares** (b) Standard error of estimate

(c) Correlation analysis (d) None of these

Q165. The line of best fit to observed data is known as:

(a) Correlation coefficient (b) **Regression line**

(c) Standard error (d) None of these

Q166.In regression of X on Y, Y is:

(a) **Explanatory variable**
(b) Explained variable
(c) Constant
(d) None of these

Q167.In regression of X on Y, X is:

(a) Explanatory variable
(b) **Explained variable**
(c) Constant
(d) None of these

Q168.For predicting the best value of X for given Y, we make use of:

(a) **Regression of X on Y**
(b) Regression of Y on X
(c) Correlation of X on Y
(d) Correlation of Y on X

Q169.If "time" is used as the independent variable in a simple linear regression analysis, then which of the following assumption could be violated

(a) The residual variation is the same for all fitted values of Y
(b) The residuals are normally distributed
(c) **Successive observations of the dependent variable are uncorrelated**
(d) There is a linear relationship between the independent and dependent variables

Q170.Relationship between the correlation coefficient and coefficient of determination is that:

(a) The coefficient of determination is the square root of the coefficient of correlation
(b) **The coefficient of determination is the coefficient of correlation squared**
(c) Both are unrelated
(d) Both are equal

Q171.The strength (degree) of the correlation between a set of independent variables X and a dependent variable Y is measured by:

(a) Standard error of estimate
(b) Coefficient of determination
(c) Coefficient of correlation
(d) **All of them**

Q172.Collinearity or Multicollinearity exists when:

(a) An independent variable is strongly correlated with a dependent variable

(b) **Independent variables are correlated less than -0.70 or more than 0.70**

(c) There is only one independent variable

(d) Relationship between the dependent and independent variable is non-linear

Q173.The percent of the total variation of the dependent variable Y explained by the set of independent variables X is measured by:

(a) Standard error of estimate

(b) **Coefficient of determination**

(c) Coefficient of correlation

(d) Coefficient of skewness

Q174.Relationship between the correlation coefficient and coefficient of determination is:

(a) Both are equal

(b) Both are unrelated

(c) Coefficient of correlation = (Coefficient of determination)^2

(d) **Coefficient of determination = (Coefficient of correlation)^2**

Q175.A coefficient of correlation is computed to be -0.95 means:

(a) The correlation coefficient cannot have this value

(b) **The relationship between the two variables is strong but negative**

(c) The relationship between the two variables is strong but positive

(d) None of these

Q176.If the coefficient of determination computed to be 0.39 in a problem involving one independent variable and one dependent variable, that means:

(a) 39% of the total variation is explained by the dependent variable

(b) **39% of the total variation is explained by the independent variable**

(c) Correlation coefficient is also 0.39

(d) None of these

Q177.The Coefficient of Correlation r is independent of:

(a) Origin only

(b) Scale of measurement

(c) **Both change of origin and scale of measurement**

(d) None of these

Q178. If $b_{yx} < 0$ and $b_{yx} \leq 0$, then:

(a) $r = 0$ (b) $\mathbf{r < 0}$

(c) $r > 0$ (d) None of these

Q179.Two regression lines are parallel to each other if their slope is:

(a) Negative (b) Different

(c) **Slope** (d) None of these

Q180.If X and Y are independent of each other, the coefficient of correlation is:

(a) -1 (b) $+1$

(c) **0** (d) ∞

Q181. If $r = 0.6$, $b_{yx} = 1$, then $b_{xy} = ?$

(a) 0.20 (b) **0.36**

(c) 0.40 (d) 0.72

Q182.In a linear regression equation, Y=a + bX, what is the b denote?

(a) **The regression coefficient, the slope of the line**

(b) The intercept with the Y-axis

(c) The correlation coefficient, the strength of the line

(d) The score on the variable X

Q183.Assumptions for partial correlation are usual assumptions of:

(a) Perfect correlation (b) **Pearson correlation**

(c) Exponential correlation (d) None of these

Q184.In multiple regression analysis, two variables may have relationship of:

(a) Exact linear relationship (b) In-exact linear relationship

(c) Exact non linear relationship (d) **Both (b) and (c)**

Q185.Linear association between two variables say x and y can be expressed through:

(a) **Correlation coefficient** (b) Ranked correlation

(c) Concurrent correlation (d) None of these

Q186.Exact linear relationship between some or all independent variables depicts:

(a) Absolute-collinearity (b) Homo-collinearity

(c) Poly-collinearity (d) **Multicollinearity**

Q187.Property of symmetry for coefficient of correlation:

(a) **Satisfies** (b) Not Satisfies

(c) Validates (d) Not validates

Q188.If multiple coefficient of determination (R^2) depicts a perfect relationship, its value should be equal to

(a) -1 (b) 0

(c) **1** (d) None of these

Q189.While performing correlation, we draw scatter plot to check assumptions of:

(a) Unbiasness (b) Multicollinearity

(c) **Linearity** (d) None of these

Q190.Regression model must be linear in:

(a) **Parameters** (b) Variables

(c) Slope coefficient (d) Variances

Q191.In coefficient of partial correlation all other variables are kept:

(a) Deviated (b) Residual mean

(c) Not fixed (d) **Constant**

Q192.Growth and trend factors are common source of

(a) Biased factoring (b) Homogeneity

(c) **Multicollinearity** (d) Interception

Q193.If multiple coefficient of determination R^2 is equals to 1, adjusted R^2 will be

(a) -1 (b) 0

(c) **1** (d) ∞

Q194.In regression model, for any value of variable x, variance of error term should be:

(a) Repeated (b) Zero

(c) **Same** (d) Not fixed

Q195.In equation [1-(1- R^2) * (n-1/n-k)] if condition is (k < 1) then which of following holds a true value?

(a) Adjusted $R^2 > R^2$ (b) **Adjusted $R^2 < R^2$**

(c) Adjusted $R^2 = R^2$ (d) Adjusted $R^2 < -R^2$

Q196.High correlation is not an indication of:

(a) Relationship measure (b) **Cause and effect**

(c) Unit-free measures (d) Scale of measurement

Q197.Computation of multiple coefficient of determination always outputs a:

(a) Zero (b) -1

(c) **Positive value** (d) Negative value

Q198. If all the values fall on the same straight line and the line has a positive slope then what will be the value of the correlation coefficient (r):

(a) $r = 0$ (b) $r = -1$

(c) **$r = +1$** (d) $r < 1$

Q199.The regression line always passes through:

(a) Origin

(b) Median point of X and Y

(c) **Mean point of X and Y**

(d) Median point of X and Mean of Y

Q200.In the least square regression line, the quantity is always:

(a) Positive (b) Negative

(c) **Zero** (d) Fractional

Q201.If the scattered diagram is drawn the scatter points lie on a straight line then it indicates:

(a) No correlation (b) **Perfect correlation**

(c) Limited correlation (d) None of these

Q202. When the regression line passes through the origin then:

(a) Regression coefficient is zero
(b) Correlation is zero
(c) **Intercept is zero**
(d) Association is zero

Q203. If $b_{yx} < 0$ and $b_{xy} < 0$, then:

(a) $r = 0$
(b) $\mathbf{r < 0}$
(c) $r > 0$
(d) None of these

8

Probability Theory

The theory of probability has its origin in the games of chance related to gambling such as throwing a die, tossing a coin, drawing cards from a pack of cards etc. **Jerame Cardon** (1501-1576), an Italian mathematician was the first man to write a book on the subject entitled "**Book on Games and Chance**", which was published after his death in 1663. The systematic and scientific foundation of the mathematical theory of probability was laid in mid-seventeenth century by two French mathematicians **B. Pascal** (1623-1662) and **Pierre de Fermat** (1601-1665).

Important terms and concepts

To understand probability, let us first get acquainted with some important terms and concepts associated with probability.

- **Trial:** An experiment is any natural or planned activity for which it is not possible to predict the particular result called a **trial.**
- **Events:** The possible outcomes of experiments of a trial are called **events.** **For example:** Tossing of a coin in an experiment and one toss of the coin is called **trial**. In tossing of a coin there are only two outcomes or **events**, i.e. head or tail.

Types of events

Events can be of following different types:

- **Exhaustive events**
- **Equally likely events**
- **Mutually exclusive events**
- **Dependent events**
- **Independent events**
- **Null or impossible events**
- **Exhaustive events:** It is the total number of all the possible outcomes of an experiment.

- **Equally likely events:** It is likelihood of the occurrence of every event in an experiment or happening is the same, it represents **equally likely event**.
- **Mutually exclusive events:** When occurrence of one event preludes the possibility of occurrence of another event, then two events are called **mutually exclusive events.** It means both events cannot occur simultaneously and the occurring of one event does not allow the occurrence of the other event.
- **Dependent events:** When the occurrence of one event influences the occurrence of the other event, the second event is said to be **dependent event.**
- **Independent events:** When the occurrence of one event does not affect the occurrence of the other event, it is said to be **independent event**.
- **Null or impossible event:** An event having no sample point i.e. no possibility of occurrence is called a **null event or impossible event**.
- **Sample space and Sample points:** A set of all possible outcomes from a random experiment is called a **sample space** and the components of sample space are known as **sample points.**

Set theory

- **Sets and Elements: A set is a** collection of objects or members. Each object in a set is called an **element** of the set. **Sets** are commonly represented using curly brackets containing a collection of **elements** of the set, separated by commas. Usually the sets are denoted by capital letters like: A, B, C,......... and the small letters like: a,b,c,....... denoted the elements of a set. **For example:** The set A has the elements 1,2,3, it may be written as: A = {1,2,3} tells us that A is a **set** whose **elements** are the first 3 natural numbers.

  Note: In set notations the symbol $\in$ means "**belongs to**" and symbol $\notin$ means "**does not belong to**". For example: For a set A = {1,2,3}; $2 \in A$ and $4 \notin A$.
- **Subset:** If every element of a set A is also an element of B, then A is called a **subset** of B. Symbolically, $A \subset B$ denotes that A is a subset of B.
- **Random experiment:** A random experiment is any attempt or operations whose results or outcomes can not be predicted with certainty.
- **Sample space:** The **sample space** of a random experiment, denoted by S, is the set of all the possible outcomes of the experiment. Each element of this set is called a **sample point** or **elementary event**.
- **Vann-Diagrams:** The relationship between subsets and the sample space S can be represented by Vann-diagrams. In this diagram the sample space

S is represented by a rectangle and the subsets by circles drawn inside the rectangles.

- **Union of two sets:** The union of two sets A and B is the set of elements that belongs to A or B or both. Symbolically, we write $A \cup B$ for the union of A and B.
- **Intersection of two sets:** The intersection of two sets A and B is the set of elements that are common to both A and B. Symbolically, we write $A \cap B$ for the intersection of A and B.
- **Disjoint sets:** The intersection of two sets is a null set; the two sets are called disjoint set. Thus, A and B are disjoint if $A \cap B = \varphi$.
- **Null set or empty set:** A set which contains no elements is called a null set or empty set, denoted by φ.
- **Complement of a set:** If A is a subset of the universal set U, then the complement of A with respect to U is the set of all elements of U that are not in A. The complement of A is denoted by A' or A^c.

Probability theory

The Probability Theory provides a means of getting an idea of the likelihood of occurrence of different events resulting from a random experiment in terms of quantitative measures between zero and one. Probability can be defined in three ways-

1) Mathematical or Classical or a-priori probability
2) Statistical or empirical or a-posteriori probability
3) Axiomatic approach to probability

1) Mathematical or Classical or a-priori probability: If an experiment has 'n' exhaustive, mutually exclusive and equally likely cases and out of which, 'm' are favourable to the occurrence(happening) of an event (A), the probability of the happening of the event A, is denoted by P(A) and is represented as:

$$P(A) = \frac{m}{n} = \frac{\textit{Number of favourable cases}}{\textit{Total number of cases}}$$

and the probability that the event A does not happen will be

$$P(\bar{A}) = \frac{n-m}{n} = \frac{\textit{Number of unfavourable cases}}{\textit{Total number of cases}}$$

Clearly,

$$P(\bar{A}) = \frac{n-m}{n} = 1 - \frac{m}{n} = 1 - P(A)$$

$$\Rightarrow P(A) = 1\text{-}P(\bar{A})$$

$$\Rightarrow P(A) + P(\bar{A}) = 1$$

Note

- The probability of an event which can occur is 0 to 1.
- The probability of an impossible event is zero, i.e. P(A) =0.
- Since $0 \leq m \leq n$, the probability of an event is a positive quantity, *i.e.*, P(A)>0.
- The probability of a sure event is 1, *i.e.* P(A) =1, therefore, the probability of happening of an event ranges between 0 and 1, i.e., $0 \leq P(A) \leq 1$.
- The sum of the probabilities of happening (success) and not happening (failure) of an event is always equal to 1.

Limitations

The classical definition of probability fails to give the probability of an event in the following cases:

- If the exhaustive number of cases (n) in a trial is infinite.
- When various outcomes of a trial are not equally likely.
- It is applicable only when sample space is finitely enumerable.

2) Statistical or empirical or a-posteriori probability: If a trial is repeated a number of times (n times) under identical conditions, the limiting value of the ratio of the number of times the event A happens to the number of trials, as the number of trials becomes infinite is called the probability of happening of the event A. Symbolically, if in n trials, an event H happens n times, then probability of event A is given by:

$$P(A) = \lim_{n \to \infty} \frac{m}{n}$$

Limitations

The statistical definition of probability fails to give the probability of an event in the following cases:

- It is difficult to prove the existence of a limit to the relative frequency.

3) Axiomatic approach to probability: The axiomatic approach, proposed by a Russian mathematician, **A. N. Kolmogrov**, in the year 1933 includes both the 'Classical' and the 'Statistical' definitions as special cases and also overcomes the limitations in each definition. According to the axiomatic approach to probability, the probability P of an event A with regard to a sample space S of an experiment satisfies the following axioms-

- $0 \leq P(A) \leq 1$
- $P(S) = 1$
- $P(A) = P(A_1)+P(A_2)+P(A_3)+\ldots\ldots\ldots\ldots+P(A_r)$

  Where $A_1, A_2, \ldots\ldots, A_r$ are the sample points comprising the event A.

Some Probability Theorems

There are some important theorems of probability-

(i) Addition theorem

(ii) Multiplication theorem

(iii) Baye's theorem

(i) **Addition theorem:** For any two events A and B are **mutually exclusive**, the probability that any one of them would happen is the sum of the probabilities of the happening of A and B. Symbolically,

$P(A \text{ or } B) = P(A \cup B) = P(A) + P(B)$

This theorem can be extended to any number of mutually exclusive events as given by:

$P(A_1 \text{ or } A_2 \text{ or } \ldots \text{ or } A_n) = P(A_1 \cup A_2 \cup \ldots\ldots \cup A_n) = P(A_1) + P(A_2) + \ldots + P(A_n)$

If two events A and B are **not mutually exclusive**, then either A or B both events can occur. In this case,

$P(A \text{ or } B) = P(A \cup B) = P(A) + P(B) - P(A \cap B)$

(ii) **Multiplication theorem:** For two events A and B are **independent** and can happen simultaneously the probability of their joint occurrence of the event is equal to the product of their separate probabilities. Symbolically,

$P(A \text{ and } B) = P(A.B) = P(A \cap B) = P(A).P(B)$

This theorem can be extended to any number of independent events as given by:

$P(A_1 \text{ and } A_2 \text{ and } \ldots \text{ and } A_n) = P(A_1 \cap A_2 \cap \ldots\ldots \cap A_n) = P(A_1) . P(A_2) \ldots\ldots P(A_n)$

For two **dependent events**: The probability of simultaneous occurrence of two events A and B is equal to the probabilities of A multiplied by the conditional probability of B given A has occurred or it is equal to the probability of B multiplied by the conditional probability of A given that B has occurred. Symbolically,

$$P\,(A \text{ and } B) = P\,(A.B) = P\,(A \cap B) = P(A).P(B/A)$$

$$= P(B).P(A/B)$$

Conditional Probability: The probability of the happening of an event A, when it is known that B has already happened, is called the conditional probability of A and is denoted by P(A/B). i.e.

$$P(A/B) = P(A \cap B)/P(B)$$

$$\text{or} P(B/A) = P(A \cap B)/P(A)$$

(iii) **Bay's theorem:** If an event B can occur only in combination with one of the n mutually exclusive and exhaustive events A_1, A_2, ..., A_n, then if actual observation B is found to have occurred, the probability that it was preceded by a particular event A, is given by-

$$P\left(\frac{A_i}{B}\right) = \frac{P(A_i).P\left(\frac{B}{A_i}\right)}{\sum_{k=1}^{n} P(A_k).P\left(\frac{B}{A_k}\right)}$$

Here, k = 1,2,3,......,n

Random variable: A real valued function defined on the sample space, is called a **random variable**. There are two types, namely-

1) Discrete Random Variable
2) Continuous Random Variable

1) **Discrete Random Variable:** If the sample space S contains a finite number of points, then it is called a discrete sample space. A random variable X defined over a discrete sample space is called a **discrete random variable.**

2) **Continuous Random Variable:** If the sample space S contains infinite number of points, over a continuous sample space, then it is called continuous sample space. A random variable X defined over a continuous sample space is called a **continuous random variable**.

Probability Distribution: The distinct values of discrete random variable X together with their associated probabilities define the **probability distribution** of the random variable X. In general, if a random variable X assumes values x_1, x_2, ..., x_n with associated probabilities p_1, p_2, ..., p_n, then the probability distribution of X is:

$$\sum_{i=1}^{n} P(X = x_i) = p_1 + p_2 + \dots\dots\dots + p_n = 1$$

Simple Expectation or Mathematical Expectation: Let X be a discrete random variable with the following probability distribution:

Probability Distribution of X

x:	x_1	x_2	x_3	....	x_n
P(X = x)	**f(x_1)**	**f(x_2)**	**f(x_3)**	....	**f(x_n)**

Then the simple expectation or mathematical expectation of the random variable X, denoted as E(X), is

$$E(X) = \sum_{i=1}^{n} x_i f(x_i) = x_1 f(x_{12}) + x_2 f(x_2) + \dots\dots\dots + x_r f(x_n)$$

Factorial Symbol: The multiplication of all positive integers, say "n", that will be smaller than or equivalent to n is known as the **factorial**. The factorial of a positive integer n, the product n(n-1)(n-2).....3.2.1 is represented by the symbol "**n!**", which read as 'n factorial'.

Permutation: It is an arrangement of all or part of a set of objects. In general, n distinct objects taken r at a time can be arranged in n(n-1)(n-2)......(n-r+1) ways. This product is represented by the symbol ^nP_r. Thus-

$$^nP_r = n(n-1)(n-2)\dots\dots\dots(n-r+1) = \frac{n!}{(n-r)!}$$

Combinations: The **combination** of n different objects taken r at a time is a selection of r out of the n objects with no attention given to the order of arrangement. The number of combinations of n objects taken r at a time is denoted by:

$$^nC_r = \frac{n!}{r!(n-r)!}$$

MCQ

Q1. Any subset of sample space is known as:

(a) Probability (b) Event

(c) Empty set (d) **Sample space**

Q2. The probability of impossible event is:

(a) 1/3 (b) 1/2

(c) 1 (d) **0**

Bold means correct answer

Q3. The probability of sure event is:

(a) 1/3 (b) 1/2

(c) **1** (d) 0

Q4. Which of the following cannot be the value of probability?

(a) **-1** (b) 0

(c) 1/2 (d) 1

Q5. What is the probability of getting a number greater than 6 on dice?

(a) 1 (b) 1/2

(c) 1/3 (d) **0**

Q6. If an event has only one sample point then it is called:

(a) Impossible event (b) Sure event

(c) **Simple event** (d) Compound event

Q7. If an event has more than one sample point then it is called:

(a) Impossible event (b) Sure event

(c) Simple event (d) **Compound event**

Q8. Event A or B is represented by:

(a) $A \cap B$ (b) $\mathbf{A \cup B}$

(c) $A \cap B'$ (d) $A' \cap B$

Q9. Event A and B is represented by:

(a) $\mathbf{A \cap B}$ (b) $A \cup B$

(c) $A \cap B'$ (d) $A' \cap B$

Q10. Event .... contains elements which are presented in A and absent in B

(a) A or B (b) A and B

(c) **A but not B** (d) B but not A

Q11. Event A but not B is represented by:

(a) $A \cap B$ (b) $A \cup B$

(c) $\mathbf{A \cap B'}$ (d) $A' \cap B$

Q12. In a simultaneous throw of two dice, what is the probability of getting a total of 7?

(a) 1/2 (b) 1/3

(c) **1/6** (d) 1/9

Q13. A dice is thrown once. Find the probability of getting an even number.

(a) **1/2** (b) 1/3

(c) 1/6 (d) 1/9

Q14. A dice is thrown once. Find the probability of getting an odd number.

(a) **1/2** (b) 1/3

(c) 1/6 (d) 1/9

Q15. If P(A) = 1/3, P(B) = 1/4 and P(A∩B) = 1/16, the value of P(A/B) is:

(a) 3/16 (b) **1/4**

(c) 1/12 (d) 7/12

Q16. If P(A ∩ B)= 0, then the event A and B are:

(a) Dependent (b) Independent

(c) Equally likely (d) **None of these**

Q17. If P(A∪B) = P(A) + P(B), then the two events A and B are:

(a) Dependent (b) Independent

(c) **Mutually exclusive** (d) None of these

Q18. If P(B/A) = P(B), then the events A and B are:

(a) Dependent (b) **Independent**

(c) Equally likely (d) None of these

Q19. Probability of a sure event is:

(a) 0 (b) 0.5

(c) **1** (d) None of these

Q20. If P(A) = 0.6, P(B/A) = 0.7, then P(A∩B) is:

(a) 0.6 (b) 0.7

(c) **0.42** (d) 0.1

Q21. If $P(A \cap B) = 0.6$, $P(A) = 0.7$, $P(B) = 0.8$, then $P(B/A)$ is:

(a) 5/6 (b) 5/7

(c) **6/7** (d) 3/4

Q22. If $P(A \cap B) = 0.6$, $P(A) = 0.7$, $P(B) = 0.8$, then $P(A/B)$ is:

(a) 5/6 (b) 5/7

(c) 6/7 (d) **3/4**

Q23. If $P(A \cap B) = 0.2$ and $P(\overline{B})=0.6$ then $P(A/B)$ is:

(a) 0.10 (b) 0.20

(c) 0.40 (d) **0.50**

Q24. Let A and B be events such that $P(A) = 1/4$, $P(B) = 1/5$ and $P(A \cap B) = 1/6$, then $P(A/B)$ is:

(a) 1/6 (b) **5/6**

(c) 1/3 (d) 2/3

Q25. Let A and B be events such that $P(A) = 1/4$, $P(B) = 1/5$ and $P(A \cap B) = 1/6$, then $P(B/A)$ is:

(a) 1/6 (b) 5/6

(c) 1/3 (d) **2/3**

Q26. Let A and B be events such that $P(A) = 1/4$, $P(B) = 1/5$ and $P(A \cap B) = 1/6$, then $P(A \cup B)$ is:

(a) 1/6 (b) 5/6

(c) 11/60 (d) **17/60**

Q27. Let A and B be events such that $P(\overline{A}) = 3/4$, $P(B) = 2/5$ and $P(A \cup B) = 7/15$, then $P(A/B)$ is:

(a) **11/24** (b) 13/24

(c) 4/15 (d) 11/15

Q28. Let A and B be events such that $P(\overline{A}) = 3/4$, $P(B) = 2/5$ and $P(A \cup B) = 7/15$, then $P(B/A)$ is:

(a) 11/24 (b) 13/24

(c) 4/15 (d) **11/15**

Q29. Two unbiased dice are thrown. Find the probability that the sum is 8 or greater, if 4 appear on the first die:

(a) **1/2** (b) 1/3

(c) 1/6 (d) 5/6

Q30. Find the probability of getting a tail in a throw of a coin:

(a) 0 (b) 1

(c) **1/2** (d) ∞

Q31. A coin is tossed three times. Find the probability of getting exactly one head.

(a) 1/2 (b) 1/3

(c) 1/8 (d) **3/8**

Q32. An unbiased die is drawn. What is the probability that digit 4 appears?

(a) 1/2 (b) 1/3

(c) **1/6** (d) 5/6

Q33. What is the probability of getting a sum 9 from two throws of a dice?

(a) 1/6 (b) 1/8

(c) **1/9** (d) 1/12

Q34. Three unbiased coin are tossed. What is the probability of getting at most two heads?

(a) 1/4 (b) 3/4

(c) 3/8 (d) **7/8**

Q35. Two dice are thrown simultaneously. What is the probability of getting two numbers whose products are even?

(a) 1/2 (b) **3/4**

(c) 3/8 (d) 5/16

Q36. A person throws three coins. Find the probability that at least one head turns up:

(a) 1/8 (b) **7/8**

(c) 1/6 (d) 5/6

Q37. A coin is tossed three times. Find the probability of getting exactly two heads.

(a) 1/2 (b) 1/3

(c) 1/8 (d) **3/8**

Q38. A coin is tossed three times. Find the probability of getting exactly one or two heads.

(a) 1/4 (b) **3/4**

(c) 1/8 (d) 3/8

Q39. Two dice are tossed. The probability that the total score is a prime number is:

(a) 1/2 (b) 1/6

(c) 7/12 (d) **5/12**

Q40. From a well shuffled of cards, one card is drawn at random. The probability of getting a queen of club or a king of heart is:

(a) 1/13 (b) 2/13

(c) **1/26** (d) 1/52

Q41. A coin is tossed three times. Find the probability of getting the head and tail alternatively:

(a) 1/2 (b) 1/3

(c) 2/3 (d) **1/4**

Q42. Two cards are drawn together from a pack of 52 cards. The probability that one is a spade and one is heart, is:

(a) 3/26 (b) 21/26

(c) 39/102 **(d) 13/102**

Q43. One card is drawn from a pack of 52 cards. What is the probability is that the card drawn is a face card?

(a) 1/4 (b) **1/13**

(c) 3/13 **(d) 9/52**

Q44. A bag contains 6 white and 8 blue balls. One ball is drawn at random from the bag. What is the probability that the ball drawn is blue?

(a) 3/7 (b) **4/7**

(c) 3/4 (d) 1/8

Q45. Two dice are rolled simultaneously. Find the probability that the number 4 comes up at least once.

(a) 1/6 (b) **5/6**

(c) 1/36 (d) 11/36

Q46. An urn contains 6 red, 4 blue, 2 green and 3 yellow marbles. If 4 marbles are picked up at random, what is the probability that at least one of them is blue?

(a) 4/15 (b) 11/15

(c) 22/91 (d) **69/91**

Q47. In a simultaneous throw of two coins, the probability of getting at least one head is:

(a) 1/2 (b) 1/3

(c) 2/3 (d) **3/4**

Q48. A card is drawn from a pack of 52 cards. The probability of getting a queen of club or a king of heart is:

(a) 1/13 (b) 2/13

(c) **1/26** (d) 1/52

Q49. If a pair of dice is drawn, find the probability that the sum is greater than 12:

(a) **0** (b) 1/2

(c) 1/3 (d) 1/4

Q50. A bag contains 3 blue, 2 green and 5 red balls. If four balls are picked at random, what is the probability that two are green and two are blue?

(a) 1/2 (b) 3/5

(c) 1/18 (d) **1/70**

Q51. A basket contains 6 blue, 2 red, 4 green and 3 yellow balls. If 5 balls are picked up at random, what is the probability that at least one is blue?

(a) 2/5 (b) 9/91

(c) **137/143** (d) 6/143

Q52. If a pair of dice is drawn, find the probability that the sum is neither 8 nor 10:

(a) 1/6 (b) 5/6

(c) **7/9** (d) 1/9

Q53. In a simultaneous throw of two dice, what is the probability of getting a total of 10 or 11?

(a) 1/4 (b) 1/6

(c) 7/12 (d) **5/36**

Q54. From a pack of 52 cards, one card is drawn at random. What is the probability that the card drawn is a nine or a club?

(a) 1/4 (b) 1/13

(c) **4/13** (d) 1/26

Q55. A man and his wife appear in an interview for two vacancies in the same post. The probability of husband's selection is 1/5 and the probability of wife's selection is 1/3. What is the probability that only one of them is selected?

(a) 1/5 (b) **2/5**

(c) 1/15 (d) 2/15

Q56. A box contains 3 green, 5 red and 8 white balls. A ball is drawn at random from the box. What is the probability that the ball drawn is either red or green?

(a) **1/2** (b) 1/5

(c) 3/5 (d) 7/16

Q57. Three unbiased coins are tossed. What is the probability of getting at most two heads?

(a) 1/4 (b) 3/4

(c) 3/8 (d) **7/8**

Q58. Two dice are tossed. The probability that the total score is a prime number is:

(a) 1/2 (b) 1/6

(c) **5/12** (d) 7/12

Q59. In which approach to probability the outcomes are equally likely to occur?

(a) Independent (b) **Mathematical probability**

(c) Statistical probability (d) Axiomatic probability

Q60. In a simultaneous throw of two dice, what is the probability of getting a total of 5?

(a) 1/3 (b) 1/6

(c) **1/9** (d) 4/9

Q61. Three unbiased coins are tossed. What is the probability of getting at least 2 heads?

(a) **1/2** (b) 1/3

(c) 1/4 (d) 1/8

Q62. Two cards are drawn from a pack of 52 cards. The probability that either both are black or both are queen is:

(a) 1/13 (b) 3/13

(c) **55/221** (d) 63/221

Q63. How many ways can 4 starting positions in a team are filled with 9 students who can play at any of the positions?

(a) 126 (b) 252

(c) **3024** (d) None of these

Q64. One card is randomly drawn from a pack of 52 cards. Find the probability that the drawn card is a king.

(a) **1/13** (b) 1/26

(c) 1/4 (d) 1/2

Q65. In a single throw of die, what is the probability of getting a number greater than 3?

(a) **1/2** (b) 1/3

(c) 2/3 (d) 1/4

Q66. Three unbiased coins are tossed. What is the probability of getting at most two heads?

(a) 1/4 (b) 3/4

(c) 3/8 (d) **7/8**

Q67. One card is randomly drawn from a pack of 52 cards. Find the probability that the drawn card is red.

(a) 1/13 (b) 1/26

(c) 1/4 (d) **1/2**

Q68. Tickets numbered 1 to 20 are mixed up and then a ticket is drawn at random. What is the probability that the ticket drawn has a number which is a multiple of 3 or 7?

(a) 1/5 (b) **2/5**

(c) 3/5 (d) 1/2

Q69. A box contains 2 red, 3 blue, 4 green and 6 yellow balls. If two balls are drawn at random from the box, what is the probability that both are blue?

(a) **1/35** (b) 2/35

(c) 1/5 (d) 2/5

Q70. One card is drawn at random from a pack of 52 cards. What is the probability that the card drawn is a face card?

(a) 1/4 (b) 3/4

(c) 1/13 (d) **4/13**

Q71. Two dice are thrown simultaneously. What is the probability of getting two numbers whose product is even?

(a) 1/4 (b) **3/4**

(c) 3/8 (d) 5/8

Q72. Two dice are thrown simultaneously. What is the probability of getting two numbers whose product is odd?

(a) **1/4** (b) 3/4

(c) 3/8 (d) 5/8

Q73. A box contains 24 electric bulbs, out of which 6 are defective. Two balls are chosen at random from this box, the probability that at least one of them is defective, is:

(a) 10/23 (b) **13/23**

(c) 1/3 (d) 2/3

Q74. One card is randomly drawn from a pack of 52 cards. Find the probability that the drawn card is red and a king.

(a) 1/13 (b) **1/26**

(c) 1/4 (d) 1/2

Q75. One card is randomly drawn from a pack of 52 cards. Find the probability that the drawn card is red or a king.

(a) 1/13 (b) **7/13**

(c) 1/4 (d) 1/2

Q76. A die is rolled. If the outcome is an odd number, what is the probability that it is a prime number?

(a) 1/6 (b) 5/6

(c) 1/3 (d) **2/3**

Q77. In measuring probability of any certain event, zero represents:

(a) **Impossible events** (b) Possible events

(c) Sure event (d) Sample event

Q78. In a class, there are 15 boys and 10 girls. Three students are selected at random. The probability that 1 girl and 2 boys are selected is:

(a) 3/25 (b) 1/50

(c) **21/46** (d) 25/117

Q79. A bag contains 4 white, 5 red and 6 blue balls. Three balls are drawn at random from the bag. The probability that all of them red, is:

(a) **2/91** (b) 3/91

(c) 2/15 (d) 1/15

Q80. Two cards are drawn together from a pack of 52 cards. What is the probability is that both the cards drawn being kings?

(a) 1/15 (b) 25/57

(c) **1/221** (d) 35/256

Q81. A bag contains 4 red balls, 6 blue balls and 8 pink balls. One ball is drawn at random and replaced with 3 pink balls. A probability that the first ball drawn was either red or blue in colour and the second drawn was pink in colour?

(a) 13/17 (b) 13/18

(c) 11/30 (d) **None of these**

Q82. One card is drawn from a pack of 52 cards. What is the probability that the card drawn is either a black card or a king?

(a) 1/2 (b) 1/4

(c) 2/13 (d) **7/13**

Q83. Tickets numbered 1 to 12 are mixed up and then a ticket is drawn at random. What is the probability that the ticket drawn has a number which is a multiple of 2 or 5?

(a) 1/2 (b) 2/5

(c) 5/12 (d) **7/12**

Q84. A basket contains 4 red, 5 blue and 3 green balls. If three balls are picked up at random what is the probability that at least one is blue?

(a) 5/12 (b) 7/12

(c) 7/44 (d) **37/44**

Q85. What is the probability that a leap year selected at random will contain either 53 Saturday or 53 Sunday?

(a) 1/8 (b) 3/8

(c) 1/7 (d) **3/7**

Q86. The probability that a student passed Economics is 2/3, the probability that he passes Statistics is 4/9. If the probability of passing at least one subject is 4/5, what is the probability that student will pass both the subjects?

(a) 1/5 (b) 1/3

(c) 1/9 (d) **14/45**

Q87. A problem in statistics is given to three students whose chances of solving are 1/2, 1/3 and 1/4, respectively. What is the probability that the problem will be solved?

(a) **1/4** (b) 3/4

(c) 1/8 (d) 7/8

Q88. Tickets numbered 1 to 20 are mixed up and then a ticket is drawn at random. What is the probability that the ticket drawn bears a number which is a multiple of 3?

(a) 1/2 (b) 2/5

(c) **3/10** (d) 3/20

Q89. Two lottery tickets are drawn from 20 for first and second prizes. Count the total number of arrangements of the tickets:

(a) **380** (b) 190

(c) 95 (d) 10

Q90. In a box, there are 8 white, 7 blue and 6 green balls. One ball is picked up randomly. What is the probability that it is neither white nor green?

(a) **1/3** (b) 3/4

(c) 8/21 (d) 9/21

Q91. A bag contains 2 red, 3 green and 2 white balls. Two balls are drawn at random. What is the probability that none of the balls drawn white?

(a) **10/21** (b) 11/21

(c) 2/7 (d) 5/7

Q92. In a lottery, there are 10 prizes and 25 blanks. A lottery is drawn at random. What is the probability of getting a prize?

(a) 2/5 **(b) 2/7**

(c) 5/7 (d) 1/10

Q93. From a pack of 52 cards, two cards are drawn together at random. What is the probability of both the cards being kings?

(a) 1/52 (b) 25/57

(c) 1/221 (d) 35/221

Q94. A, B and C are three mutually exclusive and exhaustive events. If $\mathbf{P(B)} = \frac{3}{2}P(A)$ and $P(C) = \frac{1}{2}\mathbf{P(B)}$, find P(A):

(a) 3/4 (b) 1/16

(c) **4/13** (d) 16/81

Q95. The joint probability is:

(a) Called Prior probability

(b) Based on two mutually exclusive events

(c) **The likelihood of two events happening together**

(d) The likelihood of an event happening given that another event has already happened

Q96. An urn contains 2 red, 3 green and 4 blue balls. If 2 balls are drawn at random, find the probability that no ball is blue.

(a) 2/9 (b) 5/9

(c) **5/18** (d) 5/36

Q97. A bag contains 6 green balls 7 red balls and 8 pink balls. If two balls are drawn at random from the bag, one after another what is the probability that the first ball is green and second ball is red?

(a) **1/5** (b) 3/5

(c) 1/7 (d) 5/7

Q98. An urn contains 2 red, 3 blue, 4 green and 5 yellow marbles. If two marbles are picked up at random, what is the probability that either both are blue or both are yellow?

(a) **1/7** (b) 3/7

(c) 3/7 (d) 4/7

Q99. Two coins are tossed. What is the conditional probability of getting two heads given that at least one coin shows a head?

(a) 1/2 (b) **1/3**

(c) 2/3 (d) 1/4

Q100.Let A and B be two events, such that $P(A) = 3/10$, $P(A \cup B) = 2/5$ and $P(B) = c$. Find c such that A and B are mutually exclusive.

(a) 1/5 (b) 4/5

(c) 1/7 (d) **1/10**

Q101.Let A and B be two events, such that $P(A) = 3/10$, $P(A \cup B) = 2/5$ and $P(B) = c$. Find c such that A and B are independent.

(a) 1/5 (b) 4/5

(c) **1/7** (d) 1/10

Q102. If the occurrence of one event means that another cannot happen, then the events are:

(a) **Mutually exclusive** (b) Independent

(c) Dependent (d) Exhaustive

Q103.Which of the following is not a correct statement about a probability?

(a) **It is the collection of several experiments**

(b) It must have a value between 0 and 1

(c) It can be reported as a decimal

(d) A value near 0 means that the event is not likely to occur

Q104.A listing of the possible outcomes of an experiment and their corresponding probability is called:

(a) **Probability distribution** (b) Frequency distribution

(c) Contingency table (d) Random variable

Q105.The special rule of multiplication of probability, the events must be:

(a) Mutually exclusive (b) **Independent**

(c) Favourable (d) Unfavourable

Q106.A speaks truth in 50% cases B speaks truth in 60% cases, the probability that they will way say the same thing while describing a single event is:

(a) **1/2** (b) 1/3

(c) 2/3 (d) 1/5

Q107.If you roll a pair of dice, what is the probability that (at least) one of the dice is a 4 or the sum of the dice is 7?

(a) 13/36 (b) **15/36**

(c) 21/36 (d) 1/9

Q108.If a card is chosen from a pack of cards, what is the probability of getting a diamond or a club?

(a) 1/6 (b) 1/4

(c) 1/3 (d) **1/2**

Q109.Outcomes of an experiment are classified as

(a) Probability (b) Random experiment

(c) Conditional probability (d) **Events**

Q110.The collection of one or more outcomes from an experiment is called:

(a) **Event** (b) Random experiment

(c) Probability (d) Random variable

Q111. In the special rule of addition of probability, the events are always:

(a) **Mutually exclusive** (b) Independent

(c) Favourable (d) Unfavourable

Q112.Probability of second event in situation if first event has been already occurred is classified as:

(a) Addition probability (b) **Conditional probability**

(c) Joint probability (d) Multiplication probability

Q113.In probability theories, events which can never occur together are classified as:

(a) Exclusive events (b) Complementary events

(c) **Mutually exclusive events** (d) Independent events

Q114.Joint probability of independent events A and B is equal to:

(a) **P(A) x P(B)** (b) P(A) + P(B)

(c) P(A) x P(B) + P(A - B) (d) P(A) x P(B) - P(A x B)

Q115.Types of probabilities for independent events must include:

(a) Joint events (b) Marginal events

(c) Conditional events (d) **All of above**

Q116.In measuring probability of any certain event, one represents:

(a) Impossible events (b) Possible events

(c) **Sure event** (d) Sample event

Q117. If two events A and B are considered as partially overlapping events then rule of addition can be written as:

(a) P(A or B) = P(A) - P(B) + P(A and B)

(b) P(A or B) = P(A) + P(B) * P(A - B)

(c) P(A or B) = P(A) * P(B) + P(A - B)

(d) **P(A or B) = P(A) + P(B) - P(A and B)**

Q118.If a person buys a lottery, chance of winning a Maruti car is 60%, chance of winning Mahindra car is 70% and chance of winning both is 40% then chance of winning Maruti or Mahindra is:

(a) 0.6 (b) **0.9**

(c) 0.8 (d) 0.5

Q119.According to combination rule, if total number of outcomes are 'r' and distinct outcome collection is 'n' then combinations are calculated as

(a) **n!/x!(n - x)!** (b) n!/x!(n + x)!

(c) x!/n!(n - x)! (d) x!/n!(n + x)!

Q120. An urn contains 3 green, 4 red, 5 white and 6 blue marbles. If four marbles are picked up at random, what is the probability that 2 is green, 1 are red and 1 is white?

(a) 1/13 (b) 12/13

(c) **1/51** (d) 4/51

Q121. A bag contains 10 apples out of which 4 are rotten; two apples are taken out together. If one of them is found to be good, the probability that other also good is:

(a) 7/15 (b) 8/15

(c) **1/3** (d) 2/3

Q122. A blood bank has 10 units of A+ blood in which 4 are contaminated with serum hepatitis. Three units are selected for use with three patients. What is the probability that all three patients will be exposed to hepatitis from this source?

(a) 1/10 (b) 9/10

(c) 2/5 (d) **1/30**

Q123. What is the probability of obtaining 9, 10 and 11 points with 3 dice?

(a) 25/216 (b) **79/216**

(c) 1/8 (d) 7/8

Q124. Three cards are drawn from a pack of 52 cards. What is the probability that all the three will be kings?

(a) 4/5525 (b) 3/5525

(c) **1/5525** (d) 16/5525

Q125. A peon is given 4 letters to drop into 4 letter boxes of a city. What is the probability that the peon will drop all the four letters in different letter boxes?

(a) 1/4 (b) 1/32

(c) **3/32** (d) 1/256

Q126. There are 4 letters and 4 addressed envelopes. Find the probability that all the letters are not dispatched in the right envelopes.

(a) **23/24** (b) 1/24

(c) 1/4 (d) 3/4

Q127.What is the chance that a leap year will contain 53 Sundays?

(a) 1/7
(b) **2/7**
(c) 1/365
(d) 1/366

Q128.What is the chance that a non-leap year will contain 53 Sundays?

(a) **1/7**
(b) 2/7
(c) 1/365
(d) 1/366

Q129.Mr. X kills a bird once in three shots. N this assumption he fired three shots. Find the probability that the bird is not killed.

(a) 1/3
(b) 2/3
(c) 1/27
(d) **8/27**

Q130.A die is rolled two times. What is the probability of getting an odd number both the times?

(a) 1/2
(b) 1/3
(c) **1/4**
(d) 3/4

Q131.A husband and wife both appear in an interview for two vacancies. The probability of husband selection is 1/4 and that of wife is 1/3. Find the probability of both of them selected.

(a) 7/12
(b) **1/12**
(c) 3/4
(d) 1/2

Q132.A husband and wife both appear in an interview for two vacancies. The probability of husband selection is 1/4 and that of wife is 1/3. Find the probability of none of them selected.

(a) 7/12
(b) 1/12
(c) 3/4
(d) **1/2**

Q133.A husband and wife both appear in an interview for two vacancies. The probability of husband selection is 1/4 and that of wife is 1/3. Find the probability atleast one of them selected.

(a) 7/12
(b) 1/12
(c) 3/4
(d) **1/2**

Q134.A single letter is selected at random from the word **PROBABILITY**. Find the probability that selected letter is a vowel.

(a) 1/2
(b) 1/7
(c) 1/11
(d) **4/11**

Q135.The probability of throwing more than 4 by one die is:

(a) 1/2 (b) **1/3**

(c) 2/3 (d) 1/4

Q136.If A and B are independent events then $P(A\cap B)$ is equal to:

(a) P(A) + P(B) (b) **P(A) x P(B)**

(c) P(A) x P(A/B) (d) P(B) x P(B/A)

Q137.If A and B are mutually exclusive events then $P(A\cup B)$ is equal to:

(a) P(A) + P(B) (b) **P(A) x P(B)**

(c) P(A) x P(A/B) (d) P(B) x P(B/A)

Q138.For any event A, $P\left(A+\bar{A}\right)$ is equal to:

(a) 0 (b) **1**

(c) 0.5 (d) ∞

Q139.For any event A, $P\left(A\bar{A}\right)$ is equal to:

(a) **0** (b) 1

(c) 0.5 (d) ∞

Q140.Groups are formed of 4 persons out of 12 persons. The probability that one particular player is never included is:

(a) 1/3 (b) **2/3**

(c) 1/4 (d) 3/4

Q141. A problem in Statistics is given to three students A, B and C whose probability of solving are 1/2, 1/3 and 1/4, respectively. What is the chance that the problem will be solved?

(a) 1/2 (b) 1/3

(c) 1/4 (d) **3/4**

Q142.If two events A and B are dependent, then P(A/B) is equal to:

(a) P(A)/P(B) (b) P(B)/P(A)

(c) P(AB)/P(A) (d) **P(AB)/P(B)**

Q143.If two events A and B are dependent, then P(B/A) is equal to:

(a) P(A)/P(B) (b) P(B)/P(A)

(c) **P(AB)/P(A)** (d) P(AB)/P(B)

Q144.If ‘p’ is the probability of success and ‘q’ is the probability of failure, then p x q is equal to:

(a) Less than 1
(b) 0
(c) **1**
(d) More than 1

Q145.$^6C_2$ is equal to:

(a) 10
(b) **15**
(c) 20
(d) 25

Q146.What are the chances that no two boys are sitting together for a photograph if there are 5 girls and 2 boys?

(a) 2/7
(b) 4/7
(c) **5/7**
(d) 1/21

Q147.In a drawer there are 4 white socks, 3 blue socks and 5 grey socks. Two socks are picked randomly. What is the possibility that both the socks are of same color?

(a) 2/33
(b) 13/66
(c) **19/66**
(d) 23/66

Q148.What will be the possibility of drawing a jack or a spade from a well shuffled standard deck of 52 playing cards?

(a) 1/13
(b) **4/13**
(c) 1/26
(d) 17/52

Q149. In a throw of coin what is the probability of getting head.

(a) 0
(b) **1/2**
(c) 1
(d) 2

Q150. In a throw of coin what is the probability of getting tail.

(a) 0
(b) **1/2**
(c) 1
(d) 2

Q151.From a pack of 52 cards, one card is drawn at random. Find the probability of a face card drawn:

(a) 1/4
(b) 3/4
(c) 1/13
(d) **3/13**

Q152.The probability that it will rain tomorrow is 0.85. What is the probability that it will not rain tomorrow

(a) 1/4 (b) 3/4

(c) 1/20 (d) **3/20**

Q153.What is the probability that a number selected from the numbers (1, 2, 3,..........,15) is a multiple of 4?

(a) **1/5** (b) 4/5

(c) 1/3 (d) 2/3

Q154.What are the total outcomes when we throw three coins?

(a) 4 (b) 5

(c) **8** (d) 7

Q155.The probability that a prime number selected at random from the numbers (1,2,3, ...35) is :

(a) **11/35** (b) 12/35

(c) 13/35 (d) 14/35

Q156.The probability of getting less than 3 in a single throw of a die is:

(a) 1/2 (b) **1/3**

(c) 2/3 (d) 1/4

Q157.The total number of events of throwing 10 coins simultaneously is:

(a) 10 (b) 100

(c) 512 (d) **1024**

Q158.A girl calculates that the probability of her winning the first prize in a lottery is 8/100. If 6,000 tickets are sold, how many tickets has she bought?

(a) 120 (b) 240

(c) **480** (d) 960

Q159.One card is drawn from a well-shuffled deck of 52 cards. The probability that the card will not be an ace is:

(a) 1/13 (b) 3/13

(c) 4/13 (d) **12/13**

Q160. Which of the following cannot be the probability of an event?

(a) 0.2 (b) 0.3

(c) 0.8 (d) **1.4**

Q161. Probability of getting a number x such that $1 \le x \le 4$, is:

(a) > 1 (b) 0

(c) < 1 (d) **1**

Q162. The sum of the probabilities of all elementary events of an experiment is p, then:

(a) $p = 0$ (b) $\mathbf{p = 1}$

(c) $0 \le p < 1$ (d) $0 < p < 1$

Q163. If an event cannot occur, then its probability is:

(a) **0** (b) 1/2

(c) 1/3 (d) 1

Q164. A bag contains 12 balls of which x are red, the probability of getting a red ball at random from the bag is:

(a) x/4 (b) x/8

(c) **x/12** (d) 12/x

Q165. If the probability of newborn being Rh –ve is 0.3, than the probability of newborn being Rh +ve is:

(a) 0.2 (b) 0.3

(c) **0.7** (d) 1.0

Q166. If probability of getting Rh+ve is 0.8 and probability of newborn being male is 1/4, the probability of newborn being a male with Rh+ve is equal to:

(a) **0.2** (b) 0.4

(c) 0.6 (d) 0.8

Q167. If prevalence of diabetes in a population is 10%; probability that four persons selected randomly from this population will have diabetes will be:

(a) **0.0001** (b) 0.001

(c) 0.01 (d) 0.1

Q168. If probability of a child being Rh-ve is 1/5, the probability of a child being Rh+ve in a society will be:

(a) 1/5
(b) 1/10
(c) **4/5**
(d) 9/10

Q169. If $A \subset B$ and $B \subset A$ then,

(a) P(A)>P(B)
(b) P(A)<P(B)
(c) **P(A)=P(B)**
(d) $P(\bar{A})<P(\bar{B})$

Q170. What is the probability of an impossible event?

(a) **0**
(b) 0.5
(c) 1
(d) None of these

Q171. If A and B are two events, then the probability of exactly one of them occurs is given by:

(a) $\mathbf{P(A \cap \bar{B}) + P(\bar{A} \cap B)}$
(b) P(A) + P(B)–2P(A)P(B)
(c) $P(\bar{A}) + P(\bar{B}) - 2\,P(\bar{A})P(\bar{B})$
(d) $P(A) + P(B) - P(A \cap B)$

Q172. Which of the given sets are not an event of the sample space S= {a,1,2,3,z}:

(a) {a, z}
(b) {1,2,3}
(c) {}
(d) **{a, b}**

Q173. Which among the following is a sample space obtained while tossing a coin thrice?

(a) {(H,T),(T,H),(T,T),(H,H)}
(b) {(H,H,H),(H,T,T),(T,T,T)}
(c) {(H,H),(T,T)}
(d) **{(H,H,H),(H,H,T),(H,T,T),(T,H,T),(H,T,H),(T,T,H),(T,H,H),(T,T ,T)}**

Q174. The probability of each event, when a coin is tossed for 1000 times with frequencies: Head = 480 and Tail = 520 is:

(a) 0.5 and 0.5
(b) **0.48 and 0.52**
(c) 0.04 and 0.05
(d) 0.40 and 0.60

Q175.Phenomenon of Statistical Regularity is observed when:

(a) **Number of trials of a random experiment increases**

(b) Number of trials of a random experiment is kept minimal

(c) Relative frequencies approach divergent values

(d) None of these

Q176.If P (A) = 0.3, P(B)=0.2, P(A∩B)=0.1, what is the probability that exactly one of the events occurs?

(a) **0.3** (b) 0.4

(c) 0.6 (d) 1

Q177.If P(A)=0.7, P(B)=0.2, P(A∩B))=0.2, what is the probability that neither A nor B occurs?

(a) 0.3 (b) 0.2

(c) **0.8** (d) 0.7

Q178.Which of these is the same as (A ∩ B^c) U (A∩B)?

(a) A\B (b) AUB

(c) **A** (d) A∩B

Q179.What is the probability of the event that an even number appears when tossing a fair dice?

(a) 1/6 (b) **1/2**

(c) 1/4 (d) 2/3

Q180.In a rowing race, the odds that A will win is 2 to 3 and the odds that B will win is 1 to 4.What is the probability that A or B wins the race?

(a) 1/5 (b) 2/5

(c) **3/5** (d) 4/5

Q181.Two cards are drawn from a well shuffled pack of 52 cards. Find the probability that they are both aces if the first card is replaced.

(a) 1/144 (b) **1/169**

(c) 4/169 (d) 1/36

Q182.If two events A and B are independent, then P (A/B) is equal to:

(a) **P(A)** (b) P(B)

(c) P(A∩B) (d) None of these

Q183.When a card is drawn from a well shuffled deck of cards, what is the probability of getting a black queen?

(a) 1/4 (b) 1/2

(c) **1/26** (d) 1/52

Q184.A pair of fair dice is tossed. What is the probability that the maximum of the two numbers is greater than 4?

(a) 1/9 (b) **5/9**

(c) 1/18 (d) 1/6

Q185.An urn B_1 contains 2 white and 3 black chips and another urn B_2 contains 3 white and 4 black chips. One urn is selected at random and a chip is drawn from it. If the chip drawn is found black, find the probability that the urn chosen was B_1.

(a) 3/7 (b) 4/7

(c) **21/41** (d) 20/41

Q186.At a certain university, 4% of men are over 6 feet tall and 1% of women are over 6 feet tall. The total student population is divided in the ratio 3:2 in favour of women. If a student is selected at random from among all those over six feet tall, what is the probability that the student is a woman?

(a) 2//5 (b) 3/5

(c) 2/11 (d) **3/11**

Q187.If A = {1, 2, 3, 4}, B = ϕ, then AUB is:

(a) ϕ (b) {0}

(c) **{1, 2, 3, 4}** (d) None of these

Q188.Let A = {a, b}, B = {p, q}, then A x B is:

(a) {a, b, p, q} (b) {0}

(c) ϕ (d) **{(a, p), (a, q), (b, p), (b, q)}**

Q189.P(A ∩ B) is equal to:

(a) P(A) . P(B|A) (b) P(B) . P(A|B)

(c) **Both A and B** (d) None of these

Q190.If P (A) = 0.8, P (B) = 0.5 and P (B|A) = 0.4, what is the value of P (A ∩ B)?

(a) **0.32** (b) 0.25

(c) 0.1 (d) 0.5

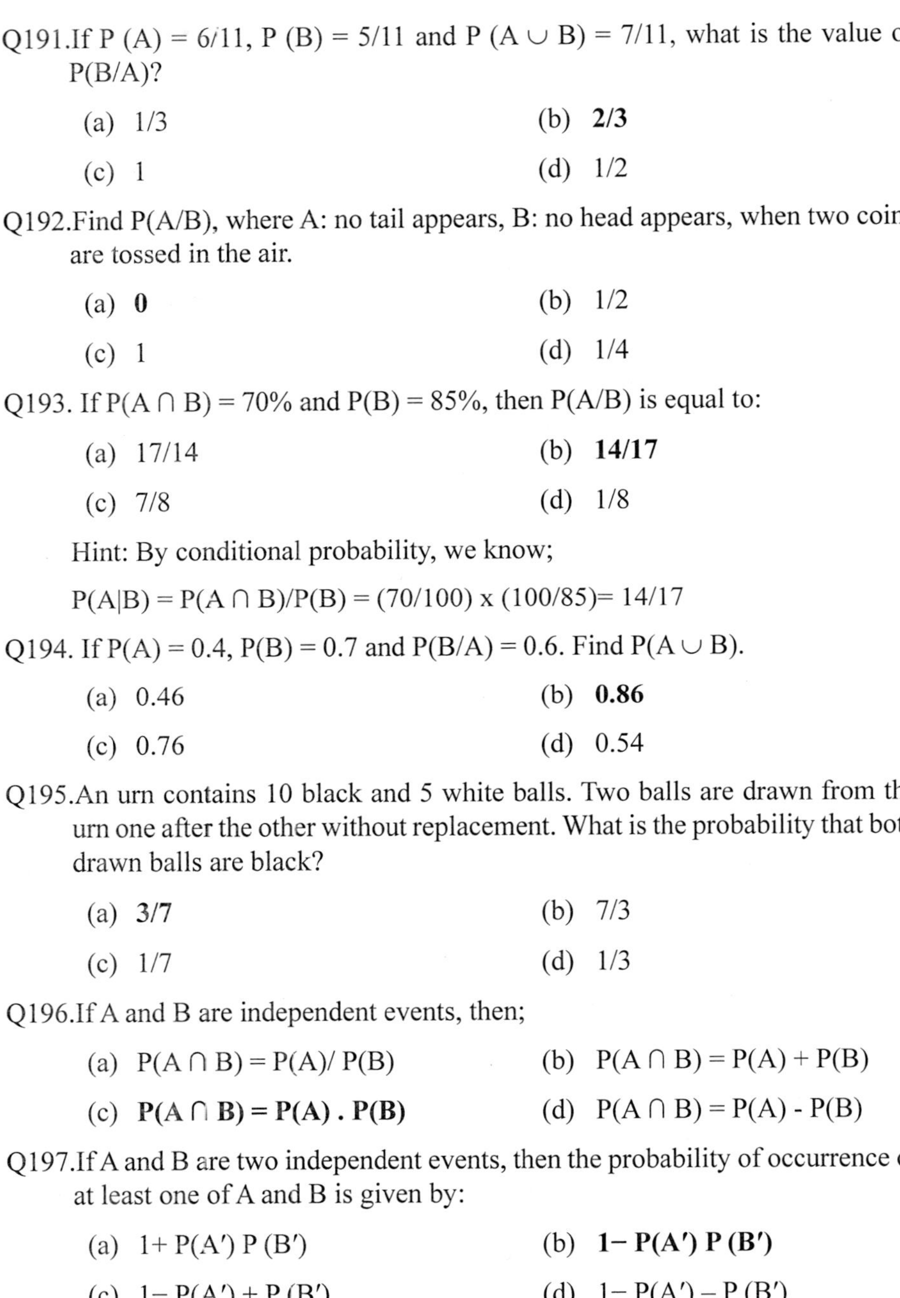

Q191.If P (A) = 6/11, P (B) = 5/11 and P (A ∪ B) = 7/11, what is the value of P(B/A)?

(a) 1/3 (b) **2/3**

(c) 1 (d) 1/2

Q192.Find P(A/B), where A: no tail appears, B: no head appears, when two coins are tossed in the air.

(a) **0** (b) 1/2

(c) 1 (d) 1/4

Q193. If P(A ∩ B) = 70% and P(B) = 85%, then P(A/B) is equal to:

(a) 17/14 (b) **14/17**

(c) 7/8 (d) 1/8

Hint: By conditional probability, we know;

P(A|B) = P(A ∩ B)/P(B) = (70/100) x (100/85)= 14/17

Q194. If P(A) = 0.4, P(B) = 0.7 and P(B/A) = 0.6. Find P(A ∪ B).

(a) 0.46 (b) **0.86**

(c) 0.76 (d) 0.54

Q195.An urn contains 10 black and 5 white balls. Two balls are drawn from the urn one after the other without replacement. What is the probability that both drawn balls are black?

(a) **3/7** (b) 7/3

(c) 1/7 (d) 1/3

Q196.If A and B are independent events, then;

(a) P(A ∩ B) = P(A)/ P(B) (b) P(A ∩ B) = P(A) + P(B)

(c) **P(A ∩ B) = P(A) . P(B)** (d) P(A ∩ B) = P(A) - P(B)

Q197.If A and B are two independent events, then the probability of occurrence of at least one of A and B is given by:

(a) 1+ P(A′) P (B′) (b) **1− P(A′) P (B′)**

(c) 1− P(A′) + P (B′) (d) 1− P(A′) – P (B′)

Hint: P(at least one of A and B) = P(A ∪ B)

= P(A) + P(B) − P(A ∩ B) = P(A) + P(B) − P(A) P(B)

$= P(A) + P(B) [1-P(A)] = P(A) + P(B). P(A')$

$= 1- P(A') + P(B) P(A') = 1- P(A') [1- P(B)]$

$= 1- P(A') P(B')$

Q198.The probability of solving the specific problems independently by A and B are 1/2 and 1/3 respectively. If both try to solve the problem independently, find the probability that exactly one of them solves the problem.

(a) 1 (b) **1/2**

(c) 1/3 (d) 1/4

Hint: P(A) = 1/2, P(B) = 1/3. Since, A and B are independent events, therefore;

$\Rightarrow P(A \cap B) = P(A). P(B)$

$\Rightarrow P(A \cap B) = \frac{1}{2} \times 1/3 = 1/6$

$P(A') = 1 - P(A) = 1 - 1/2 = 1/2$

$P(B') = 1 - P(B) = 1 - 1/3 = 2/3$

Now the probability that exactly one of them solved the problem is either the problem is solved by A and not B or vice versa.

$P(A).P(B') + P(A').P(B)$

$= 1/2 (2/3) + 1/2 (1/3)$

$= 1/3 + 1/6 = 3/6$

$\Rightarrow P(A).P(B') + P(A').P(B) = 1/2$

Q199.If A and B are two independent events, then

(a) **P(A∩B) = P(A) × P(B)** (b) P(AB) = 1 – P(A') P(B')

(c) P(AB) = 1 + P (A') P(B') P(A') (d) P (AB) = P(A')/ P(B')

Q200.The probability of an event is 3/7. Then odd against the event is:

(a) **4 : 3** (b) 7 : 3

(c) 3 : 7 (d) 3 : 4

Q201.A pair of dice is rolled. The probability of obtaining an even prime number on each die is:

(a) **1/36** (b) 1/12

(c) 1/6 (d) 0

Q202.P(A∩B) = 3/8, P(B) = 1/2 and P(A) = 1/4 then P(B′ / A′) is:

(a) 3/5 (b) 5/8

(c) 3/8 (d) **5/6**

Q203.If A and B are two events such that P(A) ≠ 0 and P(B/A) = 1 then

(a) P(A/B) = 1 (b) **P(B/A) = 1**

(c) P(A/B) = 0 (d) P(B/A) = 0

Q204.If P (A) = 3/8, P(B) = 1/2 and P(A∩B) = 1/4 then P(A′ / B′) is:

(a) 1/4 (b) **1/3**

(c) 3/4 (d) 3/8

Q205.If A and B are two events such that P(A) ≠ 0 and P(B∩A) = 1, then

(a) B ⊂ A (b) B = φ

(c) **A ⊂ B** (d) A ∩ B = φ

Q206.If A and B are any two events such that P(A) + P(B) – P(A∩B) = P(A) then

(a) P(B/A) = 1 (b) P(B/A) = 0\

(c) **P(A/B) = 1** (d) P(A/B) = 0

Q207.If A and B are events such that P (A∪B) = 3/4. P(A∩B) = 1/4, P(A) = 2/3 then P(A∩B) is

(a) 3/8 (b) **5/8**

(c) 5/12 (d) 1/4

Q208.If one card is drawn out of 52 playing cards, the probability that it is a dice is:

(a) 1/26 **(b) 1/13**

(c) 1/52 (d) 1/4

Q209.The chance of getting a double with 2 dice is

(a) 2/3 **(b) 1/6**

(c) 5/6 (d) 5/36

Q210.Two number are chosen, one by one without replacement from the set of number A = {1, 2, 3, 4, 5, 6} then the probability that minimum value of two number chosen is less than 4 is:

(a) 14/15 (b) **1/15**

(c) 1/5 (d) 8/5

Q211.Five horses are in a race. Mr. Ashok selects two of the horses at random and best on them. The probability that Mr. Ashok select the winning horses is:

(a) 4/5 (b) 3/5

(c) 1/5 (d) **2/5**

Q212.The probability of India winning a test match against West Indies is 12. Assuming independence from match to match the probability that in a match series India second win occurs at the third test is

(a) 1/6 (b) **1/4**

(c) 1/2 (d) 2/3

Q213.Three distinct numbers are selected from first 100 natural numbers. The probability divisible by 2 and 3 is:

(a) 9/25 (b) 4/35

(c) 4/55 (d) **4/1155**

Q214.The probability that A speaks truth is 4/5 while this probability for B is 3/4. The probability that they contradict each other's when asked to speak and fact is:

(a) 7/20 (b) 1/5

(c) 3/20 (d) **4/5**

Q215.Two dice are tossed once. The probability of getting an even number at the first dice are total of 8 is

(a) 1/36 (b) 3/36

(c) 11/36 (d) **5/9**

Q216.If A, B are two events associated with same random experiment such that $P(A) = 0.4$, $P(B) = 0.8$ and $P(B/A) = 0.6$ then $P(A/B)$ is

(a) **0.3** (b) 0.4

(c) 0.5 (d) 0.6

Q217.If $P(A) = 3/8$, $P(B) = 5/8$, $P(A \cup B) = 3/4$ then $P(B/A)$ is

(a) 3/47 (b) 5/49

(c) **2/3** (d) 1/4

Q218.An urn contains balls of which 3 are red, 4 are blue and 2 are green, 3 balls are drawn at random without replacement from the urn. The probability that the 3 balls have different colours is:

(a) 1/3
(b) **2/7**
(c) 1/21
(d) 2/23

Q219.If P(A) = 4/5 and P(A∩B) = 7/10, then P(B/A) is equal

(a) 1/10
(b) 1/8
(c) 7/8
(d) **17/20**

Q220.If P(A∩B) = 7/10 and P(B) = 17/20, then P(A/B) equals

(a) **14/17**
(b) 17/20
(c) 7/8
(d) 1/8

Q221.If P(A) = 7/10 P(B) = 7/10 and P(A∪B) = 7/10 then P (B/A) + P(A/B) equals

(a) 1/4
(b) 1/3
(c) 5/12
(d) **7/12**

Q222.If P(A) = 2/5, P(B) = 3/10 and P (A∩B) = 1/5, then P (A'|B'). P(B'/A') is equal to

(a) 5/6
(b) 5/7
(c) **25/42**
(d) 1

Q223.If P(A) = 0,4, P(B) = 0.8 and P(B/A) = 0.6 then P(A∪B) is equal to

(a) 0.24
(b) 0.3
(c) 0.48
(d) **0.96**

Q224.A and B are events such that P(A) = 0.4, P(B) = 0.3 and P(A∪B) = 0.5. Then P(B∩A) equals:

(a) 2/3
(b) 1/2
(c) 3/10
(d) **1/5**

Q225.Given that A and B are two events such that P(B) = 3/5, P(A/B) = 1/2 and P (A∪B) = 4/5, then P(A) equals:

(a) 3/10
(b) 1/5
(c) **1/2**
(d) 3/5

Q226.If A and B are events such that P(B) = 3/5, P(A/B) = 1/2 and P (A∪B) = then P(B/A') equals:

(a) 1/5
(b) 3/10
(c) 1/2
(d) **3/5**

Q227.Let P (A) = 7/13, P(B) = 9/13 and P (A∪B) = 9/13, Then P(A'|B) is equal to:

(a) 6/13 (b) 4/13

(c) 4/9\ (d) **5/9**

Q228.If A and B are such that events that P(A) > 0 and P(B) ≠ 1, then P (A'/ B') equal

(a) 1 – P (A|B) (b) 1 – P(A'|B)

(c) **1−P(A∪B) / P(B′)** (d) P(A') | P(B')

Q229.If two events are independent, then

(a) they must be mutually exclusive

(b) the sum of their probabilities must be equal to 1

(c) (a) and (b) both are correct

(d) **None of the above is correct**

Q230.If the event A and B are independent, then P(A∩B) is equal to

(a) P(A) + P(B) (b) P(A) – P(B)

(c) **P(A). P(B)** (d) P(A) | P(B)

Q231.Two dice are thrown once, if it is known that the sum of the numbers on the dice was less than 6 the probability of getting a sum 3 is:

(a) **1/5** (b) 2/5

(c) 5/18 (d) 1/18

Q232.If A and B are events such that P(A/B) = P (B/A), then

(a) A = B (b) A ∩B = φ

(c) A ⊂ B but A≠ B (d) **P(A) = P(B)**

Q233.If A and B are any two events such that P(A) + P(B) – P(A and B) = P(A), then

(a) **P(A/B) =1** (b) P(B/A) =1

(c) P(A/B) = 0 (d) PB/A) = 0

Q234.If A and B are independent events, then

(a) P (A ∩B) = P(A) - P(B) (b) **P (A ∩B) = P(A). P(B)**

(c) P (A ∩B) = P(A)/P(B) (d) P (A ∩B) = P(B)/P(A)

Q235.If P(A) = 1/2, P(B) =0, the P(A/B) is

(a) 0 (b) 1/2

(c) 1 (d) **Not define**

Q236.If A and B be the two events such that P(A) = 0.6, P(B) = 0.2 and P(A/B) = 0.5, the P(A'/B') is

(a) 1/8 (b) **3/8**

(c) 1/10 (d) 3/10

Q237.If P (A∩B) = 70% and P(B) = 85%, the P(A/B) is equal to:

(a) 1/8 (b) 3/8

(c) 7/8 (d) **14/17**

Q238.If A and B are two independent events, the probability of occurrence of at least one of A and B is given by:

(a) **1 – P(A'). P(B')** (b) 1 + P(A'). P(B')

(c) 1 – P(A') + P(B') (d) 1 – P(A') - P(B')

Q239.If A and B are two events such that P(A) ≠ 0 and P(B/A) =1, then

(a) A = φ (b) B = φ

(c) **A ⊂ B** (d) B ⊂ A

Q240.If P(A/B) > P(A), then which of the following is correct:

(a) P(B/A) < P(B) (b) **P(B/A) > P(B)**

(c) P(B/A) = P(B) (d) P(B/A) < P(A).P(B)

Q241.If A and B are two independent events such that 0 < P(A) < 1 and 0 < P(B) < 1, then which of the following is not correct:

(a) A' and B' are independent

(b) A' and B are independent

(c) A and B' are independent

(d) **A and B are mutually exclusive**

Q242.A child has a block in the shape of a cube with one letter written on each face as shown below:

A	B	A	D	C	E

The cube is thrown once. What is the probability of getting A?

(a) 1/2 (b) **1/3**

(c) 1/4 (d) 1/6

Q243.A die is thrown. What is the probability of getting an even prime number?

(a) 1/2 (b) 1/3

(c) 1/4 (d) **1/6**

Hint: Even prime number = 2,

∴ Probability = 1/6

Q244.A die is thrown, what is the probability of getting multiplies of 2?

(a) **1/2** (b) 1/3

(c) 1/4 (d) 1/6

Hint: Possible outcomes when a die is thrown = {1,2,3,4,5,6}

Multiplies of 2 = {2,4,6}; Therefore, total probability = 3/6 =1/2

Q245.What is the probability of picking an ace from a pack of cards?

(a) 1/2 (b) 1/4

(c) **1/13** (d) 1/52

Q246.What is the probability of picking a card of the club from a pack of cards?

(a) 1/2 (b) **1/4**

(c) 1/13 (d) 1/52

Q247.If n coins are tossed simultaneously what is the total number of outcomes?

(a) 2^{n-2} (b) 2^{n-1}

(c) $\mathbf{2^n}$ (d) 2^{n+1}

Q248.If 4 dice are rolled simultaneously then what is the total number of possible outcomes?

(a) **1296** (b) 216

(c) 36 (d) 6

Hint: If 4 dice are rolled simultaneously then total number of possible outcomes is 6^4 =1296

Q249.If 2 dice are rolled simultaneously then what is the probability of getting sum is 4?

(a) 1/2 (b) 1/4

(c) 1/6 (d) **1/12**

Hint: If 2 dice are rolled simultaneously then total number of possible outcome is $6^2 = 36$.

We can get 4 as sum in these cases {(1,3),(2,2),(3,1)}, So, probability of getting sum 4 is 3/36 =1/12.

Q250.A coin is tossed and a dice is rolled simultaneously what is total number of possible outcomes?

(a) 2 (b) 4

(c) 8 (d) **12**

Hint: If a coin is tossed and a dice is rolled simultaneously then total number of possible outcomes will be 2x6=12.

Q251.A coin is tossed and if head come then a blue dice is rolled and if tail come then a pink dice is rolled then what is the possible number of outcomes?

(a) 2 (b) 6

(c) **12** (d) 24

Hint: If head come on coin then blue dice can have 6 outcomes. So, total = 1 x 6 =6 outcomes and if tail come on coin then pink dice can have 6 outcomes, so, total = 1 x 6 = 6 outcomes. Therefore, total number of possible outcome = 6+6 =12

Q252.Method in which the previously calculated probabilities are revised with values of new probability is called:

(a) Revision theorem (b) Independent theorem

(c) Dependent theorem (d) **Baye's theorem**

Q253.A woman is known to speak truth 3 out of 4 times. She throws a die and reports that it is a six. Find the probability that it is actually a six.

(a) 1/7 (b) 2/7

(c) **3/8** (d) 5/8

Hint: Let A = event that the women report that six in the throwing of the die and

Let, E_1 = event that six occurs and E_2 = event that six does not occur.

Therefore, $P(E_1)$ = Probability that six occurs = 1/6; and

$P(E_2)$ = Probability that six does not occur = 5/6

Also, $P(A/E_1) = 3/4$ and $P(A/E_2) = 1 - 3/4 = 1/4$.

By using Baye's theorem;

$P(E_2/A) = P(E_1)P(A/E_1) / \{ P(E_1)P(A/E_1) + P(E_2)P(A/E_2) \} = (1/6 \times 3/4)/\{(1/6 \times 3/4)+(5/6 \times 1/4)\} = 3/8$

Q254. Bag 1 contains 4 white and 6 black balls while another bag 2 contains 4 white and 3 black balls. One ball is drawn at random from one of the bag and it is found to be black. Find the probability that it was drawn from bag 1.

(a) 5/12 (b) **7/12**

(c) 12/13 (d) 7/11

Hint: Let, E_1 = event of choosing bag 1 and E_2 = event choosing bag 2.

Let A be event of drawing a black ball.

$P(E_1) = 1/2$ and $P(E_2) = 1/2$; Also, $P(A/E_1)$ = P(drawing a black ball from bag 1) = 6/10 = 3/5.

$P(A/E_2)$ = P(drawing a black ball from bag 2) = 3/7.

By using Baye's theorem;

$P(E_1/A)= P(E_1)P(A/E_1)/\{ P(E_1)P(A/E_1) + P(E_2)P(A/E_2) \} = (1/2 \times 3/5) /\{(1/2 \times 3/5)+(1/2 \times 3/7)\}= 7/12$

Q255. Bag 1 contains 3 white and 7 black balls while another bag 2 contains 4 white and 6 black balls. One ball is drawn at random from one of the bag and it is found to be black. Find the probability that it was drawn from bag 2.

(a) 5/12 (b) 7/12

(c) **6/13** (d) 7/13

Hint: Let, E_1 = event of choosing bag 1 and E_2 = event choosing bag 2.

Let A be event of drawing a black ball.

$P(E_1) = 1/2$ and $P(E_2) = 1/2$; Also, $P(A/E_1)$ = P(drawing a black ball from bag 1) = 7/10.

$P(A/E_2)$ = P(drawing a black ball from bag 2) = 6/10 = 3/5.

By using Baye's theorem;

$P(E_2/A) = P(E_2)P(A/E_2)/\{P(E_1)P(A/E_1) + P(E_2)P(A/E_2)\} = (1/2 \times 3/5) / \{(1/2 \times 7/10) + (1/2 \times 3/5)\} = 6/13$

Q256.If P(A) = 0.48, then P(not A) will be:

(a) 0.48 (b) 0.50

(c) **0.52** (d) 0.60

Q257.If P(A) = 0.42, then probability of event A, not occurring is:

(a) 0.42 (b) 0.50

(c) **0.58** (d) 0.60

Q258.A dice is thrown. The probability of getting 3 and 5 is:

(a) 1/2 (b) **1/3**

(c) 1/4 (d) 1/6

Q259.If A and B are two events associated with the same sample space of a random experiment through P(A/B) is given by:

(a) P(A∩B)/ P(B) provided that P(B) = 0

(b) **P(A∩B)/ P(B) provided that P(B) ≠ 0**

(c) P(A∩B)/ P(B)

(d) P(A∩B)/ P(A)

Q260.Let A and B be events of a sample space S of an experiment, If P(S/B) = P(B/B) then value of P(S/B) is:

(a) -1 (b) 0

(c) 0.5 (d) **1**

Hint: We know that P(S/B) = P(S∩B)/P(B) (By conditional probability formula)

Which is equivalent to P(B) / P(B) =1, hence P(S/B) = 1

Q261. If P(E) = 1/5, P(F) = 0, then what will be the value of P(E/F)?

(a) 0 (b) 0.5

(c) 1 (d) **Not defined**

Hint: We know P(E/F) = P(E∩F)/P(F);

Here, P(F) =0, As the value of denominator becomes 0, the value of P(E/F) becomes undefined.

Q262.The probability of drawing an ace card from a pack of cards is:

(a) 1/4 (b) **1/13**

(c) 1/26 (d) 1/52

Q263.A batsman hits boundaries from 4 times out of 24 balls. Find the probability that he did not hit the boundaries.

(a) 1/2 (b) 1/3

(c) 1/6 (d) **5/6**

Hint: Total balls = 24; Number of boundaries = 4.

Therefore, number of balls without boundaries = 24 – 4 = 20

Probability of no boundary = 20/24 = 5/6

Q264.A card is drawn at random from a well shuffled pack of cards. Find the probability of club or queen card.

(a) 1/13 (b) 2/13

(c) **4/13** (d) 1/26

Q265.A single die is rolled. Find the probability that the number obtained is greater than 2.

(a) 1/2 (b) 1/3

(c) **2/3** (d) 1/6

Q266.Two coins are tossed simultaneously. The probability of getting at least one head is:

(a) 1/2 (b) 1/3

(c) 1/4 (d) **3/4**

Q267.Suppose A and B are two independent events with probabilities $P(A) \neq 0$ and $P(B) \neq 0$. Let A' and B' be their complements. Which one of the following statement is false?

(a) $P(A \cap B) = P(A).P(B)$ (b) **$P(A \cup B) = P(A) + P(B)$**

(c) $P(A / B) = P(A)$ (d) $P(A' \cap B') = P(A').P(B')$

Q268.A single die is thrown twice. What is the probability that the sum is neither 8 nor 9?

(a) 1/4 (b) **3/4**

(c) 1/9 (d) 5/9

Q269.A bag contains 40 balls numbered from 1 to 20. One ball is drawn at random. Find the probability that number of the drawn ball is multiple of 3 or 7.

(a) 1/10 (b) 3/10

(c) 1/5 (d) **2/5**

Hint: Let A be the event that the drawn number in a multiple of 3, then

A = {3,6,9,12,15,18}. Further, let B be the event that the drawn number in a multiple of 7, i.e.

B = {7,14}, Also, A∩ B = φ, and total balls = 20. Therefore, P(A) = 6/20 = 3/10; P(B) = 2/20 = 1/10; and P(A∩B) = 0/20 = 0. Thus, the probability of the desired event

P(A or B) = P(AUB) = P(A) + P(B) – P(A∩B) = 3/10 + 1/10 - 0 = 2/5

Q270.Two dice are rolled simultaneously. Find the probability that the number 4 comes up at least once.

(a) 1/6 (b) 5/6

(c) 13/36 (d) **11/36**

Q271.A die is rolled, if the outcome is an odd number. What is the probability that it is prime number?

(a) 1/2 (b) 1/3

(c) **2/3** (d) 1/4

Hint: Let A = the event of getting an odd number; B= the event of getting a prime number. Then,

A = {1,3,5}; B = {2,3,5}, and A∩B = { 3,5}. Now, we are required to find the conditional probability of event B when A has already occurred, i.e. P(B/A) = n(A∩B)/ n(A) = 2/3

Q272.In a class of post graduate students 40% students read Biostatistics, 25% students read Biochemistry and 15% students read both Biostatistics and Biochemistry, as a supportive course. One student is selected at random. Find the probability that he reads Biostatistics, if it is known that he reads Biochemistry.

(a) 1/5 (b) **3/5**

(c) 1/8 (d) 3/8

Hint: Let S be the set of all the 100 students. Let A and B denote the set of students reading Biostatistics and Biochemistry, respectively. Then, n(S) = 100, n(A) = 40, n(B) = 25, and n(A∩B) = 15.
The required probability is P(A/B) = n(A∩B) / n(B) = 15/25 = 3/5

Q273. A sample of 2 items is selected at random from a box containing 10 items of which 2 are defective. Find the expected number of defective items.

(a) 1/5 (b) **2/5**

(c) 3/5 (d) 4/5

Hint: Let X = the number of defective items. So, X can assume values 0, 1, or 2. Here, the sample space consists of ${}^{10}C_2$ = 10 x 9/ 2x1 = 45 distinct and equally likely sample points. Now,

P(X=0) = P(No defective occur in a sample of size 2) = ${}^{8}C_2$ x ${}^{2}C_0$ / ${}^{10}C_2$ = (28 x 1)/45 = 28/45

P(X=1) = P(One defective occur in a sample of size 2) = ${}^{8}C_1$ x ${}^{2}C_1$ / ${}^{10}C_2$ = (8 x 2)/45 = 16/45

P(X=2) = P(Two defective occur in a sample of size 2) = ${}^{8}C_0$ x ${}^{2}C_2$ / ${}^{10}C_2$ = (1 x 1)/45 = 1/45

Thus, the expected value of the number of defective X is

E(X) = $\Sigma x_i.f(x_i)$ = (0 x 28/45) + (1 x 16/45) + (2 x 1/45) = (16+2)/45 = 18/45 = 2/5

Q274. From a pack of 52 cards 2 are accidently dropped. Find the chance that they are the 2 honours of the same suit.

(a) 1/1326 (b) 5/1326

(c) **1/663** (d) 5/663

Hint: The event that they are 2 honours of the same suit shall have ${}^{2}C_1$ x ${}^{2}C_2$ elements in it. Therefore,

The required probability = ${}^{2}C_1$ x ${}^{2}C_2$ / ${}^{52}C_2$ = 2/ 1326 = 1/663

Q275. The husband and wife appeared for an interview for two vacancies. The probability of husband's selection is 1/4 and that of wife's selection is 1/5. Find the probability that only one of them will be selected.

(a) 1/5 (b) 3/20

(c) **7/20** (d) 9/20

Hint: Let p_1 and p_2 be the probabilities of husband and wife selection. Then

p_1 = 1/4 ; So, q_1 = 3/4 and p_2 = 1/5; So, q_2 = 4/5

Therefore, the required probability = $p_1q_2 + q_1p_2$ = (1/4 x 4/5) + (3/4 x 1/5) = 1/5 + 3/20 = 7/20

Q276.A bag contains 2 red, 3 green and 2 white balls. Two balls are drawn at random. What is the probability that none of the balls drawn white?

(a) 2/7 (b) 5/7

(c) **10/21** (d) 11/21

Q277.Tickets numbered 1 to 20 are mixed up and then a ticket is drawn at random. What is the probability that the ticket drawn has a number which is a multiple of 3 or 5?

(a) 8/15 (b) **9/20**

(c) 1/2 (d) 2/5

Q278.Three unbiased coin are tossed. What is the probability of getting at most two heads?

(a) 3/4 (b) **7/8**

(c) 3/8 (d) 1/8

Q279.In a lottery, there are 10 prizes and 25 blanks. A lottery is drawn at random. What is the probability of getting a prize?

(a) **2/7** (b) 5/7

(c) 2/5 (d) 3/5

Q280.Two cards are drawn together from a pack of 52 cards. The probability that one is a club and one is club, is:

(a) **13/102** (b) 39/102

(c) 3/26 (d) 21/26

Q281.In a simultaneous throw of two coins, the probability of getting at least one head is-

(a) 2/3 (b) **3/4**

(c) 1/2 (d) 1/3

Q282.Two cards are drawn from a pack of 52 cards. The probability that either both are red or both are jack is:

(a) 45/221 (b) **55/221**

(c) 1/13 (d) 3/13

Q283.One card is drawn at random from a pack of 52 cards. What is the probability that the card drawn is a face card?

(a) 1/4 (b) 3/4

(c) 1/13 (d) **4/13**

Q284.In a single throw of die, what is the probability of getting a number greater than 3?

(a) **1/2** (b) 1/3

(c) 2/3 (d) 1/4

Q285.A box contains 24 electric bulbs, out of which 6 are defective. Two balls are chosen at random from this box, the probability that at least one of them is defective, is:

(a) 10/23 (b) **13/23**

(c) 1/3 (d) 2/3

Q286.One card is drawn from a pack of 52 cards. What is the probability that the card drawn is either a black card or a queen?

(a) 1/2 (b) 1/4

(c) 2/13 (d) **7/13**

Q287.A basket contains 6 red, 2 blue, 4 green and 3 yellow balls. If 5 balls are picked up at random, what is the probability that at least one is red?

(a) 2/5 (b) 9/91

(c) **137/143** (d) 6/143

Q288.A bag contains 4 red balls, 6 blue balls and 8 pink balls. One ball is drawn at random and replaced with 3 pink balls. A probability that the first ball drawn was either red or blue in colour and the second drawn was pink in colour?

(a) 13/17 (b) 13/18

(c) 11/30 (d) **None of these**

Q289.Two dice are tossed. The probability that the total score is a prime number is:

(a) 1/2 (b) 1/6

(c) 7/12 (d) **5/12**

Q290.In a box, there are 8 blue, 7 red and 6 green balls. One ball is picked up randomly. What is the probability that it is neither blue nor green?

(a) **1/3** (b) 3/4

(c) 8/21 (d) 9/21

Q291. An urn contains 2 red, 3 green and 4 blue balls. If 2 balls are drawn at random, find the probability that no ball is blue.

(a) 2/9 (b) 5/9

(c) **5/18** (d) 5/36

Q292. A bag contains 6 green balls 7 red balls and 8 pink balls. If two balls are drawn at random from the bag, one after another what is the probability that the first ball is green and second ball is red?

(a) **1/5** (b) 3/5

(c) 1/7 (d) 5/7

Q293. An urn contains 2 red, 3 blue, 4 green and 5 yellow marbles. If two marbles are picked up at random, what is the probability that either both are blue or both are yellow?

(a) **1/7** (b) 3/7

(c) 3/7 (d) 4/7

Q294. Two coins are tossed. What is the conditional probability of getting two heads given that at least one coin shows a head?

(a) 1/2 (b) **1/3**

(c) 2/3 (d) 1/4

Q295. Let A and B be two events, such that $P(A) = 3/10$, $P(A \cup B) = 2/5$ and $P(B) = c$. Find c such that A and B are mutually exclusive.

(a) 1/5 (b) 4/5

(c) 1/7 (d) **1/10**

Q296. Let A and B be two events, such that $P(A) = 3/10$, $P(A \cup B) = 2/5$ and $P(B) = c$. Find c such that A and B are independent.

(a) 1/5 (b) 4/5

(c) **1/7** (d) 1/10

Q297. If the occurrence of one event means that another cannot happen, then the events are:

(a) **Mutually exclusive** (b) Independent

(c) Dependent (d) Exhaustive

Q298. Which of the following is not a correct statement about a probability?

(a) **It is the collection of several experiments**

(b) It must have a value between 0 and 1

(c) It can be reported as a decimal

(d) A value near 0 means that the event is not likely to occur

Q299. A listing of the possible outcomes of an experiment and their corresponding probability is called:

(a) **Probability distribution** (b) Frequency distribution

(c) Contingency table (d) Random variable

Q300. The special rule of multiplication of probability, the events must be:

(a) Mutually exclusive (b) **Independent**

(c) Favourable (d) Unfavourable

Q301. A speaks truth in 50% cases B speaks truth in 60% cases, the probability that they will way say the same thing while describing a single event is:

(a) **1/2** (b) 1/3

(c) 2/3 (d) 1/5

Q302. If you roll a pair of dice, what is the probability that (at least) one of the dice is a 4 or the sum of the dice is 7?

(a) 13/36 (b) **15/36**

(c) 21/36 (d) 1/9

Q303. If a card is chosen from a pack of cards, what is the probability of getting a diamond or a club?

(a) 1/6 (b) 1/4

(c) 1/3 (d) **1/2**

Q304. Outcomes of an experiment are classified as

(a) Probability (b) Random experiment

(c) Conditional probability (d) **Events**

Q305. The collection of one or more outcomes from an experiment is called:

(a) **Event** (b) Random experiment

(c) Probability (d) Random variable

Q306. In the special rule of addition of probability, the events are always:

(a) **Mutually exclusive** (b) Independent

(c) Favourable (d) Unfavourable

Q307.Probability of second event in situation if first event has been already occurred is classified as:

(a) Addition probability (b) **Conditional probability**

(c) Joint probability (d) Multiplication probability

Q308.In probability theories, events which can never occur together are classified as:

(a) Exclusive events (b) Complementary events

(c) **Mutually exclusive events** (d) Independent events

Q309.Joint probability of independent events A and B is equal to:

(a) **P(A) x P(B)** (b) P(A) + P(B)

(c) P(A) x P(B) + P(A - B) (d) P(A) x P(B) - P(A x B)

Q310.Types of probabilities for independent events must include:

(a) Joint events (b) Marginal events

(c) Conditional events (d) **All of above**

Q311.In measuring probability of any certain event, one represents:

(a) Impossible events (b) Possible events

(c) **Sure event** (d) Sample event

Q312. If two events A and B are considered as partially overlapping events then rule of addition can be written as:

(a) P(A or B) = P(A) - P(B) + P(A and B)

(b) P(A or B) = P(A) + P(B) * P(A - B)

(c) P(A or B) = P(A) * P(B) + P(A - B)

(d) **P(A or B) = P(A) + P(B) - P(A and B)**

Q313.If a person buys a lottery, chance of winning a Maruti car is 60%, chance of winning Mahindra car is 70% and chance of winning both is 40% then chance of winning Maruti or Mahindra is:

(a) 0.6 (b) **0.9**

(c) 0.8 (d) 0.5

Q314.According to combination rule, if total number of outcomes are 'r' and distinct outcome collection is 'n' then combinations are calculated as

(a) **n!/x!(n - x)!** (b) n!/x!(n + x)!

(c) x!/n!(n - x)! (d) x!/n!(n + x)!

Q315.An urn contains 3 green, 4 red, 5 white and 6 blue marbles. If four marbles are picked up at random, what is the probability that 2 is green, 1 are red and 1 is white?

(a) 1/13 (b) 12/13

(c) **1/51** (d) 4/51

Q316.A bag contains 10 apples out of which 4 are rotten; two apples are taken out together. If one of them is found to be good, the probability that other also good is:

(a) 7/15 (b) 8/15

(c) **1/3** (d) 2/3

Q317.A blood bank has 10 units of A+ blood in which 4 are contaminated with serum hepatitis. Three units are selected for use with three patients. What is the probability that all three patients will be exposed to hepatitis from this source?

(a) 1/10 (b) 9/10

(c) 2/5 (d) **1/30**

Q318.What is the probability of obtaining 9, 10 and 11 points with 3 dice?

(a) 25/216 (b) **79/216**

(c) 1/8 (d) 7/8

Q319.Three cards are drawn from a pack of 52 cards. What is the probability that all the three will be kings?

(a) 4/5525 (b) 3/5525

(c) **1/5525** (d) 16/5525

Q320.A peon is given 4 letters to drop into 4 letter boxes of a city. What is the probability that the peon will drop all the four letters in different letter boxes?

(a) 1/4 (b) 1/32

(c) **3/32** (d) 1/256

Q321.There are 4 letters and 4 addressed envelopes. Find the probability that all the letters are not dispatched in the right envelopes.

(a) **23/24** (b) 1/24

(c) 1/4 (d) 3/4

Q322.What is the chance that a leap year will contain 53 Sundays?

(a) 1/7 (b) **2/7**

(c) 1/365 (d) 1/366

Q323.What is the chance that a non-leap year will contain 53 Sundays?

(a) **1/7** (b) 2/7

(c) 1/365 (d) 1/366

Q324.Mr. X kills a bird once in three shots. N this assumption he fired three shots. Find the probability that the bird is not killed.

(a) 1/3 (b) 2/3

(c) 1/27 (d) **8/27**

Q325.A die is rolled two times. What is the probability of getting an odd number both the times?

(a) 1/2 (b) 1/3

(c) **1/4** (d) 3/4

Q326.A husband and wife both appear in an interview for two vacancies. The probability of husband selection is 1/4 and that of wife is 1/3. Find the probability of both of them selected.

(a) 7/12 (b) **1/12**

(c) 3/4 (d) 1/2

Q327.A husband and wife both appear in an interview for two vacancies. The probability of husband selection is 1/4 and that of wife is 1/3. Find the probability of none of them selected.

(a) 7/12 (b) 1/12

(c) 3/4 (d) **1/2**

Q328.A husband and wife both appear in an interview for two vacancies. The probability of husband selection is 1/4 and that of wife is 1/3. Find the probability atleast one of them selected.

(a) 7/12 (b) 1/12

(c) 3/4 (d) **1/2**

Q329.A single letter is selected at random from the word **PROBABILITY**. Find the probability that selected letter is a vowel.

(a) 1/2 (b) 1/7

(c) 1/11 (d) **4/11**

Q330.The probability of throwing more than 4 by one die is:

(a) 1/2 (b) **1/3**

(c) 2/3 (d) 1/4

Q331.If A and B are independent events then P(A∩ B) is equal to:

(a) P(A) + P(B) (b) **P(A) x P(B)**

(c) P(A) x P(A/B) (d) P(B) x P(B/A)

Q332.If A and B are mutually exclusive events then P(A∪ B) is equal to:

(a) P(A) + P(B) (b) **P(A) x P(B)**

(c) P(A) x P(A/B) (d) P(B) x P(B/A)

Q333.For any event A, is equal to:

(a) 0 (b) **1**

(c) 0.5 (d) ∞

Q334.For any event A, is equal to:

(a) **0** (b) 1

(c) 0.5 (d) ∞

Q335.Groups are formed of 4 persons out of 12 persons. The probability that one particular player is never included is:

(a) 1/3 (b) **2/3**

(c) 1/4 (d) 3/4

Q336. A problem in Statistics is given to three students A, B and C whose probability of solving are 1/2, 1/3 and 1/4, respectively. What is the chance that the problem will be solved?

(a) 1/2 (b) 1/3

(c) 1/4 (d) **3/4**

Q337.If two events A and B are dependent, then P(A/B) is equal to:

(a) P(A)/P(B) (b) P(B)/P(A)

(c) P(AB)/P(A) (d) **P(AB)/P(B)**

Q338.If two events A and B are dependent, then P(B/A) is equal to:

(a) P(A)/P(B)
(b) P(B)/P(A)
(c) **P(AB)/P(A)**
(d) P(AB)/P(B)

Q339.If 'p' is the probability of success and 'q' is the probability of failure, then p x q is equal to:

(a) Less than 1
(b) 0
(c) **1**
(d) More than 1

Q340.${}^{6}C_{2}$ is equal to:

(a) 10
(b) **15**
(c) 20
(d) 25

Q341.What are the chances that no two boys are sitting together for a photograph if there are 5 girls and 2 boys?

(a) 2/7
(b) 4/7
(c) **5/7**
(d) 1/21

Q342.In a drawer there are 4 white socks, 3 blue socks and 5 grey socks. Two socks are picked randomly. What is the possibility that both the socks are of same color?

(a) 2/33
(b) 13/66
(c) **19/66**
(d) 23/66

Q343.What will be the possibility of drawing a jack or a spade from a well shuffled standard deck of 52 playing cards?

(a) 1/13
(b) **4/13**
(c) 1/26
(d) 17/52

Q344. *In a throw of coin what is the probability of getting head.*

(a) 0
(b) **1/2**
(c) 1
(d) 2

Q345. *In a throw of coin what is the probability of getting tail.*

(a) 0
(b) **1/2**
(c) 1
(d) 2

Q346.From a pack of 52 cards, one card is drawn at random. Find the probability of a face card drawn:

(a) 1/4 (b) 3/4

(c) 1/13 (d) **3/13**

Q347.The probability that it will rain tomorrow is 0.85. What is the probability that it will not rain tomorrow

(a) 1/4 (b) 3/4

(c) 1/20 (d) **3/20**

Q348.What is the probability that a number selected from the numbers (1, 2, 3,..........,15) is a multiple of 4?

(a) **1/5** (b) 4/5

(c) 1/3 (d) 2/3

Q349.What are the total outcomes when we throw three coins?

(a) 4 (b) 5

(c) **8** (d) 7

Q350.The probability that a prime number selected at random from the numbers (1,2,3, ...35) is :

(a) **11/35** (b) 12/35

(c) 13/35 (d) 14/35

Q351.The probability of getting less than 3 in a single throw of a die is:

(a) 1/2 (b) **1/3**

(c) 2/3 (d) 1/4

Q352.The total number of events of throwing 10 coins simultaneously is:

(a) 10 (b) 100

(c) 512 (d) **1024**

Q353.A girl calculates that the probability of her winning the first prize in a lottery is 8/100. If 6,000 tickets are sold, how many tickets has she bought?

(a) 120 (b) 240

(c) **480** (d) 960

Q354.One card is drawn from a well-shuffled deck of 52 cards. The probability that the card will not be an ace is:

(a) 1/13 (b) 3/13

(c) 4/13 (d) **12/13**

Q355.Which of the following cannot be the probability of an event?

(a) 0.2 (b) 0.3

(c) 0.8 (d) **1.4**

Q356.Probability of getting a number x such that $1 \leq x \leq 4$, is:

(a) 1 (b) 0

(c) < 1 (d) **1**

Q357.The sum of the probabilities of all elementary events of an experiment is p, then:

(a) $p = 0$ (b) $\mathbf{p = 1}$

(c) $0 < p < 1$ (d) $0 < p < 1$

Q358.If an event cannot occur, then its probability is:

(a) **0** (b) 1/2

(c) 1/3 (d) 1

Q359.A bag contains 12 balls of which x are red, the probability of getting a red ball at random from the bag is:

(a) x/4 (b) x/8

(c) **x/12** (d) 12/x

Q360.If the probability of newborn being Rh –ve is 0.3, than the probability of newborn being Rh +ve is:

(a) 0.2 (b) 0.3

(c) **0.7** (d) 1.0

Q361.If probability of getting Rh+ve is 0.8 and probability of newborn being male is 1/4, the probability of newborn being a male with Rh+ve is equal to:

(a) **0.2** (b) 0.4

(c) 0.6 (d) 0.8

Q362.If prevalence of diabetes in a population is 10%; probability that four persons selected randomly from this population will have diabetes will be:

(a) **0.0001** (b) 0.001

(c) 0.01 (d) 0.1

Q363.If probability of a child being Rh-ve is 1/5, the probability of a child being Rh+ve in a society will be:

(a) 1/5 (b) 1/10

(c) **4/5** (d) 9/10

Q364.If $A \subset B$ and $B \subset A$ then,

(a) P(A)>P(B) (b) P(A)<P(B)

(c) **P(A)=P(B)** (d) $P(\bar{A})<P(\bar{B})$

Q365.What is the probability of an impossible event?

(a) **0** (b) 0.5

(c) 1 (d) None of these

Q366.If A and B are two events, then the probability of exactly one of them occurs is given by:

(a) $\mathbf{P(A \cap \bar{B}) + P(\bar{A} \cap B)}$ (b) P(A)+P(B)–2P(A)P(B)

(c) $P(\bar{A}) + P(\bar{B}) - 2\,P(\bar{A})P(\bar{B})$ (d) $P(A) + P(B) - P(A \cap B)$

Q367.Which of the given sets are not an event of the sample space S= {a,1,2,3,z}:

(a) {a, z} (b) {1,2,3}

(c) {} (d) **{a, b}**

Q368.Which among the following is a sample space obtained while tossing a coin thrice?

(a) {(H,T),(T,H),(T,T),(H,H)}

(b) {(H,H,H),(H,T,T),(T,T,T)}

(c) {(H,H),(T,T)}

(d) **{(H,H,H),(H,H,T),(H,T,T),(T,H,T),(H,T,H),(T,T,H),(T,H,H),(T,T ,T)}**

Q369.Classical definition of probability is applicable only when the following holds true-Sample space is finite, Events are mutually exclusive, Events are equally likely, and Events are exhaustive

(a) Sample space is finite, and Events are mutually exclusive

(b) Events are mutually exclusive, Events are equally likely, and Events are exhaustive

(c) **Sample space is finite, Events are mutually exclusive, Events are equally likely, and Events are exhaustive**

(d) Sample space is finite

Q370.Phenomenon of Statistical Regularity is observed when:

(a) **Number of trials of a random experiment increases**

(b) Number of trials of a random experiment is kept minimal

(c) Relative frequencies approach divergent values

(d) None of these

Q371.If P (A) = 0.3, P(B)=0.2, P(A∩B)=0.1, what is the probability that exactly one of the events occurs?

(a) **0.3** (b) 0.4

(c) 0.6 (d) 1

Q372.If P(A)=0.7, P(B)=0.2, P(A∩B))=0.2, what is the probability that neither A nor B occurs?

(a) 0.3 (b) 0.2

(c) **0.8** (d) 0.7

Q373.Which of these is the same as (A ∩ B^c) U (A∩B)?

(a) A\B (b) AUB

(c) **A** (d) A∩B

Q374.What is the probability of the event that an even number appears when tossing a fair dice?

(a) 1/6 (b) **1/2**

(c) 1/4 (d) 2/3

Q375.In a rowing race, the odds that A will win is 2 to 3 and the odds that B will win is 1 to 4.What is the probability that A or B wins the race?

(a) 1/5 (b) 2/5

(c) **3/5** (d) 4/5

Q376.Two cards are drawn from a well shuffled pack of 52 cards. Find the probability that they are both aces if the first card is replaced.

(a) 1/144 (b) **1/169**

(c) 4/169 (d) 1/36

Q377.If two events A and B are independent, then P (A/B) is equal to:

(a) **P(A)** (b) P(B)

(c) P(A∩B) (d) None of these

Q378.When a card is drawn from a well shuffled deck of cards, what is the probability of getting a black king?

(a) 1/4 (b) 1/2

(c) **1/26** (d) 1/52

Q379.A pair of fair dice is tossed. What is the probability that the maximum of the two numbers is greater than 4?

(a) 1/9 (b) **5/9**

(c) 1/18 (d) 1/6

Q380.An urn B_1 contains 2 white and 3 black chips and another urn B_2 contains 3 white and 4 black chips. One urn is selected at random and a chip is drawn from it. If the chip drawn is found black, find the probability that the urn chosen was B_1.

(a) 3/7 (b) 4/7

(c) **21/41** (d) 20/41

Q381.At a certain university, 4% of men are over 6 feet tall and 1% of women are over 6 feet tall. The total student population is divided in the ratio 3:2 in favour of women. If a student is selected at random from among all those over six feet tall, what is the probability that the student is a woman?

(a) 2//5 (b) 3/5

(c) 2/11 (d) **3/11**

Q382.If A = {1, 2, 3, 4}, B = ϕ, then AUB is:

(a) ϕ (b) {0}

(c) **{1, 2, 3, 4}** (d) None of these

Q383.Let A = {a, b}, B = {p, q}, then A x B is:

(a) {a, b, p, q} (b) {0}

(c) ϕ (d) **{(a, p), (a, q), (b, p), (b, q)}**

Q384.Given that E and F are events such that P (E) = 0.6, P (F) = 0.3 and P(E∩F) = 0.2. Find P(E|F)

(a) 1/2 (b) 1/3

(c) **2/3** (d) 1/4

Q385.Given that E and F are events such that P (E) = 0.6, P (F) = 0.3 and P(E∩F) = 0.2. Find P(F|E)

(a) 1/2 (b) **1/3**

(c) 2/3 (d) 1/4

Q386.If P(B)=0.4 and P (A∩B) = 0.32. Find P(A|B)?

(a) 0.32 (b) 0.64

(c) 0.72 (d) **0.80**

Q387. If P (A)=0.8, P (B)=0.5 and P(B/A)=0.4, find P(A∩B)

(a) 0.80 (b) 0.40

(c) **0.32** (d) 0.98

Q388.If 2P(A) = P(B) = 5/13 and P(A|B) = 2/5. Find P(A∪B)?

(a) 1/13 (b) 2/13

(c) **11/26** (d) 15/26

Q389.If P(A) = 6/11,P(B) =5/11 and P(A∪B) = 7/11. Find

(a) 1/5 (b) 4/5

(c) **4/11** (d) 7/11

Q390.Assume that each child born is equally likely to be a boy or a girl. If a family has two children, what is the conditional probability that both are girls given that the youngest is a girl?

(a) **1/2** (b) 1/3

(c) 1/4 (d) 3/4

Q391.Assume that each child born is equally likely to be a boy or a girl. If a family has two children, what is the conditional probability that both are girls given that at least one is girl?

(a) 1/2 (b) **1/3**

(c) 1/4 (d) 3/4

Q392.If P(A) = 12, P (B) = 0 then P (A | B) is\

(a) 0 (b) 1

(c) 12 (d) **not defined**

Q393.If A and B are events such that P(A / B) = P(B / A) then

(a) $A \subset B$ but $A \neq B$ (b) A = B

(c) $A \cap B = \varphi$ (d) **P(A) = P(B)**

Hint: P(A / B) = P(B / A). That means $P(A \cap B)/P(B) = P(A \cap B)/P(A)$

Thus P(A) =P(B)

Q394.If A and B are independent events, and if P(A) =3/5 and P(B) =1/5. Find P ($A \cap B$)?

(a) 1/15 (b) **3/25**

(c) 1/3 (d) 1/5

Q395.Two cards are drawn at random and without replacement from a pack of 52 playing cards. Find the probability that both the cards are black.

(a) 1/25 (b) 3/25

(c) 25/51 (d) **25/102**

Q396.A box of mangoes is inspected by examining three randomly selected mangoes drawn without replacement. If all the three mangoes are good, the box is approved for sale otherwise it is rejected. Find the probability that a box containing 15 mangoes out of which 12 are good and 3 are bad ones will be approved for sale.

(a) 13/44 (b) 23/44

(c) 40/91 (d) **44/91**

Hint: Total number of mangoes in a box = 15

P(a box is approved) = (12x11x10)/(15x14x13) = 44/91

Q397.Given that the events A and B are such that P(A) = 1/2, $P(A \cup B)$ = 3/5 and P(B) = k. Find k, if they are mutually exclusive:

(a) 1/2 (b) 1/3

(c) 1/5 (d) **1/10**

Hint: Let $P(A \cap B) = x$

$P(A \cup B) = P(A) + P(B) - P(A \cap B) \Rightarrow 3/5 = 1/2 + k - x \Rightarrow k - x = 3/5 - 1/2 = 1/10$

Now, when A and B are mutually exclusive, then P(A∩B) = 0

Therefore, x = 0. Hence, k = 1/10

Q398. Given that the events A and B are such that P(A) = 1/2, P(A∪B) = 3/5 and P(B) = k. Find k, if they are independent:

(a) 1/2 (b) 1/3

(c) **1/5** (d) 1/10

Hint: Let P(A∩B) = p

P(A∪B) = P(A) + P(B) – P(A∩B) ⇒ 3/5 = 1/2 + k – p

⇒ k – p = 3/5 – 1/2 = 1/10 (i)

Now, when A and B are independent, then P(A∩B) = P(A) . P(B)

⇒ p = 1/2 x k (ii)

From equation (i) and (ii), we get

k – k/2 =1/10 ⇒ k = 1/5

Q399. A die is tossed thrice. Find the probability of getting an odd number at least once.

(a) 1/2 (b) 1/6

(c) 1/8 (d) **7/8**

Hint: S = {1,2,3,4,5,6}; n(S) = 6

Let A represents an odd number. Then A = {1,3,5}; n(A) = 3; P(A) = n(A)/ n(S) = 3/6 = 1/2; n = 3

P(Atleast one success) = 1 – P(X =0) = 1 – (1/2)3 = 7/8

Q400. The probability of obtaining an even prime number on each die when a pair of dice is rolled is:

(a) 0 (b) 1/3

(c) 1/12 (d) **1/36**

Hint: n(S) =36; Let A represents an even prime number one each dice.

A = {22}; n(A) =1; P(A) = n(A)/n(S) = 1/36

Q401. Two events A and B are said to be independent, if

(a) A and B are mutually exclusive

(b) **P(A'and B') = [1 – P(A)] [1 – P(B)]**

(c) P(A) = P(B)

(d) P (A) + P (B) = 1

Q402. How many ways can 4 starting positions in a team are filled with 9 students who can play at any of the positions?

(a) 126
(b) 252
(c) **3024**
(d) None of these

Q403.One card is randomly drawn from a pack of 52 cards. Find the probability that the drawn card is a king.

(a) **1/13**
(b) 1/26
(c) 1/4
(d) 1/2

Q404.In a single throw of die, what is the probability of getting a number greater than 3?

(a) **1/2**
(b) 1/3
(c) 2/3
(d) 1/4

Q405.Three unbiased coins are tossed. What is the probability of getting at most two heads?

(a) 1/4
(b) 3/4
(c) 3/8
(d) **7/8**

Q406.One card is randomly drawn from a pack of 52 cards. Find the probability that the drawn card is red.

(a) 1/13
(b) 1/26
(c) 1/4
(d) **1/2**

Q407.Tickets numbered 1 to 20 are mixed up and then a ticket is drawn at random. What is the probability that the ticket drawn has a number which is a multiple of 3 or 7?

(a) 1/5
(b) **2/5**
(c) 3/5
(d) 1/2

Q408.A box contains 2 red, 3 blue, 4 green and 6 yellow balls. If two balls are drawn at random from the box, what is the probability that both are blue?

(a) **1/35**
(b) 2/35
(c) 1/5
(d) 2/5

Q409.One card is drawn at random from a pack of 52 cards. What is the probability that the card drawn is a face card?

(a) 1/4 (b) 3/4

(c) 1/13 (d) **4/13**

Q410.Two dice are thrown simultaneously. What is the probability of getting two numbers whose product is even?

(a) 1/4 (b) **3/4**

(c) 3/8 (d) 5/8

Q411.Two dice are thrown simultaneously. What is the probability of getting two numbers whose product is odd?

(a) **1/4** (b) 3/4

(c) 3/8 (d) 5/8

Q412.A box contains 24 electric bulbs, out of which 6 are defective. Two balls are chosen at random from this box, the probability that at least one of them is defective, is:

(a) 10/23 (b) **13/23**

(c) 1/3 (d) 2/3

Q413.One card is randomly drawn from a pack of 52 cards. Find the probability that the drawn card is red and a king.

(a) 1/13 (b) **1/26**

(c) 1/4 (d) 1/2

Q414.One card is randomly drawn from a pack of 52 cards. Find the probability that the drawn card is red or a king.

(a) 1/13 (b) **7/13**

(c) 1/4 (d) 1/2

Q415.A die is rolled. If the outcome is an odd number, what is the probability that it is a prime number?

(a) 1/6 (b) 5/6

(c) 1/3 (d) **2/3**

Q416.In measuring probability of any certain event, zero represents:

(a) **Impossible events** (b) Possible events

(c) Sure event (d) Sample event

Q417.In a class, there are 15 boys and 10 girls. Three students are selected at random. The probability that 1 girl and 2 boys are selected is:

(a) 3/25 (b) 1/50

(c) **21/46** (d) 25/117

Q418.A bag contains 4 white, 5 red and 6 blue balls. Three balls are drawn at random from the bag. The probability that all of them red, is:

(a) 2/91 (b) 3/91

(c) 2/15 (d) 1/15

Q419.Two cards are drawn together from a pack of 52 cards. What is the probability is that both the cards drawn being kings?

(a) 1/15 (b) 25/57

(c) **1/221** (d) 35/256

Q420.A bag contains 4 red balls, 6 blue balls and 8 pink balls. One ball is drawn at random and replaced with 3 pink balls. A probability that the first ball drawn was either red or blue in colour and the second drawn was pink in colour?

(a) 13/17 (b) 13/18

(c) 11/30 (d) **None of these**

Q421.One card is drawn from a pack of 52 cards. What is the probability that the card drawn is either a black card or a king?

(a) 1/2 (b) 1/4

(c) 2/13 (d) **7/13**

Q422.Tickets numbered 1 to 12 are mixed up and then a ticket is drawn at random. What is the probability that the ticket drawn has a number which is a multiple of 2 or 5?

(a) 1/2 (b) 2/5

(c) 5/12 (d) **7/12**

Q423.A basket contains 4 red, 5 blue and 3 green balls. If three balls are picked up at random what is the probability that at least one is blue?

(a) 5/12 (b) 7/12

(c) 7/44 (d) **37/44**

Q424.What is the probability that a leap year selected at random will contain either 53 Saturday or 53 Sunday?

(a) 1/8 (b) 3/8

(c) 1/7 (d) **3/7**

Q425.The probability that a student passed Economics is 2/3, the probability that he passes Statistics is 4/9. If the probability of passing at least one subject is 4/5, what is the probability that student will pass both the subjects?

(a) 1/5 (b) 1/3

(c) 1/9 (d) **14/45**

Q426.A problem in statistics is given to three students whose chances of solving are 1/2, 1/3 and 1/4, respectively. What is the probability that the problem will be solved?

(a) **1/4** (b) 3/4

(c) 1/8 (d) 7/8

Q427.Tickets numbered 1 to 20 are mixed up and then a ticket is drawn at random. What is the probability that the ticket drawn bears a number which is a multiple of 3?

(a) 1/2 (b) 2/5

(c) **3/10** (d) 3/20

Q428.Two lottery tickets are drawn from 20 for first and second prizes. Count the total number of arrangements of the tickets:

(a) **380** (b) 190

(c) 95 (d) 10

Q429.In a box, there are 8 white, 7 blue and 6 green balls. One ball is picked up randomly. What is the probability that it is neither white nor green?

(a) **1/3** (b) 3/4

(c) 8/21 (d) 9/21

Q430.A bag contains 2 red, 3 green and 2 white balls. Two balls are drawn at random. What is the probability that none of the balls drawn white?

(a) **10/21** (b) 11/21

(c) 2/7 (d) 5/7

Q431.In a lottery, there are 10 prizes and 25 blanks. A lottery is drawn at random. What is the probability of getting a prize?

(a) 2/5 (b) **2/7**

(c) 5/7 (d) 1/10

Q432.From a pack of 52 cards, two cards are drawn together at random. What is the probability of both the cards being kings?

(a) 1/52 (b) 25/57

(c) **1/221** (d) 35/221

Q433.The joint probability is:

(a) Called Prior probability

(b) Based on two mutually exclusive events

(c) **The likelihood of two events happening together**

(d) The likelihood of an event happening given that another event has already happened

Q434.If A and B be two arbitrary events, then:

(a) $P(A \cup B) = P(A) + P(B)$ (b) $P(A \cap B) = P(A) . P(B)$

(c) $\mathbf{P(A \cup B) < P(A) + P(B)}$ (d) $P(A/B) = P(B) . P(A \cap B)$

Q435. $P(\bar{A} \cup \bar{B}) = ?$

(a) 0 (b) 1/2

(c) **1** (d) None of these

Q436.All the face cards are removed from a pack of playing cards. The remaining cards are well shuffled and then a card is drawn from it. What is the probability that the drawn card is an ace?

(a) 1/4 (b) 1/5

(c) 1/6 (d) **1/10**

Q437.An experiment yields three mutually exclusive events A, B and C such that $P(A) = 2\,P(B) = 3\,P(C)$. Then $P(\bar{A}) = ?$

(a) 3/11 (b) 4/11

(c) **5/11** (d) 6/11

9

Theoretical Probability Distribution

Theoretical Probability Distribution

Many random variables associated with statistical experiments have similar properties, and as such can be described by the same probability functions. The distribution obtained from a theoretical listing of outcomes and probabilities, which can be described by a mathematical formula representing same phenomenon of interest is called a **theoretical probability distribution**. The distributions of discrete and continuous random variables are accordingly called **discrete** or **continuous probability distributions**.

Discrete Probability Distribution: It can be defined as a probability distribution of discrete random variables. Such a distribution will represent data that has a finite countable number of outcomes. If the random variable X is a discrete random variable, the probability function P(X) is called **probability mass function** and its distribution is called **discrete probability distribution**. There are two conditions that a discrete probability distribution must satisfy. These are given as follows:

(i) The probability of a discrete random variable X taking on an exact value of X, lies between 0 and 1. That means-

$0 \leq P(X = x) \leq 1$

(ii) The sum of all probabilities must be equal to 1. That means-

$\Sigma P(X = x) = 1$

Types of Discrete Probability Distributions

There are several important discrete probability distributions-

- Uniform distribution
- Binomial distribution
- Negative Binomial distribution
- Geometric distribution
- Poisson distribution
- Hyper-Geometric distribution

- **Uniform distribution:** The random variable X is said to have a uniform distribution, if its probability density function is given by-

  $f(X,n)=\frac{1}{n}$

  Here, $X = x_1, x_2, \ldots\ldots\ldots, x_n$

  This distribution has the parameter n.

- **Binomial distribution:** This distribution was developed by Swiss mathematician **James Bernoulli** (1654 – 1705). In this distribution, there is an experiment consisting of repeated independent trials, each with two possible outcomes are called '**Success**' and '**Failure**'. The probability of success is denoted by 'p' and that of failure by 'q', such that p + q = 1. Experiments of this nature are called **Bernoulli trial**. Thus, the **Bernoulli trials** have the following properties-

1) Each trial has only two possible outcomes, 'Success or Failure'.
2) The repeated trials are independent
3) The probability of success in each trial remains constant.

In case of **Binomial distribution**, if p is the probability of success in a trial, then the probability of x success in n independent trials, which is given by-

$$P(X = x) = P(x) = {}^nC_x\, p^x\, q^{n-x}$$

Where, x = 0, 1, 2, …… , n

Note

- The quantities n and p are called the parameters of the binomial distribution.
- The mean of the binomial distribution is np and variance is npq.
- The mean of the binomial distribution is always greater than its variance.
- If p = q = ½, then the binomial distribution is a symmetrical distribution.
- For p ≠ q, the binomial distribution is a skewed distribution.
- **Negative Binomial distribution:** An experiment, in which independent trials are repeated until a fixed number of success occur, with the probability of success p and a failure with probability q = 1 − p, remains constant at each trial, is called **negative binomial experiment**. The number X of trials to produce k success in a negative binomial experiment, is called a **negative binomial random variate**. The probability distribution of the random

variable X, is called the **negative binomial distribution**, denoted by B*(X, k, p), defined by-

$B^*(X, k, p) = {}^{x-1}C_{k-1}\, p^k\, q^{x-k}$; X = k, k+1, k+2, ...

Note:

- The quantities k and p are called the two parameters of the negative binomial distribution.
- **Geometric distribution:** If repeated independent trials can result in a success with probability p and a failure with probability q = 1 − p, then the probability distribution of the random variable X, the number of the trial on which the first success occurs, is-

  $g(x; p) = pq^{x-1}$, x = 1,2,3,...

Note

- The mean of the geometric distribution is 1/p and variance is (1-p)/p.
- **Poisson distribution:** This distribution was developed by French mathematician **Simon Denis Poisson** in 1837. Poisson distribution is a discrete probability distribution. It is the limiting form of the binomial distribution, when n is large (*i.e.* n→∞), p is small (i.e. p→0) and np = m (a constant). Symbolically, A random variable X taking values 0, 1, 2, ...,∞ is said to have Poisson distribution with parameter m, if for (m > 0), the probability of x successes is given by-

  $$P(X = x) = \frac{e^{-m} m^x}{x!}$$

  Where, x = 0,1,2,.........,∞

Note

- There is only one parameter in Poisson distribution, which is m. Thus, if m is known, the Poisson distribution is completely known.
- The mean and variance of the Poisson distribution are the same, which is m. That means, Mean = Variance = m
- It is a limiting case of the binomial probability distribution.
- Poisson distribution is also a discrete probability distribution.

• **Hyper-Geometric distribution:** The hyper-geometric random variables are closely related to the binomial random variables. It does not require independence and is based on sampling done **without replacement**. The hyper-geometric random variables are particularly important in quality control and in the statistical estimation of population proportions. Let probability distribution of the hyper-geometric random variable X, the number of successes in a random sample of size n selected from N items of which k are labelled success and N −k labelled failure, is

$$h(X,N,n,k)=\frac{{}^{k}C_{x}\,{}^{N-k}C_{n-x}}{{}^{N}C_{n}}$$

Where, max{0,n−(N −k)} ≤ x ≤ min{n,k}; x = 0, 1, 2, ...n

Note

- The mean of hyper-geometric distribution is np and variance is npq(N-n)/(N-1)

Continuous Probability Distribution: The probability distribution of a continuous random variable X is called continuous probability distribution. It is denoted by f(X). If the random variable X is a continuous random variable, the probability function f(X) is called **probability density function** and its distribution is called **continuous probability distribution**. For any two numbers a and b with a ≤ b is defined by

$$P(a\le X\le b)=\int_{a}^{b}f(x)\,dx$$

The probability that X is in the interval [a, b] can be calculated by integrating the probability density function of the random variable X.

Types of Continuous Probability Distributions

There are several important continuous probability distributions-

- Rectangular or Uniform Probability Distribution
- Gamma Probability Distribution
- Beta Probability Distribution
- Exponential Probability Distribution
- Normal Probability Distribution
- Cauchy Probability Distribution
- Log-Normal Probability Distribution

• **Rectangular or Uniform Probability Distribution:** The continuous random variable X is said to have a rectangular or uniform distribution, if its probability density function is defined by:

$$f(X)=\frac{1}{\beta-\alpha}$$

$\alpha \le X \le \beta$

• **Gamma Probability Distribution:** The continuous random variable X is said to have a Gamma distribution, if its probability density function, is defined by:

$$f(X,\alpha,\beta)=\frac{1}{\beta^{\alpha}\Gamma\alpha}x^{\alpha-1}e^{-\frac{x}{\beta}}$$

$$X>0$$

Here, the function represents the value of gamma function at the point α, defined by:

$$\Gamma\alpha=\int_0^{\infty}x^{\alpha-1}e^{-x}dx$$

Where, ɼ(α+1) = αɼα = α!

• **Beta Probability Distribution:** The random variable X is said to have a Beta distribution, if its probability density function, is defined by

$$f(X,\alpha,\beta)=\frac{1}{B(\alpha,\beta)}x^{\alpha-1}(1-x)^{\beta-1}$$

$$0<X<1$$

• **Exponential Probability Distribution:** It is a particular case of gamma distribution. If α=1 in gamma distribution, we have exponential distribution, which is defined as:

$$f(X)=\frac{1}{\beta}e^{-\frac{x}{\beta}}$$

$$X>0$$

• **Normal Probability Distribution:** The normal distribution was discovered by **De Moivre** in the year 1733. Normal distribution is probably the most important of all theoretical distributions for the reason that so many physical measurements and natural phenomena have observed frequency distributions which very closely resemble the normal distribution. Symbolically, A continuous random variable X is said to be normally distributed if it has the probability density function represented by the equation-

$$p(x)=\frac{1}{\sigma\sqrt{2\pi}}\,e^{-\frac{1}{2}\left(\frac{x-\mu}{\sigma}\right)^2}$$

$$-\infty<x<\infty$$

Where, μ = Mean of the normal distribution;

σ = Standard deviation of the normal distribution.

Remarks

- μ and σ are also known as the two parameters of the normal distribution.

- The graphical shape of the normal curve is the bell-shaped smooth symmetrical curve.
- The curve is symmetrical about the vertical axis through the mean. That means, if we fold the curve along with this vertical axis, the two halves of the curve would coincide.
- As a result of symmetry, the mean, median and mode of the distribution are identical. That means, Mean = Median = Mode
- The curve is asymptotic to x-axis, i.e. it becomes closer and closer to x-axis but never actually touches it.
- Since there is only one maximum point in the curve, the normal curve is uni-model, That means, it has only one mode.
- The area under the normal curve lies within $\mu \pm 1\sigma$ limits which are also called 1-sigma limits is 0.6827, $\mu \pm 2\sigma$ limits which are also called 2-sigma limits is 0.9545, $\mu \pm 3\sigma$ limits which are also called 3-sigma limits is 0.9973, respectively.
- It is possible to transform any normal random variable X with mean μ and variance σ^2 to a new normal random variable Z with mean 0 (zero) and variance 1 (one). This normal random variable Z with mean 0 and variance 1 is called **Standard Normal Variable** (S.N.V.). The transformation of X to Z is-

  $$Z = \frac{X - \mu}{\sigma}$$

The following important points should be kept in mind while computing area or probability under a standard normal curve-

- The total area under the standard normal curve is 1.
- The mean of the distribution is 0 (zero). Thus, the negative and positive values of Z will lie on the left and right of mean respectively.
- The ordinate at mean i.e., at Z = 0 divides the area under the standard normal curve into two equal parts. Thus, the area on the right and left of the ordinate at Z = 0 is 0.5. Symbolically:

  $$P(-\infty \leq Z \leq 0) = P(0 \leq Z \leq \infty) = 0.5$$

- Since the curve is symmetrical, thus

  $$P(-a \leq Z \leq 0) = P(0 \leq Z \leq a)$$

• **Cauchy Probability Distribution:** The random variable X is said to have a Cauchy distribution, if its probability density function, is defined by

$$f(X) = \frac{1}{\pi} \cdot \frac{1}{1+(X-\mu)^2}$$

$-\infty < X < \infty$

- **Log-Normal Probability Distribution:** Let X be a random variable and if $y = \log_e X$ is a normal variate, then the random variable X is said to have a Log normal distribution, if its probability density function, is:

$$f(X) = \frac{1}{X\beta\sqrt{2\pi}} e^{-\frac{1}{2\beta^2}\left[(\log X - \log\alpha)^2\right]}$$

$X > 0$

MCQ

Q1. If the values taken by a random variable are negative, the negative values will be:

(a) Negative probability

(b) **Positive probability**

(c) Insufficient data

(d) May have negative or positive probabilities

Q2. The variable that assigns a real number value to an event in a sample space is called:

(a) Static variable (b) Defined variable

(c) **Random variable** (d) Uncertain variable

Q3. In binomial probability distribution, the success and failure generated by the trial is respectively denoted by

(a) **p and q** (b) n and p

(c) p + q (d) p - q

Q4. A random variable that assumes a finite or a countable infinite number of value is called:

(a) Uncertain random variable (b) Discrete random variable

(c) **Continuous random variable** (d) Irregular random variable

Q5. If P(x) = 0.6 and x = 5, then E(x) is:

(a) 0.12 (b) 1

(c) 0.6 (d) **3**

Hint: Because E(x) = x.P(x)

Bold means correct answer

Q6. In a discrete probability distribution, the sum of all probabilities is always:

(a) 0 (b) **1**

(c) Infinite (d) None of these

Q7. The covariance of two independent random variables is:

(a) -1 (b) **0**

(c) 1 (d) None of these

Q8. In random experiment, observations of random variable are classified as:

(a) Trials (b) Functions

(c) **Events** (d) Composition

Q9. If X is a discrete random variable taking values x_1, x_2, x_3, where P(X=0) =0, P(X=1) =1/3, P(X=2) = 1/3. Find the value of P(X=3)

(a) 0 (b) 1/2

(c) **1/3** (d) 1

Hint: Since $\Sigma P(x_i) = 1$. That means P(X=0)+P(X=1)+P(X=2)+P(X=3) = 1.

Therefore P(X=3) = 1- {P(X=0)+P(X=1)+P(X=2)} = 1 – (0+1/3+1/3) = 1/3

Q10. Let X be the random variable, P(X=x) is the probability mass function is given by:

X	:	0	1	2	3
P(X=0)	:	0	2k	3k	4k

Determine the value of K?

(a) 0 (b) 1/2

(c) 1/3 (d) **1/9**

Hint: Since $\Sigma P(x_i) = 1$. That means P(X=0)+P(X=1)+P(X=2)+P(X=3) = 1.

Therefore 1 = 0 + 2k + 3k + 4k

1 = 9k K=1/9

Q11. Let X be the random variable, P(X=x) is the probability mass function is given by:

X	:	0	1	2	3
P(X=0)	:	1/8	1/2	1/4	1/8

Find the value of $P(X \geq 1)$.

(a) 3/8 (b) 5/8

(c) 3/4 (d) **7/8**

Hint: Since $P(X \geq 1) = P(X=1)+P(X=2)+P(X=3) = 1/2 + 1/4 + 1/8 = 7/8$

Q12. Let X be the random variable, P(X=x) is the probability mass function is given by:

X	:	0	1	2	3
P(X=0)	:	1/8	1/2	1/4	1/8

Find the value of f(1).

(a) 3/8 (b) **5/8**

(c) 3/4 (d) 7/8

Hint: Since $f(1) = P(X \leq 1) = P(X=0)+P(X=1) = 1/8 + 1/2 = 5/8$

Q13. What is the other name for Bernoulli trial?

(a) Two-way experiment (b) Three-way experiment

(c) **Dichotomous experiment** (d) Nucleo experiment

Q14. How many outcomes can a Bernoulli trial?

(a) **2** (b) 3

(c) 2^n (d) 2^{n+1}

Hint: In Bernoulli trial, there are two possible outcomes i.e. success and failure.

Q15. If the value of success in binomial probability distribution is 0.40 and failure is 0.60 and the number of values in distribution is 5 then the moment coefficient of skewness is:

(a) **0.167** (b) 0.267

(c) 0.367 (d) 0.467

Q16. Probability of failure in binomial distribution is denoted by:

(a) $p = q + 1$ (b) $p = q - 1$

(c) $q = 1 + p$ (d) $\mathbf{q = 1 - p}$

Q17. In binomial distribution, the formula of calculating mean is:

(a) $\mu = pq$ (b) $\mu = npq$

(c) $\mu = p + q$ (d) $\boldsymbol{\mu = np}$

Q18. If the value of p is smaller or lesser than 0.5 then the binomial distribution is classified as:

(a) **Skewed to right** (b) Skewed to left

(c) Skewed to infinity (d) Skewed to integers

Q19. The standard deviation of the binomial distribution is:

(a) $\sqrt{p}$ (b) $\sqrt{p}\,q$

(c) $\sqrt{np}$ (d) $\sqrt{\mathbf{npq}}$

Q20. If n is large, neither p nor q is small, the limiting form of the binomial distribution is:

(a) **Normal** (b) Poisson

(c) Binomial (d) None of these

Q21. In binomial distribution n=6 and p=0.9, then the value of P(X=7) is:

(a) Less than zero (b) **Zero**

(c) More than zero (d) One

Q22. In which distribution successive trials are without replacement:

(a) **Hypergeometric distribution** (b) Binomial distribution

(c) Sampling distribution (d) Geometric distribution

Q23. The successive trials are with replacement in:

(a) Hypergeometric distribution (b) **Binomial distribution**

(c) Sampling distribution (d) Geometric distribution

Q24. Binomial distribution is symmetrical when:

(a) $\mathbf{p = q}$ (b) $p < q$

(c) $p > q$ (d) $np > npq$

Q25. Hypergeometric distribution has parameters:

(a) 1 (b) 2

(c) **3** (d) 4

Q26. The variance of the binomial distribution is always:

(a) Equal to mean (b) **Less than mean**

(c) More than mean (d) None of these

Q27. Each trial in Binomial distribution has:

(a) One outcome (b) **Two outcome**

(c) Three outcome (d) Four outcome

Q28. A variable that can assume any value between two given points is called:

(a) **Continuous random variable** (b) Discrete random variable

(c) Irregular random variable (d) Uncertain random variable

Q29. A table with all possible value of a random variable and its corresponding probabilities is called:

(a) Probability mass function (b) Probability density Function

(c) Cumulative distribution function (d) **Probability distribution**

Q30. Which of the following mentioned standard Probability density functions is applicable to discrete Random Variables?

(a) Gaussian distribution (b) **Poisson distribution**

(c) Rayleigh distribution (d) Exponential distribution

Q31. If a variable can certain integer values between two given points is called:

(a) Continuous random variable (b) **Discrete random variable**

(c) Irregular random variable (d) Uncertain random variable

Q32. The expected value of a discrete random variable 'x' is given by:

(a) P(x) (b) $\sum P(x)$

(c) **$\sum x P(x)$** (d) 1

Q33. If 'X' is a continuous random variable, then the expected value is given by:

(a) P(X) (b) $\sum x P(x)$

(c) $\int X P(X)$ (d) None of these

Q34. Out of the following values, which one is not possible in probability?

(a) P(x) = 1 (b) $\sum x P(x) = 3$

(c) P(x) = 0.5 **(d) P(x) = – 0.5**

Q35. If the values taken by a random variable are negative, the negative values will have:

(a) **Positive probability**

(b) Negative Probability

(c) May have negative or positive probabilities

(d) Insufficient data

Q36. If f(x) is a probability density function of a continuous random variable, then =?

(a) 0
(b) 1
(c) Undefined
(d) None of these

Q37. The variable that assigns a real number value to an event in a sample space is called:

(a) **Random variable**
(b) Defined variable
(c) Uncertain variable
(d) Static variable

Q38. A random variable that assumes a finite or a countable infinite number of values is called:

(a) Continuous random variable
(b) **Discrete random variable**
(c) Irregular random variable
(d) Uncertain random variable

Q39. A random variable that assumes an infinite or a uncountable infinite number of values is called:

(a) **Continuous random variable**
(b) Discrete random variable
(c) Irregular random variable
(d) Uncertain random variable

Q40. If $\Sigma P(x) = k^2 - 3$ then, the value of k is?

(a) 0
(b) 1
(c) 2
(d) 3

Q41. If P(x) = 0.5 and x = 4, then E(x) = ?

(a) 0.5
(b) 1
(c) **2**
(d) 4

Q42. In a discrete probability distribution, the sum of all probabilities is always?

(a) 0
(b) **1**
(c) -1
(d) Infinite

Q43. The weight of persons in a state is:

(a) Discrete random variable

(b) **Continuous random variable**

(c) Irregular random variable

(d) Not a random variable

Q44. The mean of the binomial distribution is:

(a) n (b) **np**

(c) npq (d) $\sqrt{npq}$

Q45. Successive trials in binomial distribution are:

(a) Mutually exclusive (b) Equally likely

(c) Dependent (d) **Independent**

Q46. The probability of success changes from trial to trial in:

(a) **Hypergeometric distribution** (b) Binomial distribution

(c) Sampling distribution (d) Geometric distribution

Q47. Binomial distribution is negatively skewed when

(a) $p < 1/2$ (b) $\mathbf{p > 1/2}$

(c) $p = 0$ (d) $p = 1$

Q48. Which of the following is not the property of binomial distribution?

(a) Trials are independent

(b) **Probability of success varies from trial to trial**

(c) Trials are independent

(d) n is fixed

Q49. The mean, median and mode for binomial distribution will be equal when:

(a) $\mathbf{p = 0.5}$ (b) $p < 0.5$

(c) $p > 0.5$ (d) $p = 1$

Q50. The mean of the hypergeometric distribution is:

(a) nN/k (b) **nk/N**

(c) (N-k)/n (d) (N+k)/n

Q51. If in a binomial distribution n=1 then E(X) is:

(a) 0 (b) 1

(c) **p** (d) q

Q52. Binomial distribution has parameters:

(a) 1 (b) **2**

(c) 3 (d) 4

Q53. A random variable X has binomial distribution with n=10 and p=0.3 then variance of X is:

(a) **2.1** (b) 21

(c) 12 (d) 10

Q54. In a binomial probability distribution it is impossible to find:

(a) $P(X = 0)$ (b) $\mathbf{P(X < 0)}$

(c) $P(X > 0)$ (d) $P(0 < X < n)$

Q55. Binomial distribution is symmetrical when

(a) **1/2** (b) 1/5

(c) 3/5 (d) 7/16

Q56. If n is large, p is small and np = m (a finite positive constant), the limiting form of the binomial distribution is:

(a) Poisson (b) **Normal**

(c) Binomial (d) None of these

Q57. If X is a normal variable with mean $\mu = 50$ and variance $\sigma^2 = 10$, then $P(X > 50)$ is:

(a) 0 (b) 0.2

(c) **0.5** (d) 1

Q58. If X is a normal variable with mean $\mu = 100$ and variance $\sigma^2 = 25$, then $P(90 < X < 120)$ is the same as:

(a) $\mathbf{P(-2 < Z < 4)}$ (b) $P(-1 < Z < 1)$

(c) $P(2 < Z < 4)$ (d) $P(1 < Z < 2)$

Q59. If the mean of binomial probability distribution is 25 then the mean of Poisson probability distribution is:

(a) 70 (b) 50

(c) **25** (d) 50

Q60. If the value of failure in binomial probability distribution is 0.70 and success is 0.30 and the number of values in distribution is 7 then the moment coefficient of kurtosis is:

(a) **0.18** (b) 0.28

(c) 0.48 (d) 0.58

Q61. Variance of binomial probability distribution is larger in value if:

(a) **p and q are equal** (b) p and q are greater than 0.5

(c) p is greater than 0.5 (d) q is greater than 0.5

Q62. In an experiment, positive and negative values are equally likely to occur. The probability for obtaining at most one negative value in five trials is:

(a) 1/16 (b) **3/16**

(c) 1/32 (d) 3/32

Hint: In binomial distribution, n = 5; p = 1/2; Let X represents number of negative values in 5 trials.

Therefore, P (At most one negative value) = P(X < 1) = P(X = 0) + P(X = 1)

$$P(X = 0) = {}^5C_0 \text{ x } \left(\frac{1}{2}\right)^0 x\left(\frac{1}{2}\right)^5 = \frac{1}{32};\ P(X = 1) = {}^5C_1 \text{ x } \left(\frac{1}{2}\right)^1 x\left(\frac{1}{2}\right)^{5-1} = \frac{5}{32}$$

P(X<1) = 1/32 + 5/32 = 6/32 = 3/16

Q63. Let X~ N(μ, σ^2), If $\mu^2 = \sigma^2$, Here μ > 0, then the value of P(X < - μ | X < μ) in terms of cumulative function N(0,1) is:

(a) [1 – P(Z < 1)] (b) 2[1 – P(Z < 1)]

(c) [1 – P(Z < 2)] (d) **2[1 – P(Z < 2)]**

Q64. Let X be a random variable that follows binomial distribution with expectation E(X) =7 and Var(X) = 6. Then the probability of success p is:

(a) 1/49 (b) 36/49

(c) **1/7** (d) 6/7

Hint: Given E(X) = 7 = npq; Var(X) = 6 = np. Therefore q = 6/7

So, p = 1-6/7 = 1/7

Q65. Consider a binomial random variable X. If x_1, x_2, ……,x_n are independent and identically distributed samples from the distribution of X with sum Y = $\sum^n_{i=1} X_i$ then the distribution of Y as n→∞ can be approximated as:

(a) Binomial (b) Bernoulli

(c) **Normal** (d) Poisson

Q66. Consider an unbiased cubic dice with opposite faces coloured identically and each face coloured red, blue or green such that each colour appears only two times on the dice. If the dice is thrown thrice, the probability of obtaining red colour on top face of the dice at least twice is:

(a) 1/27 (b) **7/27**

(c) 8/27 (d) 13/27

Hint: P(each colour) = 2/6 = 1/3; q = 1-p = 2/3; n = 3; Using Binomial distribution,

Probability of getting red on the top face at least twice is-

$P(X > 2) = {}^nC_2\, p^2\, q^{n-2} + {}^nC_3\, p^3\, q^{n-3} = {}^3C_2\, (1/3)^2\, (2/3)^{3-2} + {}^3C_3\, (1/3)^3\, (2/3)^{3-3} = 6/27 + 1/27 = 7/27$

Q67. The first moment about origin of binomial distribution is:

(a) **np** (b) npq

(c) np(q-1) (d) nq(p-1)

Q68. A lot has 10% defective items. Five items are chosen randomly from this lot. The probability that exactly 2 of the chosen items are defective is:

(a) 0.0279 (b) **0.0729**

(c) 0.0927 (d) 0.0972

Hint: n = 5; p = 1/10; q = 9/10; Using Binomial distribution,

P(Exact 2 of the chosen items are defective) = ${}^5C_2\, (1/10)^2\, (9/10)^{5-2} = 0.0729$

Q69. In a binomial distribution, the mean is 40 and variance is 20, then parameter n is:

(a) 4 (b) 8

(c) 40 (d) **80**

Q70. In a binomial distribution, the mean is 4 and variance is 3, then its mode is:

(a) 3 (b) **4**

(c) 5 (d) 6

Q71. Let X denotes the number of times heads occur in n tosses of a fair coin, If P(X = 4), P(X=5) and P(X =6) are in AP, then the value of n is:

(a) 5 (b) **7**

(c) 9 (d) 12

Q72. The mean of binomial distribution 552.5 and the probability is 65% then the number of values of binomial distribution:

(a) 1258 (b) 1050

(c) **850** (d) 700

Q73. If the value of p is smaller than 0.5 then the binomial distribution is classified as:

(a) Skewed to left (b) **Skewed to right**

(c) Skewed to integers (d) Skewed to infinity

Q74. If P(1) = P(4) in Poisson distribution, find the value of mean:

(a) 0.88 (b) 1.88

(c) **2.88** (d) 3.88

Hint: For Poisson distribution, P(1) =P(4); That means

$$\frac{e^{-m}m^1}{1!} = \frac{e^{-m}m^4}{4!} \rightarrow m^3 = 24 \rightarrow m = 2.88$$

Since, For Poisson distribution, Mean = Variance = m. Hence, Mean = 2.88

Q75. If X is a Poisson random variable with mean 3, Then P(|X – 3| <1) will be:

(a) $e^{-3}/2$ (b) $3e^{-3}$

(c) **$9\ e^{-3}/2$** (d) $99\ e^{-3}/8$

Hint: For Poisson distribution, m = 3,

For only X = 3 value of |X – 3| <1; Therefore, P(|X – 3| <1) = P(X = 3)

$$\text{Using, } P(X) = \frac{e^{-m}m^x}{x!} \rightarrow P(X=3) = \frac{e^{-3}3^3}{3!} \rightarrow P(X=3) = \rightarrow P(X=3) = \frac{9.e^{-3}}{2}$$

Q76. If a random variable X is satisfies the Poisson distribution with a mean value of 3, then the probability that the X > 2 is:

(a) $4e^{-3}$ (b) **$1 - 4e^{-3}$**

(c) $1 + 4e^{-3}$ (d) $1 - 2e^{-3}$

Hint: For Poisson distribution, m =3

$$P(X > 2) = 1 - P(X = 0) - P(X = 1) = 1 - \frac{e^{-3}3^0}{0!} - \frac{e^{-3}3^1}{1!} = 1 - e^{-3}(1 + 3) =$$
$1 - 4\ e^{-3}$

Q77. If X is a Poisson random variate with mean 3, then P(|X – 3| < 1) will be:

(a) $(1/3)e^{-3}$ (b) $(2/3)e^{-3}$

(c) **$(9/2)e^{-3}$** (d) $(7/2)e^{-3}$

Hint: Given, Mean (m) =3, X = 3,

$$P(X=x)=\frac{e^{-m}m^{x}}{x!}\rightarrow P(X=3)=\frac{e^{-3}3^{3}}{3!}=\frac{27e^{-3}}{6}=\frac{9e^{-3}}{2}$$

Q78. If calls arrive at a telephone exchange such that the time of arrival of any call is independent of the time of arrival of earlier or future calls, the probability distribution function of the total number of calls in a fixed time interval will be:

(a) Binomial (b) Bernaulli

(c) **Poisson** (d) Exponential

Q79. In a Poisson distribution, if n is the number of trials and p is the probability of success, then the mean value is given by:

(a) **m = np** (b) m = npq

(c) $m = (npq)^{1/2}$ (d) m = p

Q80. If 'm' is the mean of a Poisson distribution, then variance is given by:

(a) $m^{1/2}$ (b) m^2

(c) m^3 (d) **m**

Q81. Poisson distribution is applied for:

(a) **Discrete random variable** (b) Continuous random variable

(c) Uncertain random variable (d) Irregular random variable

Q82. If 'm' is the mean of a Poisson distribution, the P(0) is given by:

(a) m (b) m^2

(c) e^m (d) **e^{-m}**

Q83. For a Poisson distribution, If mean =1, then P(1) is:

(a) e (b) e/2

(c) e^2 (d) **1/e**

Q84. Normal distribution is symmetric about:

(a) Variance (b) Covariance

(c) Standard deviation (d) **Mean**

Q85. Standard normal distribution has which of the following properties:

(a) Mean = Standard deviation
(b) Mean = Variance
(c) **Mean = 0, Variance = 1**
(d) Mean = 1, Variance = 0

Q86. Let X be a normal random variable with mean zero and variance 9. If a = P(X > 3), the P(|X| < 3) equals:

(a) 1 – a
(b) **1 – 2a**
(c) a
(d) 2a

Hint: Given, Mean (μ) = 0; Variance (σ^2) =9; Therefore, S.D.(σ) = 3; a = P(X > 3)

$Z = \frac{X-\mu}{\sigma} = \frac{3-0}{3} = 1$; Hence, for normal distribution with Z =1

P(X > 3) = P(X < 3) = a; Maximum probability of any event = 1; That means, P(-∞ < X < ∞) =1

P(-∞ < X < ∞) = P(X < 3) + P(-3 < X < 3) + P(X > 3) → 1 = P(X < 3) + P(|X| < 3) + P(X > 3)

1 = a + P(|X| < 3) + a → P(|X| < 3) = 1 – 2a

Q87. For a random variable X, let $\bar{X}$ be the sample mean. The sample size is n. The mean and the standard deviation of X are μ and σ, respectively. The standard deviation of $\bar{X}$ is:

(a) σ/n
(b) σ/n^2
(c) **$\sigma/n^{1/2}$**
(d) σ

Q88. Approximate what area is covered under the normal distribution curve between + 3 standard deviation?

(a) 68.27%
(b) 92.54%
(c) 95.45%
(d) **99.73%**

Q89. Approximate what area is covered under the normal distribution curve between + 2 standard deviation?

(a) 68.27%
(b) 92.54%
(c) **95.45%**
(d) 99.73%

Q90. Approximate what area is covered under the normal distribution curve between + 1 standard deviation?

(a) **68.27%**
(b) 92.54%
(c) 95.45%
(d) 99.73%

Q91. The annual rainfall of a city is normally distributed with mean and standard deviation is 1100 mm and 150mm, respectively. The probability that the annual rainfall will be more than 1250mm is:

(a) 100% (b) 75%

(c) 60% (d) **< 50%**

Hint: Given, σ = 220 mm; Mean (μ) = 1100 mm; Now, for X = 1250

$Z = \frac{X-\mu}{\sigma} = \frac{1250-1100}{150} = 1$; P(X > 1250) = P(Z > 1); Here, Z is a standard normal variable

We know that, P(-1 < Z < 1) = 0.68; Therefore, P(0 < Z < 1) = 0.68/2 = 0.34;

So, P(Z > 1) = 0.50 – 0.34 = 0.16; That means, < 50%

Q92. The area under a standard normal curve is:

(a) 0 (b) -1

(c) **1** (d) ∞

Q93. The standard normal curve is symmetric about the value:

(a) **0** (b) -1

(c) 1 (d) ∞

Q94. Foe a standard normal variate, the value of standard deviation is:

(a) 0 (b) -1

(c) **1** (d) ∞

Q95. In Normal distribution, the highest value of ordinate occurs at:

(a) Standard deviation (b) **Mean**

(c) Variance (d) Extremes

Q96. In standard normal distribution, the value of median is:

(a) **0** (b) -1

(c) 1 (d) ∞

Q97. The mode of a normal distribution is 50 with S.D. 10. Then its median will be:

(a) 10 (b) 40

(c) **50** (d) 60

Q98. 5th docile of a normal distribution is 80. Then its median will be:

(a) 40 (b) **80**

(c) 120 (d) 160

Q99. 2ed quartile of a normal distribution is 90. Then its mode will be:

(a) 30 (b) **90**

(c) 120 (d) 150

Q100.The probability density function of a random variable X is given by:

$$f_x(x) = \begin{cases} k & a \leq x \leq b \\ 0 & Otherwise \end{cases}$$

Where k is a constant. If a = -1 and b = 2, then P(|X| < c) for c = 1/2 is:

(a) 1/2 (b) **1/3**

(c) 2/3 (d) 3/4

Hint: $f_x(x) = \begin{cases} k & a \leq x \leq b \\ 0 & Otherwise \end{cases}$ with a = -1 and b = 2, we have

P(|X| < 1/2) = P(- ½ < X < 1/2) = $\int_{-\frac{1}{2}}^{\frac{1}{2}} f_x(x)\,dx = \int_{-\frac{1}{2}}^{\frac{1}{2}} \frac{1}{3}\,dx = \frac{1}{3}$

Q101.In an experiment, positive and negative values are equally likely to occur. The probability of obtaining at most one negative value in five trials is:

(a) **6/32** (b) 3/32

(c) 2/32 (d) 1/32

Hint: Given, n = 5, p = ½; Let X represents number of negative values in 5 trials, at most one negative value means, k can be 0 and 1.

P(At most one negative value) = P(X < 1) = P(X =0) + P(X = 1)

= $^5C_0\,(1/2)^0\,(1/2)^{5-0} + {}^5C_1\,(1/2)^1\,(1/2)^{5-1}$ = 1/32 + 5/32 = 6/32

Q102.The mean and variance for Gamma distribution are:

(a) E(X) = α/λ, Var(X) = $1/\lambda^2$ (b) **E(X) = α/λ, Var(X) = α/λ^2**

(c) E(X) = 1/λ, Var(X) = $1/\lambda^2$ (d) E(X) = 1/λ, Var(X) = α/λ^2

Q103.Putting α = 1 in Gamma distribution results in:

(a) Binomial distribution (b) Poisson distribution

(c) Normal distribution (d) **Exponential distribution**

Q104.If X is a uniformly distributed random variable that takes value between 0 and 1. The value of $E(X^3)$ will be:

(a) 0 (b) 1/2

(c) **1/4** (d) 1/8

Hint: Given, a = 0, b = 1; f(x) = 1/(b – a) = 1/(1-0) = 1;

$$E\left(X^3\right)=\int_0^1 X^3 f(x)\,dx=\left[\frac{x^4}{4}\right]_0^1=\frac{1}{4}$$

Q105.Consider a binomial random variable X. If $x_1, x_2, x_3, \ldots\ldots, x_n$ are independent and identically distributed samples from the distribution of X with sum $Y=\sum_{i=1}^{n} X_i$ then the distribution of Y as $n\to\infty$ can be approximated as:

(a) Binomial (b) Bernoulli

(c) **Normal** (d) Exponential

Q106.Let x be a random variable that follows Binomial distribution with expectation E(X) = 7 and Var(X)=6. Then the probability of success p is:

(a) 1/49 (b) 36/49

(c) **1/7** (d) 6/7

Hint: Given, E(X) = 7 = np; Var(X) = 6 = npq

Therefore, q = 6/7.

So, p = 1/7

Q107. The mean and variance of geometric distribution are:

(a) p/q and p/q (b) p/q and p^2/q

(c) q/p and q/p (d) **q/p and q/p^2**

Q108.Find the mean of x, if x is a Poisson variate satisfying the condition P(3) = P(4) ?

(a) 2 (b) 3

(c) **4** (d) 8

Hint: Here, x = 3 and 4 → $(e^{-m} \times m^3)/3! = (e^{-m} \times m^4)/4! \to m = 4!/3! \to m = 4$

Q109.Let X and Y are two independent normal variables with mean 50 and 80, respectively and standard deviation as 3 and 4, respectively. What is the distribution of X + Y?

(a) N(130, 3) (b) N(130, 4)

(c) **N(130, 5)** (d) N(130, 6)

Hint: Since X and Y are independent then

$Z = X + Y \sim N [(X + Y), (\sigma_x^2 + \sigma_y^2)] \rightarrow P(Z) = P(X+Y) \sim N [130, (4^2 + 3^2)] \rightarrow N(130, 25)$

Therefore, $N [\mu, \sqrt{(\sigma_x^2 + \sigma_y^2)}]$ is N(130, 5)

Q110.The hyper geometric distribution has:

(a) Two parameters (b) **Three parameters**

(c) Four parameters (d) Five parameters

Q111.Nature of hyper geometric random variable is:

(a) **Discrete** (b) Continuous

(c) Qualitative (d) Quantitative

Q112.In hyper geometric distribution, the successive trials are:

(a) Very large (b) Very small

(c) **Dependent** (d) Independent

Q113.If in a hyper geometric distribution, N = 10, k =5 and n = 4, then the probability of failure is:

(a) 0.10 (b) 0.25

(c) **0.50** (d) 1.00

Q114.The range of hyper geometric distribution is:

(a) 0 to n

(b) 0 to N

(c) 0 to k

(d) **0 to k or n (whichever is less)**

Q115.The number of trials in hyper geometric distribution is:

(a) Small (b) Large

(c) **Fixed** (d) Not fixed

Q116.The probability of success changes from trial to trial in:

(a) Binomial distribution (b) Poisson distribution

(c) **hyper geometric distribution** (d) Normal distribution

Q117.In hyper geometric distribution, the relation between mean and variance is:

(a) Mean = Variance
(b) Mean < Variance
(c) **Mean > Variance**
(d) Mean = 2 Variance

Q118.Hyper geometric distribution reduces to Binomial distribution when:

(a) $n \to \infty$
(b) $\mathbf{N \to \infty}$
(c) N = n
(d) N < n

Q119.In a hyper geometric distribution N = 6, k = 3 and n = 4, then the mean is equal to:

(a) 1
(b) **2**
(c) 4
(d) 8

Q120.The mean of hyper geometric distribution is:

(a) **n x k/N**
(b) n x k/N – 1
(c) n-1 x k/N
(d) n x k -1/N

Q121.A ball is thrown at a circular bin such that it will land randomly over the area of the bin. Find the probability that it lands closer to the center than to the edge?

(a) 51%
(b) 34%
(c) **25%**
(d) 12%

Hint: The set of outcomes are all of the points on the bin, which make up an area of where is the radius of the circle. The points which are closer to the center than to the edge are those that lie within the circle of radius around the center. Hence, the area of the success outcomes is $\pi(r/2)^2 = \pi r^2/4$. Thus,

P(Closer to center than edge) = (Area of the desired outcome)/(Area of the total outcome) = $(\pi r^2/4)/\ \pi r^2 = 1/4 = 25\%$

Q122.What is the variance of a geometric distribution having parameter p = 0.72?

(a) 13%
(b) 26%
(c) **54%**
(d) 68%

Q123.Consider a Poisson distribution for the tossing of an unbiased coin. The mean for the distribution is μ. The standard deviation for this distribution given by:

(a) μ
(b) μ^2
(c) $\boldsymbol{\mu^{1/2}}$
(d) $1/\mu$

Q124. Sum of n independent exponential random variable results in:

(a) Binomial distribution (b) **Gamma distribution**

(c) Poisson distribution (d) Normal distribution

Q125. Find the value of $\Gamma(5/2)$.

(a) $1/4 . \pi^{1/2}$ (b) **$3/4 . \pi^{1/2}$**

(c) $5/4 . \pi^{1/2}$ (d) $7/4 . \pi^{1/2}$

Q126. A table with all possible value of a random variable and its corresponding probabilities is called:

(a) Probability density function

(b) Probability mass function

(c) **Probability distribution**

(d) Cumulative distribution function

Q127. A variable that can assume any value between two given points is called:

(a) Discrete random variable

(b) **Continuous random variable**

(c) Uncertain random variable

(d) Irregular random variable

Q128. If a variable can certain integer values between two given points is called:

(a) **Discrete random variable** (b) Continuous random variable

(c) Uncertain random variable (d) Irregular random variable

Q129. The expected value of a discrete random variable X is given by:

(a) $\sum P(x)$ (b) **$\sum X.P(X)$**

(c) $P(x)$ (d) 1

Q130. If X is a continuous random variable, then the expected value is given by:

(a) $\sum P(X)$ (b) $\sum X.P(X)$

(c) **$\int X.P(X)$** (d) $P(X)$

Q131. Out of the following values, which one is not possible in probability?

(a) **$P(X) = -0.5$** (b) $P(X) = 0.5$

(c) $P(X) = 1$ (d) $P(X) = 2$

Q132.A variable X is exponentially distributed for $X > 0$ with a mean of one. The probability that the value of X will lie between 1 and $1 < X < 2$ is:

(a) $e^{1} - e^{-2}$ (b) $\mathbf{e^{-1} - e^{-2}}$

(c) $e^{-1} - e^{2}$ (d) $e^{1} - e^{2}$

Q133.A random variable P follows exponential distribution with mean value 0.5. The expectation of P^2 will be:

(a) 0.10 (b) 0.30

(c) **0.50** (d) 0.60

Hint: For an exponential distribution, $f(p) = \theta\, e^{-\theta p}$

Where, Mean = $1/\theta$ and Standard deviation = $1/\theta$

Given that, Mean = 0.5, than variance = 0.25

Mean = E(P), Variance = $E(P^2) - \{E(P)\}^2 = 0.25 \rightarrow E(P^2) = 0.25 + \{E(P)\}^2$ $= 0.25 + 0.25 = 0.50$

Q134.Probability distribution of X bar is called its:

(a) Standard deviation (b) Sampling distribution

(c) Standard error (d) **Expected value**

Q135.The probability distribution of a statistic is called:

(a) Standard error (b) **Sampling error**

(c) Sampling distribution (d) Parameter

Q136.If the mean of the sampling distribution is equal to the parameters then estimators will be:

(a) Biased (b) **Unbiased**

(c) Consistent (d) Sufficient

Q137.If X is a Poisson random variable and $E(X^2) = 6$. Then V(X) is:

(a) -3 (b) 1

(c) **2** (d) 3

Hint: $V(X) = (X^2) - [E(X)]^2 = E(X) = \lambda$

$\lambda = 6 - \lambda^2 \rightarrow \lambda = -3$ or 2

Therefore, V(X) = 2, because V(X) can't be negative.

Q138.In case of symmetrical distribution:

(a) $\mu_1 = \mu_2$
(b) $\beta_1 = \beta_2$
(c) $\mu_1 = \mu_4$
(d) **Mean = Median = Mode**

Q139.The normal curve is asymptotic to:

(a) **X – axis**
(b) Y – axis
(c) Along Y = X
(d) None of these

Q140.If Y = 5X + 10 and X is N(10, 25), then mean of Y is:

(a) 50
(b) **60**
(c) 70
(d) 135

Q141.Area under the normal curve on either side of mean is:

(a) -1
(b) 0
(c) **0.5**
(d) 1

Q142.Which of the following parameter control the relative flatness of normal distribution?

(a) Mean
(b) Median
(c) Mode
(d) **Standard deviation**

Q143.The shape of the normal curve dep[ands upon:

(a) Mean
(b) Mode
(c) Range
(d) **Standard deviation**

Q144.In normal distribution $\beta_1 = 0$ and β_2 is equal to:

(a) 0
(b) 0.5
(c) 1
(d) **3**

Q145.Area to the right of (Z =0) is the area of....................values of Z?

(a) Negative
(b) Zero
(c) **Positive**
(d) None of these

Q146.The standardized normal variate Z means ....away from the mean?

(a) **S.D.**
(b) Variance
(c) Mean
(d) Mode

Q147.In a normal distribution, what is the Z score associated with the mode?

(a) -1 (b) **0**

(c) 0.5 (d) 1

Q148.If X~ N(μ, σ^2) then Z follows:

(a) N(0, σ^2) (b) N(μ, 0)

(c) **N(0, 1)** (d) N(1, 0)

Q149.Normal distribution is the distribution of:

(a) Qualitative variable (b) Quantitative variable

(c) **Continuous random variable** (d) Discrete random variable

Q150.Standard normal variable Z is:

(a) X - μ (b) **(X - μ)/ σ**

(c) (μ - X)/ σ (d) (X - σ)/ μ

Q151.In a normal distribution E(X - μ)^2 is:

(a) Mean deviation (b) Standard deviation

(c) **Variance** (d) Range

Q152.We use normal distribution when n is:

(a) Fixed (b) **Large**

(c) Small (d) None of these

Q153.If X~ N(55, 16) then σ is:

(a) 55 (b) 16

(c) **4** (d) 0

10

Time Series Analysis

Time Series

A series of observations recorded over time is known as a **time series**. In other words, a **time series** is a set of observations taken at specified times, usually at equal intervals. For example: Annual retail sales, rainfall measurements, heartbeats per minute, etc. Such studies are to be based on the analysis of time series data collected over time. Thus, the analysis of time series plays an important role in an investigation of economic, social, commercial and even biological phenomenon.

Components of time series

In a long time series, generally, there are following four components:

- Secular trend or long term movements
- Seasonal variations
- Cyclic variations
- Random or irregular movements
- **Secular trend or long term movements:** It is that characteristic of a time series which extends consistently throughout the entire period of time under consideration. It shows a long term tendency of an activity to grow or to decline. The term **long period of time** in a relative phenomenon and cannot be defined exactly. For some cases a period as small as a weak may be fairly long while in other cases, a period as long as two years may not be assumed long.
- **Seasonal variations:** The component responsible for the regular rise and fall in the magnitude of the time series is called **seasonal variation**. In other words, it refers to identical, or almost identical, patterns which a time series appears to follow during corresponding months of successive years. Such variations are due to recurring events which take place annually, quarterly, monthly, weakly or even daily, depending on the type of data available. But in no case this period is to exceed one year.

- **Cyclic variations:** It refers to the long term oscillations or swings about a trend line. These cycles may or may not be periodic, i.e., they may or may not follow exactly similar patterns after equal interval of time. Such variations are of longer duration that a year and they do not show the type of regularity as observed in the case of seasonal variations. An important example of cyclical variations is the so-called business cycles representing intervals of prosperity, recession, depression and recovery. Each phase changes gradually into the phase which follows it in the given order. In a business activity, these phases follows each other with steady regularity and the period from the peak of one boom to the peak of the next boom is called a **complete cycle**. The usual periods of a business cycle may be ranging between 5 – 11 years.
- **Random or irregular movements:** It refers to such variations in a time series do not repeat in a definite pattern. Irregular movements in a time series may be of two types:
    - Random or chance variations
    - Episodic variations
- **Random or chance variations:** It is a real phenomenon is inevitable by nature. It does affect a series in a random way, and as such, the effect of change or random variations on a series is small.
- **Episodic variations:** It is a time series arise due to specific events or episodes like epidemic, fire, strike or natural calamities like flood, earthquake or late monsoon etc. In some cases, irregular variations may not have a significant importance while in others these may be so intense as to result in new cyclical variations.

Analysis of Time Series

The main objective of time series analysis is to give a mathematical description of component variations present. In order to meet this objective, it is necessary to break down the series into its main components namely, the secular trend, the seasonal periodic variations, cyclical variations and the irregular variations. Thus, the main problems in the analysis of time series are to identify the components at work and to isolate study and measure them independently. The following are the two mathematical models commonly used for the decomposition of a time series into its components.

- Additive Model
- Multiplicative Model

- **Additive Model:** The analysis of time series greatly depends on the hypothesis as to how the components interact. In the simplest case, we can assume that the effects of various components in a time series are independent and additive. Thus a time series can be expressed as:

  $Y = T + S + C + I$

  Where, Y = the value of the variable at time t; T = Trend value; S = Seasonal variation:

  C = Cyclical variation; and I = Irregular variation.

  In the above additive model, it should be noted that

  - The component S will be positive or negative according to the season of the year.
  - The component C too will be positive or negative according to above normal or below normal phase of the cycle.
  - The component I will also have positive or negative values, however, for a long series, the sum of irregular variations, is assumed to be zero.
  - The term S will not appear in the model when the series consists of annual data.

- **Multiplicative Model:** As an alternative to the **additive model,** we formulate a more useful model called **multiplicative model**. The **multiplicative model** of the time series may be formulated as:

  $\mathbf{Y = T \times S_1 \times C_1 \times I_1}$

  Where, $S_1 = \dfrac{T+S}{T}; C_1 = \dfrac{T+S+C}{T+S}; I_1 = \dfrac{T+S+C+I}{T+S+C}$

In the multiplicative model, only the trend component is viewed as an absolute value whereas, the remaining three components (S_1, C_1 and I_1) are expressed as rates or percentages. Therefore, instead of having positive or negative values, S_1, C_1 and I_1 will have indexing values greater than or less than unity. Further, it may be pointed out that the multiplicative decomposition of a time series is the same as additive decomposition if we consider logarithms, i.e.

$$\log Y = \log T + \log S_1 + \log C_1 + \log I_1$$

which is again an additive model. This shows that there is no difference, in principle, between the two models.

Remark

- The presence of all the four components is not necessary in the time series. For example; in case of annual data, as the seasonal component in annual data is not there, the multiplicative decomposition of the time series will be:

$$Y = T \times C \times I$$

Similarly, cyclical component may be absent in relatively short period data and thus, the model takes the form

$$Y = T \times S \times I$$

Measurement of Trend

The main objectives behind the study of the trend of a time series are:

- To describe the long term growing or declining trend in a phenomenon under study.
- To eliminate the trend component in order to bring into focus the remaining components in the time series.

In order to meet these objectives, some statistical methods of estimation or determination of trend are as follows:

- Free hand, graphic or inspection method
- Semi-average method
- Moving average method
- Method of least squares

- **Free hand, graphic or inspection method:** This is the simplest method of trend determination. According to this method, we graph the series and then draw a free hand curve through the points on the graph. The method does not involve complex mathematical calculations and can be used to describe all types of trend, linear or non-linear.
- **Semi-average method:** The method consists of dividing the data into two parts and averaging the data in each part. In this way, we obtain two points on the graph of the time series. The line obtained by joining these two points is the required trend line and may be extended in both the directions for estimating the trend values.
- **Moving average method:** The method of moving averages attempts to smooth out the irregularities in a series by a process of averaging. By using averages of appropriate orders, cyclical, seasonal and irregular variations may be eliminated, thus leaving only the trend component. Moving averages of extent m (or period) is series of successive averages of m terms at a time, starting from 1st, 2nd, 3rd term and so on until we exhaust the whole time series. If m is odd, say equal to (2k+1), then the moving average is put

against the mid-value of the period it covers, i.e., against t = k+1. On the other hand, if m is even, say equal to 2k, it is placed between two middle values of the period it covers. Thus, when an even number of years is taken in moving average, the average does not coincide with an original time period. For overcoming this situation, moving average of extent two of these moving averages are taken and the first of such values is put against t = k + 1. This procedure of centring puts the moving averages against the time points of the series rather than between these points.

- **Method of least squares:** This method is very useful for fitting mathematical functions to a given set of data. The method is objective, and therefore, gives correct and accurate estimation of trend, once the form of equation representing trend is determined. An examination of graphical plot of the time series often provides an adequate basis for deciding the functional form of the trend. Some of the common used for representing trend are:

(i) $X = a + bX$ (Linear or straight line trend)

(ii) $Y = a + bX + cX^2$ (Parabolic or Quadratic trend)

(iii) $Y = ab^X$ (Exponential trend)

MCQ

Q1. A time series consists of:

(a) Long-term variations (b) Short-term variations

(c) Irregular variations (d) **All of them**

Q2. Secular trend can be measured by:

(a) Five methods (b) **Four methods**

(c) Three methods (d) Two methods

Q3. A long run increase or decrease in data is known to be:

(a) Cyclical variation (b) Seasonal variation

(c) **Trend variation** (d) Irregular variation

Q4. Seasonal variations repeat with in:

(a) **A year** (b) Two years

(c) Five years (d) Ten years

Q5. In an additive model, the component measurements are:

(a) Negative (b) Positive

(c) **Absolute** (d) None of these

Bold means correct answer

Q6. An overall tendency of rise or fall in a time series is called:

(a) **Trend** (b) Seasonal

(c) Irregular (d) Cyclical

Q7. The component useful for long-term forecasting is:

(a) **Trend** (b) Seasonal

(c) Irregular (d) Cyclical

Q8. The component useful for short term forecasting is:

(a) Trend (b) **Seasonal**

(c) Irregular (d) Cyclical

Q9. Increase in the number of patients in the hospital due to heat stroke is:

(a) Secular trend (b) **Seasonal variation**

(c) Irregular variation (d) Cyclical variation

Q10. Mustard crop badly damaged on account of rains is:

(a) Secular trend (b) Seasonal movement

(c) **Random movement** (d) Cyclical movement

Q11. The following are the movement(s) in the secular trend:

(a) Steady (b) Regular

(c) Smooth (d) **All of them**

Q12. A complete cycle consists of a period of:

(a) Prosperity and recovery (b) Recession and recovery

(c) Prosperity and depression (d) **Prosperity and recession**

Q13. A complete cycle passes through:

(a) Five stages (b) **Four stages**

(c) Three stages (d) Two stages

Q14. Most frequency used mathematical model of a time series is:

(a) Mixed model (b) Additive model

(c) **Multiplicative model** (d) Regression model

Q15. In semi-average method, we decide the data into:

(a) **Two equal parts** (b) Three equal parts
(c) Four equal parts (d) Five equal parts

Q16. Moving average method is used for measurement of trend when:

(a) Trend is non-linear (b) **Trend is linear**
(c) Trend is curvilinear (d) None of these

Q17. The long term trend of a time series graph appears to be:

(a) Upward (b) Downward
(c) Straight line (d) **Parabolic curve**

Q18. Indicate which of the following an example of seasonal variations is:

(a) Recovery in business
(b) Sudden causes by wars
(c) Death rate increases due to COVID
(d) **Increases the sale of woollens during winter**

Q19. The most commonly used mathematical method for measuring the trend is:

(a) Semi-average method (b) Moving average method
(c) **Method of least square** (d) Graphic method

Q20. A trend is the better fitted trend for which the sum of squares of residuals is:

(a) Positive (b) Negative
(c) Maximum (d) **Minimum**

Q21. Decomposition of time series is called:

(a) Detrending (b) Histogram
(c) Historigram (d) **Analysis of time series**

Q22. The fire in a showroom of readymade garment is an example of:

(a) Secular trend (b) Seasonal variation
(c) **Irregular variation** (d) Cyclical variation

Q23. Increased demand of admission in the subject of computer in Uttar Predesh is:

(a) **Secular trend** (b) Seasonal trend
(c) Irregular trend (d) Cyclical trend

Q24. Damages due to floods, droughts, strikes, fires and political disturbances are:

(a) Trend
(b) Seasonal
(c) **Irregular**
(d) Cyclical

Q25. The general pattern of increase or decrease in economics or social phenomena is shown by:

(a) **Secular trend**
(b) Seasonal trend
(c) Irregular trend
(d) Cyclical trend

Q26. In moving average method, we cannot find the trend values of some:

(a) Starting period
(b) Middle period
(c) End period
(d) **Between extreme periods**

Q27. The best fitting trend is one which the sum of squares of residuals is:

(a) Negative
(b) Zero
(c) Least
(d) Maximum

Q28. The rise and fall of a time series over periods longer than one year is called:

(a) Secular trend
(b) Seasonal variation
(c) Irregular variation
(d) **Cyclical variation**

Q29. The multiplicative time series model is:

(a) $Y = a + bX$
(b) $Y = T + S + C + I$
(c) $Y = a + bX + cX^2$
(d) $\mathbf{Y = T \times S \times C \times I}$

Q30. The additive model of the time series is:

(e) $Y = a + bX$
(f) $\mathbf{Y = T + S + C + I}$
(g) $Y = a + bX + cX^2$
(a) $Y = T \times S \times C \times I$

Q31. Variations that do not follow any specific pattern is called:

(a) Trend variation
(b) Cyclical variation
(c) Seasonal variation
(d) **Irregular variation**

Q32. A pattern that is repeated throughout a time series and has a recurrence period of at most one year is called:

(a) Long-term variation
(b) **Seasonal variation**
(c) Irregular variation
(d) Cyclical variation

Q33. A business cycle has:

(a) One stage (b) Two stages

(c) Three stages (d) **Four stages**

Q34. When the production of a thing is maximum, this stage is called:

(a) Recovery (b) Recession

(c) Depression (d) **Boom**

Q35. The straight line is fitted to the time series when the movements in the time series are:

(a) Cyclical (b) Irregular

(c) Non-linear (d) **Linear**

Q36. When the production of thing is decreasing, this stage is called:

(a) Recovery (b) **Recession**

(c) Depression (a) Prosperity

Q37. When the production of thing is increasing towards prosperity, this stage is called as:

(a) **Recovery** (b) Recession

(c) Depression (d) Boom

Q38. When the production of a thing is minimum, this stage is called:

(d) Recovery (e) Recession

(f) **Depression** (g) Prosperity

Q39. A second degree parabola has:

(a) No constant (b) One constant

(c) Two constant (d) **Three constant**

Q40. If an annual time series consisting of even number of years is coded, then each coded interval is equal to:

(a) One year (b) Half year

(c) Two years (d) **Both (a) & (b)**

Q41. The normal equations in fitting a second degree parabola are:

(a) One (b) Two

(c) **Three** (d) Four

Q42. For even number of years when origin is in the centre and the unit of X being one year, then X can be coded as:

(a) X = year – 0.5 average of years
(b) **X = year – average of years**
(c) X = (year – average of years)/2
(d) X = average of years - year

Q43. In fitting a second degree parabola, the value of c is not affected with the change of:

(a) **Origin**
(b) Scale
(c) Both (a) & (b)
(d) Neither (a) nor (b)

Q44. The second degree parabola is fitted to the time series when the variations are:

(a) Random
(b) Downward
(c) **Non-linear**
(d) Linear

Q45. For even number of years when origin is in the centre and the unit of X being half year, then X can be coded as:

(a) X = year – 0.5 average of years
(b) X = year – average of years
(c) **X = 2(year – average of years)**
(d) X = average of years - year

Q46. If r is the simple correlation coefficient, the quantity r^2 is known as:

(a) Coefficient of alienation
(b) **Coefficient of determination**
(c) Coefficient of non-determination
(d) None of these

Q47. For odd number of year, formula to code the values of X by taking origin at centre is:

(a) X = year – first year
(b) **X = year – average of years**
(c) X = year – last year
(d) X = year – 0.5 average of years

Q48. A pattern of variation of a time series that repeats every year is called:

(a) Trend
(b) Secular variation
(c) Cyclical variation
(d) **Seasonal variation**

Q49. Removing the highest and lowest actual to moving average values when the computing seasonal index for annual data reduces:

(a) Seasonal variation (b) Irregular variation

(c) **Extreme cyclical variations** (d) Secular trend

Q50. The repetitive movement around a trend line in a four month period is best described by:

(a) Secular trend (b) **Seasonal variation**

(c) Irregular variation (d) Cyclical variation

Q51. The trend values in freehand curve method are obtained by:

(a) **Graph** (b) Equation of straight line

(c) Second degree parabola (d) Linear equation

Q52. In semi averages method, if the number of values is odd then we drop:

(a) Last value (b) First value

(c) Middle two values (d) **Middle value**

Q53. Which of the following is not a component of time series?

(a) Trend (b) **Variance**

(c) Seasonal variation (d) Cyclical variation

Q54. The method of moving average is used to find the:

(a) **Secular trend** (b) Seasonal variation

(c) Irregular variation (d) Cyclical variation

Q55. The best fitted trend line is one for which sum of squares of residuals is:

(a) **Minimum** (b) Maximum

(c) Positive (d) Negative

Q56. Prosperity, recession and depression in a business is an example of:

(a) Seasonal trend (b) **Cyclical trend**

(c) Secular trend (d) Irregular trend

Q57. In the measurement of the secular trend, the moving averages:

(a) **Smooth out the time series**

(b) Measure the seasonal variation

(c) Measure the cyclical variation

(d) Give the trend in a straight line

Q58. In the theory of time series, shortage of certain consumer goods before the annual budget is due to:

(a) **Seasonal trend** (b) Cyclical trend

(c) Secular trend (d) Irregular trend

Q59. A set of observations recorded at an equal interval of time is called:

(a) Trend (b) **Time series data**

(c) Data (d) Array data

Q60. The unsystematic sequence which follows irregular pattern of variations is called:

(a) Linear (b) Non-linear

(c) **Noise** (d) Signal

Q61. A time series of annual data will contain which of the following components:

(a) Secular variation (b) Cyclical variation

(c) Seasonal variation (d) **Both (a) & (b)**

Q62. What variation does the moving average method eliminate?:

(a) Secular trend (b) Cyclical

(c) Irregular (d) **Seasonal**

Q63. In a multiplicative model, C, S and I have:

(a) **Indexing values** (b) Absolute values

(c) Positive values (d) Positive or negative values

Q64. The difference between the actual value of the time series and the forecasted value is called:

(a) Sum of variation (b) Sum of squares of residual

(c) **Residual** (d) None of these

Q65. Pool data comprises of columns that is used for the representation of:

(a) Sectional data (b) Cross-sectional data

(c) **Time series data** (d) Time-interval data

Q66. A time series may show gradual shifts or movements over a long period of time, which exhibits the existence of:

(a) Variability trend | **(b) Trend pattern**
(c) Stationary pattern | (d) Seasonal pattern

Q67. The most important factors causing seasonal variations are:

(a) Change in fashions | (b) Technological improvement
(c) Weather and social customs | (d) Growth in population

Q68. The best fitted trend line is one for which sum of squares of errors is:

(a) **Minimum** | (b) Maximum
(c) Positive | (d) Negative

Q69. Shortage of certain consumer goods before annual budget is due to:

(a) Secular trend | (b) Irregular variation
(c) **Seasonal variation** | (d) Cyclical variation

Q70. A time series is a set of data recorded:

(a) Periodically | (b) At time interval
(c) At successive points of time | (d) **All of them**

Q71. Simple average method is used to find:

(a) **Seasonal variation** | (b) Cyclical variation
(c) Irregular variation | (d) Secular variation

Q72. Complete cycle consists of a period of:

(a) Prosperity of depression | (b) Prosperity and recovery
(c) **Prosperity and recession** | (d) Recession and recovery

Q73. In a straight line equation Y = a + bX, a is the:

(a) **Y - intercept** | (b) X - intercept
(c) Slope | (d) None of these

Q74. In a straight line equation Y = a + bX, constant ‘a’ is the:

(a) Y - intercept | (b) X - intercept
(c) **Slope** | (d) None of these

Q75. An increase in harvesting time is associated in the:

(a) Trend (b) **Seasonal component**

(c) Cyclical component (d) Irregular component

Q76. After detrending, the time series can be represented as:

(a) Y = T x I (b) Y = T x C x I

(c) **Y = S x C x I** (d) Y = T x S x C x I

Q77. After seasonalization, the time series can be represented as:

(a) Y = T x I (b) **Y = T x C x I**

(c) Y = S x C x I (d) Y = T x S x C x I

Q78. After detrending, the time series multiplicative model is:

(a) Y = T x I (b) Y = T x C x I

(c) **Y = S x C x I** (d) Y = T x S x C x I

Q79. The seasonal indices are obtained by using:

(a) Moving average method (b) Free hand curve method

(c) Link relative method (d) Semi-average method

Q80. The first step in the analysis of time series is:

(a) **To edit the series** (b) To obtain trend

(c) To obtain seasonal variation (d) To obtain irregular variation

Q81. The method used in the study of cyclical variations is:

(a) Link relative method (b) **Residual method**

(c) Method of least square (d) Ratio-to-trend method

Q82. Decline in mortality rate due to improvement in science is associated to the:

(a) Irregular component (b) Cyclical component

(c) Seasonal component (d) **Trend**

Q83. Decline in cement production due to worker's strike is associated to the:

(a) Irregular component (b) Cyclical component

(c) **Seasonal component** (d) Trend

Q84. Loss of working hours due to fire in the power station in a factory is associated to the:

(a) **Random component** (b) Cyclical component

(c) Seasonal component (d) Trend

Q85. The trend is determined by:

(a) Ratio-to-trend method

(b) Ratio-to-moving average method

(c) Link relative method

(d) None of these

Q86. If Y = 2 - 0.2X, then the value of Y intercept is equal to:

(a) -0.2 (b) **2**

(c) 0.2X (d) All of the above

Q87. If regression line of = 5, then value of regression coefficient of Y on X is:

(a) **0** (b) 0.5

(c) 1 (d) 5

Q88. If one regression coefficient is greater than one, then other will be:

(a) **Less than one** (b) More than one

(c) Equal to one (d) Equal to zero

Q89. In the regression equation X = a + bY, the X is called:

(a) Independent variable (b) **Dependent variable**

(c) Qualitative variable (d) None of the above

Q90. In the regression equation Y = a + bX, the Y is called:

(a) Independent variable (b) **Dependent variable**

(c) Qualitative variable (d) None of the above

Q91. In the regression equation X = a + bY, the Y is called:

(a) **Independent variable** (b) Dependent variable

(c) Qualitative variable (d) None of the above

Q92. In the regression equation Y = a + bX, the X is called:

(a) **Independent variable** (b) Dependent variable

(c) Qualitative variable (d) None of the above

Q93. In the regression equation Y = a +bX, here a is called:

(a) Independent variable (b) Dependent variable

(c) **Y-intercept** (d) X-intercept

Q94. In the regression equation X = a +bY, here a is called:

(a) Independent variable (b) Dependent variable

(c) Y-intercept (d) **X-intercept**

Q95. In the regression equation X = a + bY, here b is called:

(a) **Regression coefficient of X on Y**

(b) Regression coefficient of Y on X

(c) Correlation coefficient of X and Y

(d) Correlation coefficient of Y and X

Q96. In the regression equation Y = a + bX, here b is called:

(a) Regression coefficient of X on Y

(b) **Regression coefficient of Y on X**

(c) Correlation coefficient of X and Y

(d) Correlation coefficient of Y and X

Q97. The regression equation always passes through:

(a) (X, Y) (b) (a, b)

(c) $\left(\bar{X}, \bar{Y}\right)$ (d) $(\bar{X}, Y)$

Q98. The graph showing the paired points of (X_i, Y_i) is called:

(a) **Scatter diagram** (b) Histogram

(c) Bar diagram (d) Pie diagram

Q99. The independent variable in a regression line is:

(a) Random variable (b) **Non-random variable**

(c) Qualitative variable (d) None of the above

Q100. When b_{XY} is positive, then byx will be:

(a) Zero (b) One

(c) Negative (d) **Positive**

Q101.When regression line passes through the origin, then:

(a) Association is zero (b) **Intercept is zero**

(c) Regression coefficient is zero (d) Correlation is zero

Q102 The correlation coefficient is the............... of two regression coefficients:

(a) Arithmetic mean (b) **Geometric mean**

(c) Harmonic mean (d) Median

Q103.It is possible that two regression coefficients have:

(a) Opposite signs (b) **Same signs**

(c) No sign (d) None of these

Q104.Regression coefficient is independent of:

(a) Scale (b) Origin

(c) **Both (a) and (b)** (d) Nether (a) nor (b)

Q105.In the regression line Y = a+ bX:

(a) $\sum X = \sum Y$ (b) $\sum X = \sum \hat{X}$

(c) $\boldsymbol{\sum Y = \sum \hat{Y}}$^ (d) $X = Y$

Q106.In the regression line Y = a + bX, the following is always true:

(a) $\sum(X-\hat{X}) = \sum(Y-\hat{Y})$ (b) $\sum(X-\hat{X}) = 0$

(c) $\boldsymbol{\sum\left(Y-\hat{Y}\right) = 0}$ (d) $\sum(Y-\hat{Y})^2 = 0$

Q107.In the regression line X = a + bY, the following is always true:

(a) $\sum(X-\hat{X}) = \sum(Y-\hat{Y})$ (b) $\boldsymbol{\sum\left(X-\hat{X}\right) = 0}$

(c) $\sum(Y-\hat{Y}) = 0$ (d) $\sum (Y-\hat{Y})^2 = 0$

Q108.When the ratio of variations in the related variables is constant, it is called:

(a) **Linear correlation** (b) Nonlinear correlation

(c) Positive correlation (d) Negative correlation

Q109.The sum of the difference between the actual values of Y and its values obtained from the fitted regression line is always:

(a) **Zero** (b) Positive

(c) Negative (d) Minimum

Q110.If all the actual and estimated values of Y are same on the regression line, the sum of squares of error will be:

(a) **Zero** (b) Minimum

(c) Maximum (d) Unknown

Q111.A measure of the strength of the linear relationship that exists between two variables is called:

(a) Slope (b) Intercept

(c) **Correlation coefficient** (d) Regression equation

Q112.If both variables X and Y increase or decrease simultaneously, then the coefficient of correlation will be:

(a) **Positive** (b) Negative

(c) Zero (d) One

Q113.If one variable X increase or decrease and Y decrease or increase, then the coefficient of correlation will be:

(a) Positive (b) **Negative**

(c) Zero (d) One

Q114.If one item is fixed and unchangeable and the other item varies, the correlation coefficient will be:

(a) Positive (b) Negative

(c) **Zero** (d) Undecided

Q115.In scatter diagram, if most of the points lie in the first and third quadrants, then coefficient of correlation is:

(a) Negative (b) **Positive**

(c) Zero (d) All of the above

Q116.If the two series move in reverse directions and the variations in their values are always proportionate, it is said to be:

(a) Limited negative correlation (b) Limited positive correlation

(c) **Perfect negative correlation** (d) Perfect positive correlation

Q117.If both the series move in the same direction and the variations are in a fixed proportion, correlation between them is said to be:

(a) Perfect negative correlation (b) **Perfect positive correlation**

(c) Linear correlation (d) Nonlinear correlation

Q118.If X is measured in kg and Y is measured in inches, then correlation coefficient has the unit:

(a) kg (b) Inches

(c) Both (a) and (b) (d) **No unit**

Q119.The signs of regression coefficients and correlation coefficient are always:

(a) Different (b) **Same**

(c) Positive (d) Negative

Q120.The range of regression coefficient is:

(a) -1 to +1 (b) 0 to 1

(c) **$-\infty$ to $+\infty$** (d) 0 to ∞

Q121.The arithmetic mean of the two regression coefficients is greater than or equal to:

(a) -1 (b) +1

(c) 0 (d) **r**

Q122.If the value of regression coefficient is zero, then the two variables are called:

(a) Independent (b) Dependent

(c) **Both (a) and (b)** (d) Neither (a) nor (b)

Q123.The term regression was used by:

(a) Fisher (b) Pearson

(c) Spearman (d) **Galton**

Q124.When the two regression lines are parallel to each other, then their slopes are:

(a) Zero (b) Different

(c) **Same** (d) Positive

Q125.In correlation problem both variables are:

(a) Unknown (b) Fixed

(c) **Random** (d) Equal

Q126.In the regression equation Y = a + bX, where a and b is called:

(a) Constants (b) Estimates

(c) **Both (a) and (b)** (d) Neither (a) nor (b)

Q127.The correlation coefficient between X and -X is:

(a) **- 1** (b) 0

(c) 1 (d) ∞

Q128.If $b_{yx} = b_{xy} = r_{xy}$, then:

(a) $\sigma_x \neq \sigma_y$ (b) $\boldsymbol{\sigma_x = \sigma_y}$

(c) $\sigma_x > \sigma_y$ (d) $\sigma_x < \sigma_y$

Q129.If $r_{xy} = 0.2$, then $r_{(2x, 2y)}$ is equal to:

(a) **0.2** (b) 0.4

(c) 0 (d) 1

Q130.If byx = -4 and rxy= -1, then bxy is equal to:

(a) **-1/4** (b) -4

(c) -1 (d) 1/4

Q131.If $b_{yx} = -0.8$ and $b_{xy} = -0.2$, then r_{yx} is equal to:

(a) **-0.4** (b) 0.4

(c) -0.8 (d) 0.8

Q132.If $\hat{Y} = 4 - X$, then r will be:

(a) 0 (b) 1

(c) **-1** (d) 1/4

Q133.If $\hat{Y} = X + 5$, then r equal to:

(a) **1** (b) -1

(c) 1/2 (d) -1/2

Q134.If Y = 10X and X = 0.1Y, then r is equal to:

(a) 0.1 (b) **1**

(c) -1 (d) 10

Q135.If $r_{xy} = 1$, then:

(a) $b_{yx} = b_{xy}$ (b) $b_{yx} > b_{xy}$

(c) $b_{yx} < b_{xy}$ (d) $\boldsymbol{b_{yx} \cdot b_{xy} = 1}$

Q136. The relation between the regression coefficient b_{yx} and correlation coefficient r is:

(a) σ_y/σ_x (b) σ_x/σ_y

(c) **$r.\sigma_x/\sigma_y$** (d) $r.\sigma_y/\sigma_x$

Q137. The relation between the regression coefficient b_{xy} and correlation coefficient r is:

(a) σ_y/σ_x (b) σ_x/σ_y

(c) **$r.\sigma_x/\sigma_y$** (d) $r.\sigma_y/\sigma_x$

Q138. If the sum of the product of the deviation of X and Y from their means is zero, the correlation coefficient between X and Y is:

(a) **Zero** (b) Maximum

(c) Minimum (d) Undecided

Q139. If the coefficient of correlation between the variables X and Y is r, the coefficient of correlation between X^2 and Y^2 is:

(a) r (b) **r^2**

(c) 0 (d) 1

Q140. If $r_{xy} = 0.4$, then r_{yx} will be:

(a) **0.4** (b) - 0.4

(c) 1/0.4 (d) -1/0.4

Q141. If $\hat{Y} = a$, then b_{yx} is equal to:

(a) **0** (b) 1

(c) a (d) None of these

Q142. If $\hat{Y} = bX$, then a is equal to:

(a) **0** (b) 1

(c) b (d) None of these

Q143. If $\hat{X} = bY$, then a is equal to:

(a) **0** (b) 1

(c) b (d) None of these

Q144. When $r_{xy} < 0$, then b_{yx} and b_{xy} will be:

(a) Zero (b) Not equal to zero

(c) **< 0** (d) > 0

Q145. When $r_{xy} > 0$, then b_{yx} and b_{xy} are both:

(a) Zero (b) Not equal to zero

(c) < 0 (d) **> 0**

Q146. If $r_{xy} = 0$, then:

(a) $b_{yx} = 0$ (b) $b_{xy} = 0$

(c) $b_{yx} \neq b_{xy}$ (d) **Both (a) and (b)**

Q147. A regression model may be:

(a) Linear (b) Non-linear

(c) **Both (a) and (b)** (d) Neither (a) and (b):

Q148. In regression equation $y=\alpha+\beta x+e$, both x and y variables are:

(a) Y is fixed and X is random (b) **X is fixed and Y is random**

(c) Fixed (d) Random

Q149. If the regression equation is equal to Y=23.0+43.0 X, then 23.0 is the _____ while 43.0 is the ____ of the regression line:

(a) Slope, Intercept (b) **Intercept, slope**

(c) Slope, regression coefficient (d) Regression coefficient, slope

Q150. If the equation of regression line is y=3, then what result will you take out from it?

(a) **The line is parallel to X-axis**

(b) The line is parallel to Y-axis

(c) The line is perpendicular to X-axis

(d) The line is perpendicular to Y-axis

Q151. The product of two regression coefficients is:

(a) 1 (b) Greater than 1

(c) **Less than 1** (d) None of these

Q152. For minimizing the error sum of squares, we obtain:

(a) Standard error of estimate (b) Correlation coefficient

(c) **Normal equations** (d) None of these

Q153.The standard form of the line of regression of Y on X is:

(a) **Y = a + bX** (b) X = a + bX

(c) aX + bY + c = 0 (d) X + Y + c = 0

Q154.The standard form of the line of regression of X on Y is:

(a) Y = a + bX (b) **X = a + bX**

(c) aX + bY + c = 0 (d) X + Y + c = 0

Q155.If 3X + 2Y = 6 is the regression of Y on X, then b_{yx} is,

(a) 2/3 (b) 3/2

(c) **– 3/2** (d) -2/3

Q156.If 3X + 2Y = 6 is the regression of X on Y, then b_{xy} is,

(a) 2/3 (b) 3/2

(c) – 3/2 (d) **-2/3**

Q157.In a bivariate study, the line of regression of Y on X is 4X – 3Y = 7, the correlation between X and Y will be:

(a) **Positive** (b) Negative

(c) Zero (d) One

Q158.In a bivariate study, the line of regression of X on Y is 4X – 3Y = 7, the correlation between X and Y will be:

(a) **Positive** (b) Negative

(c) Zero (d) One

Q159.In a bivariate study, the line of regression of X on Y is 3Y + 4X = 5, the correlation between Y and X will be:

(a) Positive (b) **Negative**

(c) Zero (d) One

Q160.If b_{yx} = - 3/2 and b_{xy} = - 1/6, the value of r will be:

(a) 1/4 (b) -1/4

(c) 1/2 (d) **-1/2**

Q161.If b_{yx} = 1/3 and b_{xy} = 3/4, the value of r will be:

(a) **1/2** (b) -1/2

(c) 1/4 (d) -1/4

Q162.If b_{xy} is positive, then b_{yx} is:

(a) Negative (b) **Positive**

(c) Undetermined (d) None of these

Q163.The change in the variable X corresponding to a unit change in the variable Y is measured by:

(a) b_{yx} (b) $\mathbf{b_{xy}}$

(c) r_{yx} (d) r_{xy}

Q164.The line of best fit is obtained by:

(a) **Method of least squares** (b) Standard error of estimate

(c) Correlation analysis (d) None of these

Q165.The line of best fit to observed data is known as:

(a) Correlation coefficient (b) **Regression line**

(c) Standard error (d) None of these

Q166.In regression of X on Y, Y is:

(a) **Explanatory variable** (b) Explained variable

(c) Constant (d) None of these

Q167.In regression of X on Y, X is:

(a) Explanatory variable (b) **Explained variable**

(c) Constant (d) None of these

Q168.For predicting the best value of X for given Y, we make use of:

(a) **Regression of X on Y** (b) Regression of Y on X

(c) Correlation of X on Y (d) Correlation of Y on X

Q169.If "time" is used as the independent variable in a simple linear regression analysis, then which of the following assumption could be violated

(a) The residual variation is the same for all fitted values of Y

(b) The residuals are normally distributed

(c) **Successive observations of the dependent variable are uncorrelated**

(d) There is a linear relationship between the independent and dependent variables

Q170.Relationship between the correlation coefficient and coefficient of determination is that:

(a) The coefficient of determination is the square root of the coefficient of correlation

(b) **The coefficient of determination is the coefficient of correlation squared**

(c) Both are unrelated

(d) Both are equal

Q171.The strength (degree) of the correlation between a set of independent variables X and a dependent variable Y is measured by:

(a) Standard error of estimate

(b) Coefficient of determination

(c) Coefficient of correlation

(d) **All of them**

Q172.Collinearity or Multicollinearity exists when:

(a) An independent variable is strongly correlated with a dependent variable

(b) **Independent variables are correlated less than -0.70 or more than 0.70**

(c) There is only one independent variable

(d) Relationship between the dependent and independent variable is non-linear

Q173.The percent of the total variation of the dependent variable Y explained by the set of independent variables X is measured by:

(a) Standard error of estimate

(b) **Coefficient of determination**

(c) Coefficient of correlation

(d) Coefficient of skewness

Q174.Relationship between the correlation coefficient and coefficient of determination is:

(a) Both are equal

(b) Both are unrelated

(c) Coefficient of correlation=(Coefficient of determination)^2

(d) **Coefficient of determination=(Coefficient of correlation)^2**

Q175.A coefficient of correlation is computed to be -0.95 means:

(a) The correlation coefficient cannot have this value

(b) **The relationship between the two variables is strong but negative**

(c) The relationship between the two variables is strong but positive

(d) None of these

Q176.If the coefficient of determination computed to be 0.39 in a problem involving one independent variable and one dependent variable, that means:

(a) 39% of the total variation is explained by the dependent variable

(b) **39% of the total variation is explained by the independent variable**

(c) Correlation coefficient is also 0.39

(d) None of these

Q177.The Coefficient of Correlation r is independent of:

(a) Origin only

(b) Scale of measurement

(c) **Both change of origin and scale of measurement**

(d) None of these

Q178. If $b_{xy} < 0$ and $b_{yx} \leq 0$, then:

(a) $r = 0$ (b) $\mathbf{r < 0}$

(c) $r > 0$ (d) None of these

Q179.Two regression lines are parallel to each other if their slope is:

(a) Negative (b) Different

(c) **Slope** (d) None of these

Q180.If X and Y are independent of each other, the coefficient of correlation is:

(a) -1 (b) $+1$

(c) **0** (d) ∞

Q181. If $r=0.6$, $b_{yx}=1$, then $b_{xy}=?$

(a) 0.20 (b) **0.36**

(c) 0.40 (d) 0.72

Q182.In a linear regression equation, Y=a + bX, what is the b denote?

(a) **The regression coefficient, the slope of the line**

(b) The intercept with the Y-axis

(c) The correlation coefficient, the strength of the line

(d) The score on the variable X

Q183.Assumptions for partial correlation are usual assumptions of:

(a) Perfect correlation
(b) **Pearson correlation**
(c) Exponential correlation
(d) None of these

Q184.In multiple regression analysis, two variables may have relationship of:

(a) Exact linear relationship

(b) In-exact linear relationship

(c) Exact non linear relationship

(d) **Both (b) and (c)**

Q185.Linear association between two variables say x and y can be expressed through:

(a) **Correlation coefficient**
(b) Ranked correlation
(c) Concurrent correlation
(d) None of these

Q186.Exact linear relationship between some or all independent variables depicts:

(a) Absolute-collinearity
(b) Homo-collinearity
(c) Poly-collinearity
(d) **Multicollinearity**

Q187.Property of symmetry for coefficient of correlation:

(a) **Satisfies**
(b) Not Satisfies
(c) Validates
(d) Not validates

Q188.If multiple coefficient of determination (R^2) depicts a perfect relationship, its value should be equal to

(a) -1
(b) 0
(c) **1**
(d) None of these

Q189.While performing correlation, we draw scatter plot to check assumptions of:

(a) Unbiasness
(b) Multicollinearity
(c) **Linearity**
(d) None of these

Q190.Regression model must be linear in:

(a) **Parameters**
(b) Variables
(c) Slope coefficient
(d) Variances

Q191.In coefficient of partial correlation all other variables are kept:

(a) Deviated
(b) Residual mean
(c) Not fixed
(d) **Constant**

Q192.Growth and trend factors are common source of

(a) Biased factoring
(b) Homogeneity
(c) **Multicollinearity**
(d) Interception

Q193.If multiple coefficient of determination R^2 is equals to 1, adjusted R^2 will be

(a) -1
(b) 0
(c) **1**
(d) ∞

Q194.In regression model, for any value of variable x, variance of error term should be:

(a) Repeated
(b) Zero
(c) **Same**
(d) Not fixed

Q195.In equation [1-(1- R^2) * (n-1/n-k)] if condition is ($k < 1$) then which of following holds a true value?

(a) Adjusted $R^2 > R^2$
(b) **Adjusted $R^2 < R^2$**
(c) Adjusted $R^2 = R^2$
(d) Adjusted $R^2 < -R^2$

Q196.High correlation is not an indication of:

(a) Relationship measure
(b) **Cause and effect**
(c) Unit-free measures
(d) Scale of measurement

Q197.Computation of multiple coefficient of determination always outputs a:

(a) Zero
(b) -1
(c) **Positive value**
(d) Negative value

11

Index Numbers

An **index number** is a statistical measure designed to show changes in a variable or group of related variables with respect to time, geographical location or other characteristics like: income, profession, etc. A collection of index numbers for different years, months, locations, etc. is called an **index series**. Since index number are usually used to measure the level of business and economic activities over a period of time, so they are rightly called **barometers of economic activities** or simple called **economic barometers**. The various problems faced in the construction of an index number are to known the specific purpose of index number, selection of items, data for index numbers, choice of base periods, choice of an average, selection of weights and choice of a suitable formula.

Methods of Constructing Index Numbers

A number of formulae have been developed for constructing index numbers, which may be grouped into the following categories:

- Unweighted index numbers
- Weighted index numbers
- **Unweighted index numbers:** In this case, weights are not expressly assigned. They can be further divided into two methods:
  - (i) Simple Aggregate Method
  - (ii) Simple Average of Price Relatives Method
  - (i) **Simple Aggregate Method:** This is the simplest method of computing index numbers. In this method, the total of the current year prices for various commodities is expressed as a percentage of the total of the base year prices for these commodities. Symbolically,

$$\text{Simple Aggregate Price Index}\left(P_{0n}\right) = \frac{\Sigma p_n}{\Sigma p_0} \text{ x}100$$

Here, P_{0n} = Current year index number;

$\Sigma\, p_n$ = Total of commodity prices in the current year

$\Sigma\, p_0$ = Total of commodity prices in the base year

(ii) **Simple Average of Price Relatives Method:** In this method, we first obtain price relatives for all commodities included in the index numbers. As usual, the price relative of the current year expressed as a percentage of the price of the base year is-

$$\frac{p_n}{p_0} x 100$$

Now these commodity price relatives may be averaged by using averages like arithmetic mean, median, mode, geometric mean or harmonic mean. However, if we use arithmetic mean, the formula for computing index number becomes-

$$P_{0n} = \frac{\sum \frac{p_n}{p_0}}{N} x 100$$

Here, $\sum \frac{p_n}{p_0}$ = Sum of commodity price relatives; N = number of commodities.

Similarly, the geometric mean of the relatives is used to find the index number of the current year as-

$$P_{0n} = Antilog\left[\frac{log\left(\frac{p_n}{p_0} x 100\right)}{N}\right] = Antilog\left[\frac{\sum \log R}{N}\right]$$

Where, $R = \frac{p_n}{p_0} x 100$

- **Weighted index numbers:** In **unweighted index numbers** assign equal importance to all the commodities or items included in the index. However, various commodities included are not of equal importance. In this way, the importance of the commodities is also reflected in the index number. Weighted index numbers are of two types:

(i) Weighted aggregate index numbers

(ii) Weighted average of relatives

(i) **Weighted aggregate index numbers:** In this method, appropriate weights are assigned to various commodities which reflect their relative importance in the group. A reasonable assumption is to consider the quantities consumed or produced as weights. If W is the weight attached to commodity then a general weighted price index can be formulated as –

$$\text{Weighted aggregate price index}\left(P_{0n}\right)=\frac{\Sigma p_n W}{\Sigma p_0 W}x100$$

By using different types of weights, a number of formulas have been developed for the construction of index numbers. These methods are named after the persons who have suggested them, are-

(a) Laspeyres' index or base year method

(b) Paasche's index or given year method

(c) Marshall-Edgeworth's method

(d) Walsh's index number method

(e) Bowley's index number method

(f) Fisher's ideal index number method

(a) **Laspeyres' index or base year method:** In this method, taking the quantity of the base year. The formula for constructing the index is:

$$\text{Laspeyres' index}\left(\mathrm{P}_{0n}\right)=\frac{\Sigma p_n q_0}{\Sigma p_0 q_0}x100$$

(b) **Paasche's index or given year method:** In this method, the weights are determined by quantities in the given year. The formula for constructing the index is:

$$\text{Paasche's Price index}\left(\mathrm{P}_{0n}\right)=\frac{\Sigma p_n q_n}{\Sigma p_0 q_n}x100$$

(c) **Marshall-Edgeworth's method:** In this case, the average of the base year and given year quantities is taken as weight. Thus the formula is:

$$\text{Marshall – Edgeworth's Price index}\left(\mathrm{P}_{0n}\right)=\frac{\Sigma p_n(q_0+q_n)}{\Sigma p_0(q_0+q_n)}x100$$

(d) **Walsh's index number method:** In this case, the average of the base year and given year quantities is taken as weight. Thus the formula is:

$$\text{Walsh's Price index}\left(\mathrm{P}_{0n}\right)=\frac{\Sigma p_n\left[\sqrt{(q_0 q_n)}\right]}{\Sigma p_0\left[\sqrt{(q_0 q_n)}\right]}x100$$

(e) **Bowley's index number method:** In this case, taking the arithmetic mean of Laspeyres' and Paasche's index numbers, then we get:

$$\text{Bowley's Price index}\left(\mathrm{P}_{0n}\right)=\frac{1}{2}\left[\frac{\Sigma p_n q_0}{\Sigma p_0 q_0}+\frac{\Sigma p_n q_n}{\Sigma p_0 q_n}\right]x100$$

(f) **Fisher's ideal index number method:** It is the geometric mean of

Laspeyres' and Paasche's index numbers, then we get:

$$\text{Fisher's Price index}\left(P_{0n}\right)=\sqrt{\left[\frac{\sum p_n q_0}{\sum p_0 q_0} x \frac{\sum p_n q_n}{\sum p_0 q_n}\right]} x100$$

(ii) Weighted average of relatives: It is used to overcome the limitations of the simple average of relatives. Weighted arithmetic mean is the most common weighted average for computing index number, although weighted geometric mean can also be used. The general formula for weighted average of relatives can be written as:

$$P_{0n}=\frac{\sum\left(\frac{p_n}{p_0} x100\right)W}{\sum W}=\frac{\sum RW}{\sum W}$$

Where,

$$R=\left(\frac{p_n}{p_0} x100\right)$$

Similarly, the index number based on geometric mean of price relative is given by-

In this method, each price relative is weighted by the total value of the commodity in terms of some monetary value, say in Rs. etc. The value of the commodity is obtained by multiplying the price p of the commodity by the quantity q. Thus, the weights are given by the value, i.e., v = pq. Depending on whether the base year values p_0q_0 or the given year values p_nq_n are taken as weights, the price index formulae obtained by using weighted average of relatives' are-

(i) Weighted arithmetic mean of price relatives using base year value weights is-

$$P_{0n}=\frac{\sum p_n q_0}{\sum p_0 q_0} x100$$

(ii) Weighted arithmetic mean of price relatives using given year value weights is-

$$P_{0n}=\frac{\sum\left(\frac{p_n}{p_0}\right)W}{\sum W} x100$$

Where,

W = p_nq_n = Value in the given current year

Quantity or Volume index numbers: Instead of comparing price changes over a period of time, we may be interested in analyzing changes in quantity of production or consumption of certain commodities over a given period of time. In order to study such changes, we construct quantity index numbers. In this regard, a few quantity index number (Q_{0n}) formulae may be:

(i) **Simple aggregate quantity index:**

$$Q_{0n} = \frac{\sum \frac{q_n}{q_0}}{N} x100$$

(ii) **Simple average of relatives method:**

$$Q_{0n} = \frac{\sum \frac{q_n}{q_0}}{N} x100$$

(iii) **Laspeyres' quantity index:** Using base year prices as weights, Laspeyres' quantity index is given by-

$$Q_{0n} = \frac{\sum q_n p_0}{\sum q_0 p_0} x100$$

(iv) **Paasche's quantity index:** Using given year prices as weights, Paasche's' quantity index is given by-

$$Q_{0n} = \frac{\sum q_n p_0}{\sum q_0 p_0} x100$$

(v) **Fisher's quantity index:**

$$Q_{0n} = \sqrt{\left[\frac{\sum q_n p_0}{\sum q_0 p_0} x \frac{\sum q_n p_n}{\sum q_0 p_n}\right]} x100$$

(vi) **Marshall-Edgeworth's index:**

$$Q_{0n} = \frac{\sum q_n (p_0 + p_n)}{\sum q_0 (p_0 + p_n)} x100$$

Test of adequacy of index numbers

We observed that relatives follow certain properties. When these properties are true for an individual commodity, these should also hold good for a group of commodities. As such, the index number as an aggregative relative should also satisfy these properties. In view of this, we discuss these properties in the light of various index number formulae. A good index number should satisfy the following tests of properties-

(i) Unit test

(ii) Time reversal test

(iii) Factor reversal test

(iv) Circular test

(i) **Unit Test:** According to this test, an index number should be independent of the unit in which prices and quantities of various commodities are quoted.

This test is satisfied by all the index number formulae except the simple aggregative index.

(ii) **Time reversal test:** According to **Prof. Fisher**, the formula for calculating an index number should be such that it gives the same ratio between one point of comparison and the other, no matter which of the two is taken as base. In other words, if the two periods, the base and the reference period are interchanged, the product of the two index numbers should be unity, i.e., they should be reciprocal of each other. Symbolically,

$$P_{0n} \times P_{n0} = 1$$

An index formula satisfying the criteria is said to satisfy the **time reversal test**.

(iii) **Factor reversal test:** We know that if the two factors p and q are interchanged in a price index formula (P_{0n}). We get a quantity index formula (Q_{0n}). Then we expect that the product of P_{0n} and Q_{0n} should be equal to the true value ratio $\frac{\sum p_n q_n}{\sum p_0 q_0}$. In other words, the product of the price index and the corresponding quantity index should give the true ratio of value in the given year to the value in the base year. Symbolically, the factor reversal test is satisfied, if

$$P_{0n} \times Q_{0n} = \frac{\sum p_n q_n}{\sum p_0 q_0} = \frac{\text{Value in the given year}}{\text{Value in the base year}}$$

Now, in case of Laspeyres' index

$$P_{0n} = \frac{\sum p_n q_0}{\sum p_0 q_0}$$

and

$$Q_{0n} = \frac{\sum q_n p_0}{\sum q_0 p_0}$$

Thus,

$$P_{0n} \times Q_{0n} = \frac{\sum p_n q_0}{\sum p_0 q_0} \times \frac{\sum q_n p_0}{\sum q_0 p_0} \neq \frac{\sum p_n q_n}{\sum p_0 q_0}$$

Thus, **Laspeyre's index** does not satisfy factor reversal test. However, in the case of Fisher's index, we have

$$P_{0n} = \sqrt{\left[\frac{\sum p_n q_0}{\sum p_0 q_0} \times \frac{\sum p_n q_n}{\sum p_0 q_n}\right]}$$

and

$$Q_{0n} = \sqrt{\left[\frac{\Sigma q_n p_0}{\Sigma q_0 p_0} \times \frac{\Sigma q_n p_n}{\Sigma q_0 p_n}\right]}$$

Therefore,

$$P_{0n} \times Q_{0n} = \sqrt{\left[\frac{\Sigma p_n q_0}{\Sigma p_0 q_0} \times \frac{\Sigma p_n q_n}{\Sigma p_0 q_n}\right] x \sqrt{\left[\frac{\Sigma q_n p_0}{\Sigma q_0 p_0} \times \frac{\Sigma q_n p_n}{\Sigma q_0 p_n}\right]}} = \frac{\Sigma p_n q_n}{\Sigma p_0 q_0}$$

Thus, Fisher's index also satisfies the factor reversal test. Now, since Fisher's index number is one which satisfies both the time reversal and the factor reversal test, so it is called **Fisher's ideal index number**.

(iv) **Circular test:** This is based on the shifting of the base period and thus is an extension on the time reversal test. According to this test, an index number should also work in a circular way, i.e., the test requires that

$$P_{01} \times P_{12} \times P_{20} = 1$$

For, four time points, we have

$$P_{01} \times P_{12} \times P_{23} \times P_{30} = 1$$

In particular, for two time points, 0 and n, one gets

$$P_{0n} \times P_{n0} = 1$$

which is nothing but time reversal test. It can be observed that none of the weighted index numbers satisfies this test. However, for three years 0, 1 and 2, if the test is applied to index obtained by simple aggregate method, one gets-

$$P_{01} \times P_{12} \times P_{20} = \frac{\Sigma p_1}{\Sigma p_0} \times \frac{\Sigma p_2}{\Sigma p_1} \times \frac{\Sigma p_0}{\Sigma p_2} = 1$$

Similarly, if we apply circular test to index numbers obtained by fixed weight aggregative method, we get-

$$P_{01} \times P_{12} \times P_{20} = \frac{\Sigma p_1 q}{\Sigma p_0 q} \times \frac{\Sigma p_2 q}{\Sigma p_1 q} \times \frac{\Sigma p_0 q}{\Sigma p_2 q} = 1$$

Thus, for index obtained by simple aggregate method or by fixed weight aggregative method, the circular test is satisfied.

Link and Chain Relatives

Let p_1, p_2, p_3, ... denote prices of a commodity during successive intervals of time 1, 2, 3, ..., respectively. Then the price relatives of each time interval with respect to the preceding time interval as base, i.e., P_{12}, P_{23}, P_{34}, ... etc., are called link relatives. Now using this series of price relatives and the circular property of relatives, one can get the price relative for a given period with respect to any other period as base. Now, let us consider the modified circular property of price relatives which states-

$$P_{13} = P_{12} \,.\, P_{23}$$

or $$P_{14} = P_{12} \,.\, P_{23} \,.\, P_{34}$$

That means, the price relative of period 3 with respect to period 1 as base can be obtained from the link relatives P_{12} and P_{23}. Similarly, link relatives P_{12}, P_{23} and P_{34} can be used to get P_{14}, i.e., the price relative of period 4 with respect to period a as base. Thus, the price relatives with respect to a fixed base period can be obtained by using link relatives. These link relatives of one period with respect to proceeding as base are chained together by successive multiplication to give a chain price relative of a period with respect to a fixed base period. In view of this, it is called a **chain relative** or **chain index**. The concept of **link** and **chain relatives** can be similarly used in the case of quantity and value relatives. Thus, the construction of chain indices involves the following steps-

Steps

1. Express the figures for each period as percentage of the preceding period to get the link relatives (L.R.).
2. These link relatives are chained together by successive multiplication to get the chain indices or relatives of a period with respect to a fixed base period. Symbolically, the chain relatives or indices are computed as-

$$P_{01} = \text{First link relative}$$

$$P_{02} = P_{01} \text{ x } P_{12}$$

$$P_{03} = P_{01} \text{ x } P_{12} \text{ x } P_{23} = P_{02} \text{ x } P_{23}$$

$$P_{04} = P_{01} \text{ x } P_{12} \text{ x } P_{23} \text{ x } P_{34} = P_{03} \text{ x } P_{34}$$

$$P_{0k} = P_{01} \text{ x } P_{12} \text{ x } \ldots\ldots\ldots\ldots \text{ x } P_{(k-1)k} = P_{0(k-1)} \text{ x } P_{(k-1)k}$$

or use the following formula for computing chain index (C.I.)

$$\text{C.I.} = \frac{\text{Current year L.R. x Preceding year C.I.}}{100}$$

Further, in some specific situations, we also need to convert chain base index numbers (C.B.I.) to fixed base index (F.B.I.) for which the following formula is applicable-

$$\text{Current year F.B.I.} = \frac{\text{Current year C.B.I. x Previous year F.B.I.}}{100}$$

Base shifting, splicing and Deflating of index numbers

Base Shifting: It means to prepare a new index series with a new base period in place of an old one. Thus, base shifting is nothing but a procedure of recasting an index series by shifting its base period to some recent or more relevant base period. It is necessary in the following situations-

1. When the base period of the index series is too old or too distant from the current period.
2. When we wish to compare two or more index series with different base periods then, for making valid comparisons, it is necessary that the given index series be expressed with a common base period.

In base shifting procedure, the index number of the new base year is taken as 100 and then the remaining index numbers in the series are expressed as percentage of the index number selected as new base. Thus, the index series may be recast by using the following formula-

$$\text{Recent index number of any year} = \frac{\text{Old index number of the year}}{\text{Index number of new base year}} \text{x}100$$

Splicing: It means combining two or more index series. For retaining continuity in comparison between two or more index series, we splice or combine them into a new single index series. For clarity, suppose there is an index series 'A' with base period 2010 which was discontinued in 2015 and then a new index series 'B' is prepared with base period 2015. Then for comparing the two index series, we can splice them into a new continuous index series in the following manner-

1. Splicing the index series 'B' to 'A' to obtain a new continuous index series with base 2010. This splicing procedure is known as **forward splicing**. In fact, in this splicing the base 2015 of index series 'B' has to be shifted to base 2010.
2. Splice the index series 'A' to 'B' to get a new continuous index series with base 2015. This splicing procedure is known a **backward splicing**. In other words, in this type of splicing the base period 2010 of index series 'A' is to be shifted to base 2015. Thus, the procedure is splicing is very much alike to that involved in base shifting. The formula used for **forward splicing** is-

$$\text{Required index} = \frac{\text{Old index number on the existing base x Index number to be spliced}}{100}$$

Also the formula used for **backward splicing** is-

$$\text{Needed index} = \frac{100}{\text{Old index number on the existing base x Index number to be spliced}}$$

Deflating: It means "making allowance for the effect of changing price levels". An increase in price level of consumer goods over a period means a reduction in the purchasing power of the people. For example, if the price of wheat rises from Rs. 1200 per quintal in 2017 to Rs 2400 per quintal in 2022. This simply means that in 2022 the person can buy only half the amount of wheat in 2022 if he decides to spent the same amount which he was spending in 2017. In other words, the value of the rupees is 50 paisa in 2022 as compared to that in 2017, i.e., the purchasing power of the money in 2022 is half as compared to that in 2017. Therefore, the purchasing power is given by the reciprocal of the index number and consequently the real income is obtained by dividing the nominal income of the period by the corresponding index number and expressing the ratio in percentage. Thus,

$$\text{Real wages} = \frac{\text{Money or nominal wages}}{\text{Price index}} \text{x}100$$

The real wages is also known as **deflated wages (or income)**. In this way, we observe that, in the deflating procedure, the relevant price index is the deflator and a deflated value can be computed by using the formula-

$$\text{Deflated value} = \frac{\text{Current value}}{\text{Deflator}} \text{x}100 = \frac{\text{Current value}}{\text{the relevant index number}} \text{x}100$$

MCQ

Q1. The index number for base year is always:

(a) 10 (b) **100**

(c) 1000 (d) 10000

Q2. The weights used in Paasche's price index are donated as:

(a) p_n (b) $\mathbf{q_n}$

(c) p_0 (d) q_0

Q3. The weights used in Paasche's quantity index are donated as:

(a) $\mathbf{p_n}$ (b) q_n

(c) p_0 (d) q_0

Q4. The weights used in Laspeyre's price index are donated as:

(a) p_n (b) q_n

(c) p_0 (d) $\mathbf{q_0}$

Q5. The weights used in Laspeyre's quantity index are donated as:

(a) p_n (b) q_n

(c) $\mathbf{p_0}$ (d) q_0

Bold means correct answer

Q6. Which of the following are the major characteristics of index numbers?

(a) It measures changes over a period of time

(b) Which of the following are the major characteristics of index numbers?

(c) It measures the net or relative changes in variables

(d) **All of them**

Q7. Index number is a type of:

(a) Correlation (b) Regression

(c) Dispersion (d) **Average**

Q8. The time period for which an index number is determined is known as:

(a) Normal period (b) **Current period**

(c) Base period (d) None of these

Q9. Which of the following are limitations of using index numbers?

(a) It is useful only for short term comparison

(b) It ignores the quality of commodities

(c) The use of each index number is restricted to a specific object

(d) **All of the above**

Q10. Which of the following statements is/are correct about average prices if the price index is 110?

(a) The prices have decreased by 10 per cent

(b) **The prices have increased by 10 per cent**

(c) The prices have increased by 110 per cent

(d) The prices have decreased by 110 per cent

Q11. The weights used in a quantity index are:

(a) **Price** (b) Quantity

(c) Values (d) None of these

Q12. Which of the following is the variation within two or more variable studies by the index?

(a) Simple index (b) Price index

(c) **Composite index** (d) None of these

Q13. An index number that can serve many purposes is known as:

(a) Special purpose index (b) **General purpose index**

(c) Both (a) & (b) are correct (d) Both (a) & (b) are incorrect

Q14. Which of the following methods is used to calculate the Consumer Price Index?

(a) Paasche's formula

(b) Fisher's formula

(c) **Laspeyre's formula**

(d) Marshall-Edgeworth's formula

Q15. Which of the following is the index that is used to measure changes in total money value?

(a) **Value index** (b) Quantity index

(c) Price index (d) None of these

Q16. Fisher's method of calculating the index number is based on

(a) **Arithmetic mean** (b) Geometric mean

(c) Harmonic mean (d) None of these

Q17. What is the name of the monthly price index that takes price changes in consumer goods and services and uses it to determine changes in the price of those products over a period of time?

(a) Paasche's index (b) Wholesale price index

(c) **Consumer price index** (d) None of these

Q18. Commodities that show considerable price fluctuations can be measured by:

(a) Value index (b) **Quantity index**

(c) Price index (d) None of these

Q19. Which of the following is the index that is used to measure changes in total money value?

(a) **Value index** (b) Quantity index

(c) Price index (d) None of these

Q20. The purchasing power of money is:

(a) Equal to the price index number

(b) Not equal to the price index number

(c) **Reciprocal of the price index number**

(d) None of these

Q21. Which index in India is used to evaluate inflation?

(a) Commodity price index (b) **Wholesale price index**

(c) Productive price index (d) GDP deflator

Q22. What does the consumer price index (CPI) calculate?

(a) Price of services only (b) Price of goods only

(c) **Price of both goods and services** (d) None of these

Q23. Who releases the figures of the wholesale price index?

(a) Ministry of finance

(b) Ministry of Home affairs

(c) **Ministry of Commerce and Industry**

(d) Ministry of Statistics and Program Implementation

Q24. How often is the data of the consumer price index (CPI) released?

(a) Every day (b) Every week

(c) **Every month** (d) Every year

Q25. Who issues the data of the consumer price index (CPI)?

(a) Ministry of finance

(b) Ministry of Home affairs

(c) Ministry of Commerce and Industry

(d) **Ministry of Statistics and Program Implementation**

Q26. What is the base year of the consumer price index (CPI)?

(a) 2001 (b) 2005

(c) 2010 (d) **2012**

Q27. Index number is always expressed in:

(a) Ratio (b) Proportion

(c) Percentage (d) None of these

Q28. Index number is also called as:

(a) Constant (b) Parameter

(c) Estimate (d) **Economic barometer**

Q29. Which index number is called as ideal index number:

(a) Paasche's
(b) **Fisher's**
(c) Laspeyre's
(d) Marshall-Edgeworth's

Q30. In Lasperys price index number weight is considered as:

(a) Price in base year
(b) **Quantity in base year**
(c) Price in current year
(d) Quantity during current year

Q31. In Paasche's price index number weight is considered as:

(a) Price in base year
(b) Quantity in base year
(c) Price in current year
(d) **Quantity during current year**

Q32. Fishers price index number is:

(a) Arithmetic mean of Lasperys and Paasche's index number
(b) **Geometric mean of Laspeyre's and Paasche's index number**
(c) Difference between Lasperys and Paasche's index number
(d) None of these

Q33. A period for which index number is determined is called as:

(a) Base period
(b) **Current period**
(c) Normal period
(d) None of these

Q34. The period of comparison or with respect to which index number is determined is called as:

(a) **Base period**
(b) Current period
(c) Normal period
(d) None of these

Q35. If all the values are of equal importance, the index numbers are called:

(a) Value index
(b) Composite
(c) Weighted
(d) **Unweighted**

Q36. An index number is called a simple index when it is computed from:

(a) **Single variable**
(b) Bi-variable
(c) Tri-variable
(a) Multi-variable

Q37. When the prices of wheat are to be compared, we compute:

(a) **Price index**
(b) Value index
(c) Volume index
(d) Aggregative index

Q38. Index numbers can be used for:

(a) Fixed prices
(b) Constant prices
(c) **Forecasting**
(d) Different prices

Q39. Price relatives are a percentage ratio of current year price and:

(a) Current year quantity
(b) Previous year quantity
(c) Base year quantity
(d) **Base year price**

Q40. An index number that can serve many purposes is called:

(a) Special purpose index
(b) **General purpose index**
(c) Cost of living index
(d) None of these

Q41. Another name of consumer's price index number is:

(a) Composite
(b) Sensitive
(c) **Cost of living index**
(d) Whole-sale price index number

Q42. Cost of living at two different cities can be compared with the help of:

(a) Volume index
(b) **Consumer price index**
(c) Value index
(d) Unweighted index

Q43. Purchasing power of money can be accessed through:

(a) Volume index
(b) Simple index
(c) Fisher's index
(d) **Consumer price index**

Q44. Consumer price index numbers are obtained by:

(a) Paasche's formula
(b) Fisher's formula
(c) **Laspeyre's formula**
(d) Marshall-Edgeworth's formula

Q45. Laspeyre's index = 110, Paasche's index = 108, then Fisher's Ideal index is equal to:

(a) 100 (b) 108

(c) **109** (d) 110

Q46. Most commonly used index number is:

(a) Value index number (b) **Price index number**

(c) Volume index number (d) Simple index number

Q47. Consumer price index is obtained by:

(e) Paasche's formula

(f) Fisher's formula

(g) **Family budget method formula**

(a) Marshall-Edgeworth's formula

Q48. The aggregative expenditure method and family budget method always give:

(a) Approximate results (b) Different results

(c) **Same results** (d) None of these

Q49. If all the values are not of equal importance the index number is called:

(a) **Weighted** (b) Unweighted

(c) Simple (d) Composite

Q50. Which of the following formula satisfy the time reversal test?

(a) Paasche's (b) **Fisher's**

(c) Laspeyre's (d) Marshall-Edgeworth's

Q51. When the price of a divided by the price of the preceding year, we get:

(a) **Link relative** (b) Value index

(c) Simple relative (d) None of these

Q52. The most appropriate average in averaging the price relatives is:

(a) Arithmetic mean (b) **Geometric mean**

(c) Harmonic mean (d) Median

Q53. The most suitable average in chain base method is:

(a) Arithmetic mean (b) **Geometric mean**

(c) Harmonic mean (d) Median

Q54. In constructing index number geometric mean relatives are:

(a) Reciprocal (b) **Reversible**

(c) Non-reversible (d) None of these

Q55. Base year quantities weights are used in:

(a) Paasche's method

(b) Fisher's method

(c) **Laspeyre's method**

(d) Marshall-Edgeworth's method

Q56. The Federal Bureau of Statistics prepares:

(a) Consumer price index (b) Sensitive price indicator

(c) Wholesale price index (d) **All of them**

Q57. While computing a weighted index, the current period quantities are used in the:

(a) **Paasche's method**

(b) Fisher's method

(c) Laspeyre's method

(d) Marshall-Edgeworth's method

Q58. Fisher's ideal index number is the geometric mean of the:

(a) **Laspeyre's and Paasche's indices**

(b) Paasche's and Marshall-Edgeworth's indices

(c) Laspeyre's and Marshall-Edgeworth's indices

(d) All of them

Q59. When the base year values are used as weights, the weighted average of relatives price index number is the same as the:

(a) Paasche's index (b) quantity index

(c) **Laspeyre's index** (d) Simple aggregative index

Q60. Paasche's price index number is also called:

(a) Simple aggregative index (b) Base year weighted

(c) **Current year weighted** (d) Consumer price index

Q61. Laspeyre's price index number is also called:

(a) Simple aggregative index (b) **Base year weighted**

(c) Current year weighted (d) Cost of living index

Q62. Index number calculated by Fisher's formula is ideal because it satisfy:

(a) Time reversal test (b) Factor reversal test

(c) Circular test (d) **All of them**

Q63. Paasche's Formula is the type of:

(a) **Weighted aggregative formula** (b) Weighted average relative

(c) Unweighted average relative (d) None of these

Q64. Price Index Numbers measures the changes in?

(a) Single variable

(b) Physical quantity of goods produced

(c) **Relative changes in prices of commodities between two periods**

(d) None of these

Q65. Whose formula is considered ideal for the construction of index number?

(a) Paasche's formula

(b) **Fisher's formula**

(c) Laspeyre's formula

(d) Marshall-Edgeworth's formula

Q66. In usual notations, p_{01}, 1 stands for:

(a) **Current year** (b) Base year

(c) Reference year (d) None of these

Q67. Fisher's index number is considered ideal because:

(a) It satisfies time reversal test

(b) It satisfies factor reversal test

(c) It is based on variable reversal test

(d) All of them

Q68. Indices calculated by the chain base method are free from:

(a) **Seasonal variations** (b) Ratio

(c) Percentages (d) Errors

Q69. Index numbers helps in:

(a) Deflating values (b) Studying the trends

(c) Policy formation (d) **All of them**

Q70. The index used to measure changes in total money value is called:

(a) Quantity index (b) Price index

(c) **Value index** (d) All of them

Q71. We use price index numbers:

(a) To compare prices (b) To measure prices

(c) **To measure and compare prices** (d) None of these

Q72. Complete cycle consists of a period of:

(a) Prosperity of depression (b) Prosperity and recovery

(c) **Prosperity and recession** (d) Recession and recovery

Q73. The variation in two or more variables studies by the index is called:

(a) Simple index (b) Price index

(c) **Composite index** (d) None of these

Q74. A relative obtained by dividing the price in a given year price in the base year and expressed as a percentage is called:

(a) Link relative (b) **Price relative**

(c) Volume relative (d) None of these

Q75. The weights used in a quantity index are:

(a) Values (b) **Prices**

(c) Quantities (d) None of these

Q76. Index Numbers may be categorized in terms of:

(a) Numbers (b) Constants

(c) **Variables** (d) All of them

Q77. Which is the one of the problem in the construction of index number?

(a) Big formula
(b) **Purpose of index number**
(c) Tedious job
(d) None of these

Q78. Which of the following is useful to calculate Dearness allowances of employee?

(a) Regression
(b) Correlation
(c) **Index number**
(d) None of these

Q79. Which is the best suitable measure of central tendency to construct index number?

(a) Arithmetic mean
(b) Geometric mean
(c) **Harmonic mean**
(d) Median

Q80. Which one of the following are not the problems in the construction Index Number:

(a) Selection of weight
(b) Selection of base period
(c) **Purpose of index number**
(d) Selection of proper scale

Q81. Base period always:

(a) Base period should be far ago
(b) **Normal period**
(c) Abnormality should be present in the base period
(d) None of these

Q82. Which of the following is used to denote prices in the base period?

(a) 100
(b) 10
(c) 1
(d) **0**

Q83. A relative obtained by dividing the price in a given year price in the base year and expressed as a percentage is called:

(a) Volume index
(b) Link relatives
(c) **Price relatives**
(d) None of these

Q84. The weights used in a quantity index are:

(a) **Prices**
(b) Values
(c) Quantities
(d) None of these

Q85. An index number that can serve many purposes is called:

(a) Consumer price
(b) Special purpose index
(c) Simple index number
(d) General purpose index

Q86. ....Index numbers in which different items of the series are recorded different which is depending upon their importance:

(a) Simple index
(b) **Complex index**
(c) Linear index
(d) Price index

Q87. If a sub-index is zero, then the total index must:

(a) **Not be zero**
(b) Be zero
(c) Be infinity
(d) indeterminate

Q88. If all the sub-indices are equal to P, then the total index will be:

(a) Less than P
(b) More than P
(c) Equal to one
(d) **Equal to P**

Q89. The index number for 2015 to the base 2010 is 150 and for 2010 to the base 2015 is 70. The given indices satisfy:

(a) Circular test
(b) **Time reversal test**
(c) Factor reversal test
(d) All of them

Q90. Consumer price index reflects on the price changes experienced by:

(a) A particular family
(b) **All families of a population**
(c) An individual
(d) None of the above

Q91. The wholesale price index is different from consumer price index in respect of:

(a) Weights
(b) Collection of price quotations
(c) Inclusion of services
(d) **All of them**

Q92. The quantities of items consumed or produced taken as weights come under the category of:

(a) Fixed weights
(b) **Explicit weights**
(c) Implicit weights
(d) None of the above

Q93. Deflation of index numbers is meant for calculating:

(a) Money income index (b) Real wages

(c) Real income index number (d) **All of them**

Q94. The difference between the quantity indices due to Laspeyre and Paasche is called:

(a) Sampling error (b) Homogeneity error

(c) **Formula error** (d) None of them

Q95. The range of homogeneity error in reference to index numbers is:

(a) -1 to +1 (b) **0 to +1**

(c) 0 to ∞ (d) - ∞ to + ∞

Q96. Data for index numbers should be collected from:

(a) Selected group of persons (b) Wholesale dealers

(c) Retailer (d) **None of them**

Q97. The error(s) involved in the construction of index numbers is/are:

(a) Formula error (b) Error in collected data

(c) Error of sampling (d) **All of them**

Q98. Index number is a:

(a) Percentage relative (b) Measure of relative changes

(c) A special type of an average (d) **All of them**

Q99. Index numbers help:

(a) For adjusting national income

(b) In farming of economic policies

(c) Assessing the purchasing power of money

(d) **All of them**

Q100.Diffusion index reveals the changes in:

(a) **A group of time series** (b) Industrial production

(c) Elite (d) None of these

Q101.Index number plays an important role in the field of:

(a) Economic and education
(b) Economic and science
(c) Economic and intelligence
(d) **Economic and business**

Q102.The period for which the index number is determined is known as:

(a) Base period
(b) Normal period
(c) Current period
(d) **None of these**

Q103.Purchasing power of money is stated as _____ price index?

(a) **Reciprocal**
(b) Equal
(c) Unequal
(d) None of these

Q104.The weighted system changes with:

(a) Month
(b) Quarter
(c) Year
(d) **Time**

Q105.Which of the below is another object to determine the index number?

(a) Reliable
(b) **Scope**
(c) Index number
(d) Special purpose

Q106.Index number for base year is always considered as:

(a) 200
(b) 201
(c) **100**
(d) 101

Q107.Index number is a special type of:

(a) Correlation
(b) Regression
(c) **Average**
(d) Dispersion

Q108.A period for which index number is determined is called as:

(a) Base period
(b) Normal period
(c) **Current period**
(d) None of these

12

Statistical Inference and Tests of Hypothesis

Statistical Inference

It is the process of drawing valid inference or conclusion about the population on the basis of a sample drawn from the population. In other words, **Statistical inference** is the process of using a sample to infer the properties of a population. It requires studies pertaining to three aspects:

- Point Estimation
- Interval Estimation
- Testing of Hypothesis
- **Point Estimation:** It is to find a single value, which is used as an estimator of unknown parameter. **For example**, sample mean may be used as an estimate of the population mean μ.
- **Interval Estimation:** It is to find confidence limits based on sample values, within which the unknown parameter lies with confidence coefficient described in probability term. In other words, we estimate an unknown parameter using an **interval of values** that is likely to contain the true value of that parameter and state how confident we are that this interval indeed captures the true value of the parameter.
- **Testing of Hypothesis:** The tests used to ascertain whether the differences between estimator and the parameter or between two estimators are real or due to chance are called **tests of hypothesis** or **tests of significance**. In other words, the procedures, which enable us to decide whether to accept or reject hypothesis, are called **tests of hypothesis** or **tests of significance**. Two types of problems mainly arise in tests of significance:
    - Testing the significance of the difference between estimator and the parameter.
    - Testing the significance of the difference between two estimators.

Statistical Hypothesis

It may be defined as a statement about one or more populations. The following terms are commonly used in testing of hypothesis-

- Null hypothesis
- Alternative hypothesis
- **Null hypothesis:** The hypothesis, which is tested for possible rejection under the assumption that it is true, called **null hypothesis**. It is denoted as H_0.
- **Alternative hypothesis:** Any hypothesis, which is complementary to the null hypothesis, is called an **alternative hypothesis**. It is also called as **research hypothesis**. It is denoted as H_1.

Example: If we wish to set up the null hypothesis that the mean of the population is μ_0 (A specified value of μ), then these two hypothesis can be written as-

$H_0: \mu = \mu_0$ (Null hypothesis)

$H_1: \mu \neq \mu_0$ (Alternative hypothesis)

i.e. $\mu > \mu_0$ or $\mu < \mu_0$

Errors in Hypothesis testing

The decision to accept or reject the null hypothesis (H_0) is made on the basis of sample observations. There is always possibility of committing two kinds of error in testing of a hypothesis.

(i) **Type-I error:** Rejecting a null hypothesis (H_0) when it is true is called **type-I error**. It is denoted by 'α'. Thus-

α = P (Type-I error) = P (Rejecting H_0 when H_0 is true)

(ii) **Type-II error:** Accepting a null hypothesis (H_0) when it is false is called **type-II error**. It is denoted by 'β'. Thus-

β = P (Type-II error) = P (Accepting H_0 when H_0 is false)

Note: (1-β) is called the **power of a test.**

Level of significance

In testing of hypothesis, we wish to minimize sizes of both types of errors. However, with fixed size testing procedure, both the errors cannot be minimized simultaneously. Thus, we keep the size or the probability of committing type-I error

(α) fixed at certain level, called the **level of significance.** The level of significance is also known as the **size of rejection region** or the **size of the critical region.** The level of significance, which are usually employed in tests of significance are 5% and 1%. If the level of significance is chosen as 5 per cent, it means that the probability of accepting a true hypothesis is 95 per cent.

Confidence Interval

The **confidence interval** is a range of values that's likely to include a population value with a certain degree of confidence. It is often expressed as a percentage (%) whereby a population mean "μ" lies between an upper and lower interval. Symbolically, suppose $x_1, x_2, \ldots\ldots, x_n$ is a random sample drawn from a population having a population mean 'μ' and variance 'σ^2'. The mean $\overline{x} = \frac{\sum x}{n}$ is an estimate for the corresponding population mean 'μ'. Generally, sample mean (is not equal to population mean (μ). Therefore, it is desirable to find out an interval in which value of the population mean μ may be accepted to lie and that interval is known as **confidence interval.** The limits of confidence interval are known as **confidence limit** of population mean 'μ'. Mostly, the confidence level is selected before examining the data. The commonly used confidence level is 95% confidence level and 99% confidence level.

Critical region

A region in the sample space S in which if the computed value of the test statistic lies, we reject the null hypothesis, are called the **critical region** or **rejected region.** When the rejection region consists of two regions each associated with probability $\alpha/2$, called **two tailed test.** On the other hand, when the rejection region consists of only one region, either on the right or left, associated with probability 'α', called **one tailed test.**

Degree of freedom

It is the number of independent observation used in the making of the statistic. In general, the number of degree of freedom is the total number of observations minus the number of independent constrains imposed on the observations. Thus, if k is the number of independent constraints in a set of data of n observations, then the degree of freedom will be (n - k). The number of degree of freedom for a statistic is usually denoted by 'λ'. That means,

$\lambda = n - k$

Procedure of Hypothesis Testing

Following steps can carry out hypothesis testing-

- Set up the null hypothesis (H_0)
- Select the level of significance
- Decide about an appropriate test statistic
- Find out degree of freedom
- Find the rejection region and locate the position of the computed test statistic in it. If the test statistic lies in the rejection region, H_0 is rejected otherwise it is not rejected and we conclude that the data do not provide sufficient evidence to cause rejection of null hypothesis.
- Finally, we conclude the testing problem with a statistical decision stating clearly the level of significance.

Tests of Significance

Tests of significance can be broadly classified into two categories:

(i) Large sample tests

(ii) Small sample tests

(i) Large sample tests: When the samples are of size n >30, almost all distributions are closely approximated by normal distribution, therefore normal distribution form the statistical basis of all the **large sample tests.** Thus, in the entire large sample test we compute the test statistic Z under H_0 where Z is a standard normal distribution with mean **zero** and variance **one.** Symbolically, $Z \sim N(0, 1)$. The test is usually performed at 5% and 1% level of significance in two tail test ($\alpha = 0.05$ and $\alpha = 0.01$) at which the critical values of Z are 1.96 and 2.58 respectively.

(ii) Small sample tests: When the samples are of size $n < 30$, the assumptions on which analysis of large samples is done generally do not hold well in case of **small samples.** In case of large samples, we had presumed that the random sampling distributions of statistics are approximately normal and further that the values obtained by sampling study are close to the population values and can be used in their place for the calculation of the standard error of the estimate. These assumptions do not hold well, if the size of the samples may or may not be normally distributed and similarly, it is not possible to substitute the mean or the standard deviation of a small sample in place of the parameter mean or standard deviation for the calculation of standard errors. Under such circumstances the analysis of small samples has to be done by techniques which are different from those applicable in case of large samples.

This distribution applicable to small samples was developed by **W. S. Gossett** who was employed by the Guinness & Son., Dublin bravery, Ireland. The t-distribution is commonly called **Student's t-distribution**. Symbolically, If $x_1, x_2, \ldots\ldots, x_n$ is a random sample of size n drawn from a normal population with unknown mean μ and variance σ^2. The t-statistic is defined as-

$$t = \frac{\overline{x} - \mu}{S/\sqrt{n}} \sim t_{(n-1)d.f.}$$

Condition for applying t-distribution

- The t-distribution depends on n, the size of the sample.
- The sample is drawn by simple random from a normal population.
- The sample size is small i.e. less than 30.
- Population variance is unknown.
- The limit of t-distribution is - ∞ to + ∞.

Chi-square distribution

It is a non-parametric test. Such non-parametric tests have assumed great importance in statistical analysis and statistical inference because they are easy to compute and can be used without making assumptions about parameters as they are distribution-free tests. It is a test which describes the magnitude of difference between observed frequencies and the frequencies expected under certain assumptions. The chi-square test was first used by **Prof. Karl Pearson** in the year 1900. The Chi-square test is denoted by the Greek letter 'c^2'. The quantity c^2 describes the magnitude of the discrepancy between theory and observation. Symbolically, Let $O_1, O_2, \ldots\ldots, O_n$ be a set of observed frequencies in classes 1, 2, 3,................,n and $E_1, E_2, \ldots\ldots, E_n$ be the corresponding expected frequencies, then a chi-square statistic for testing the agreement between the observed and expected frequencies is defined as-

$$Chi\text{-}square\ (\chi^2) = \Sigma\frac{(O\text{-}E)^2}{E} \sim \chi^2_{(n-1)d.f.}$$

Condition for applying Chi-square distribution

- The sample size is large i.e. more than 50.
- The cell frequencies should be independent.
- None of the class should have frequency less than 5, if it is so, then rearrangement or regrouping of the data is done.
- The range of c^2- test is from 0 to ∞.

- It has a uni-modal curve.
- The distribution of Chi-square lies in the first quadrant.

F- Distribution

The F-test was first developed by **Sir Ronald A. Fisher.** The F-statistic is simply a ratio of two variances. Variances are a measure of dispersion, or how far the data are scattered from the mean. Larger values represent greater dispersion. The object of the F-test is to find out whether the two independent estimates of population variance differ significantly or whether the two samples are drawn from the normal populations having the same variance. Suppose, $x_1, x_2, \ldots, x_{n1}$ be a random sample of size n_1 drawn from a normal population with mean μ_1 and variances σ_1^2 and another random sample $y_1, y_2, \ldots, y_{n2}$ of size n_2 drawn from a normal population with mean μ_2 and variance σ_2^2 respectively. Then, we define the F-distribution as-

$$F = \frac{S_1^2}{S_2^2} \sim F_{(n_1-1)(n_2-1)\, d.f.}$$

Where, $S_1^2 > S_2^2$

$$S_1^2 = \frac{\Sigma(x-\bar{x})^2}{(n_1-1)} \quad and \quad S_2^2 = \frac{\Sigma(y-\bar{y})^2}{(n_2-1)}$$

Here, the F-distribution has two degree of freedoms i.e. (n_1-1) degree of freedom for **numerator** and (n_2-1) degree of freedom for **denominator**.

Since F-distribution is based on the ratio of two variances, it is also known as **variance ratio distribution**.

Assumption in F-test

The F-distribution is based on the following assumptions:

- The values in each group should be normally distributed.
- The larger variance should always be placed in the numerator.
- The samples must be independent.

Analysis of Variance

The t-test is used for testing the hypothesis of equality of two normal population means when sample sizes are small. However, In this case we should have a technique by which significance of the difference amongst more than two sample means is carried out at the same time. The **analysis of variance (ANOVA)** technique enables us to perform this simultaneous test. The analysis of variance technique, developed by **R. A. Fisher** in 1920, is capable of fruitful application to a diversity of practical problems.

There is always variation in the data recorded during an experiment. This variation is called **total variation**. The total variation is the result of so many factors. These factors are of two types:

- Assignable factors
- Non-assignable factors

The **assignable factors** are those which can be identified or traced easily while the **non-assignable** factors are a large number of small independent factors which cannot be traced easily. These non-assignable factors are also known as **chance factors or experimental error**. The analysis of variance estimates the variation due to assignable factors and non-assignable factors separately and then compares them.

Thus, '**Analysis of Variance**' is a technique of partitioning or splitting the total variation i.e. assignable factors and the chance factors and then their comparison. The analysis of variance frequently referred to by the contraction ANOVA is a statistical technique specially designed to test whether the means of more than two sample means is carried out at the same time.

Assumptions used in Analysis of Variance

The following assumptions are made in the analysis of variance techniques;

- Samples are independently drawn
- Samples are randomly drawn from the population
- Samples are drawn from normally distributed populations
- Population variances are equal.

Types of Analysis of Variance: There are three types of analysis of variance-

- When there is only one assignable factor present in the body of experiment. Then it is called **one way classification** of analysis of variance.
- When there are two assignable factors present in the experimental data. Then it is called **two way classification** of analysis of variance.
- When there are three assignable factors present in the experimental data. Then it is called **three way classification** of analysis of variance.

The analysis of variance frequently referred to by the contraction ANOVA is a statistical technique specially designed to test whether the means of more than two sample means is carried out at the same time.

MCQ

Q.1. The probability of Type 1 error is referred as?

(a) 1-α (b) β

(c) **α** (d) 1-β

Q2. A statement made about a population for testing purpose is called?

(a) Test-Statistic (b) Level of Significance

(c) **Hypothesis** (d) Statistic

Q3. A statement whose validity is tested on the basis of a sample is called?

(a) Simple hypothesis (b) Composite hypothesis

(c) Null hypothesis (d) **Statistical hypothesis**

Q4. If the assumed hypothesis is tested for rejection considering it to be true is called?

(a) Simple hypothesis (b) Composite hypothesis

(c) **Null hypothesis** (d) Statistical hypothesis

Q5. The rejection probability of null hypothesis when it is true is called as?

(a) Level of rejection (b) Level of confidence

(c) **Level of significance** (d) Level of margin

Q6. If the null hypothesis is false then which of the following is accepted?

(a) Null hypothesis (b) Positive hypothesis

(c) Negative hypothesis (d) **Alternative hypothesis**

Q7. A hypothesis which defines the population distribution is called?

(a) **Simple hypothesis** (b) Statistical hypothesis

(c) Null hypothesis (d) Composite hypothesis

Q8. The point where the Null Hypothesis gets rejected is called as?

(a) **Critical value** (b) Rejection value

(c) Acceptance value (d) Significant value

Q9. If the critical region is evenly distributed then the test is referred as?

(a) One tailed (b) **Two tailed**

(c) Three tailed (d) Zero tailed

Bold means correct answer

Q10. The type of test is defined by which of the following?

(a) Null hypothesis (b) Simple hypothesis

(c) **Alternative hypothesis** (d) Composite hypothesis

Q11. Which of the following is defined as the rule or formula to test a null hypothesis?

(a) **Test statistic** (b) Population statistic

(c) Variance statistic (d) Null statistic

Q12. Consider a hypothesis where H_0 where $\phi_0 = 23$ against H_1 where $\phi_1 < 23$. The test is?

(a) Right tailed (b) **Left tailed**

(c) Center tailed (d) Cross tailed

Q13. Consider a hypothesis H_0 where $\phi_0 = 5$ against H_1 where $\phi_1 > 5$. The test is?

(a) **Right tailed** (b) Left tailed

(c) Center tailed (d) Cross tailed

Q14. The purpose of statistical inference is:

(a) To draw conclusions about the known value of population parameter

(b) To collect sample data and use them to formulate hypotheses about a population

(c) **To draw conclusions about populations from sample data**

(d) To draw conclusion about populations and then collect sample data to support the conclusions

Q15. Type-1 error occurs when?

(a) **We reject H_0 if it is true** (b) We reject H_0 if it is false

(c) We accept H_0 if it is true (d) We accept H_0 if it is false

Q16. Alternative hypothesis is also called as?

(a) Null hypothesis (b) **Research hypothesis**

(c) Simple hypothesis (d) Composite hypothesis

Q17. If a hypothesis is rejected at 0.6 level of significance then:

(a) It will be rejected at any level

(b) It must be rejected at 0.5 level

(c) **It may be rejected at 0.5 level**

(d) It cannot be rejected at 0.5 level

Hint: (c), If the hypothesis is rejected at 0.6 Level of Significance then p < 0.6. Hence p can be less than 0.5 also. Therefore, it may be rejected at 5 percent level of significance.

Q18. In a two tailed test when a null hypothesis is rejected for a true alternative hypothesis then it has:

(a) Type-I error
(b) Type-II error
(c) **No error**
(d) Many errors

Q19. In a hypothesis test, what does the p value signify?

(a) **Smallest level of significance for rejection of null hypothesis**

(b) Largest level of significance for rejection of null hypothesis

(c) Smallest level of significance for acceptance of null hypothesis

(d) Smallest level of significance for acceptance of null hypothesis

Q20. A null hypothesis has Level of Significance 9%. For what values of level of significances, it will be rejected?

(a) 0.99
(b) 0.099
(c) **0.009**
(d) 0.9

Hint: (c), The Level of Significance of Null Hypothesis is 0.09. Hence the hypothesis will be rejected at values less than 0.09.

Q21. Consider a trial of a criminal. If a type-1 error has occurred in the judgement then which of the following statement is true?

(a) a guilty person is set free

(b) **an innocent person is convicted**

(c) a guilty person is convicted

(d) an innocent person is set free

Hint: (b); Type-1 error occurs when the null hypothesis is true but rejected. Considering null hypothesis as innocent person and setting free as acceptance an innocent person getting convicted is a type-I error.

Q22. The range of level of significance lies between:

a) $-\infty$ and 0
b) $-\infty$ and ∞
c) 0 and ∞
d) **0 and 1**

Q23. The test statistic for a hypothesis testing is given by the formula:

(a) Sample-Population/Standard Error

(b) **Sample statistic-Parameter/Standard Error**

(c) Sample mean-Population mean/Population standard deviation

(d) Statistic-E(statistic)/Variance

Hint: (b); The Test Statistic of a hypothesis is given as the Sample statistic-Parameter/Standard Error.

The test statistic provides a basis for testing a null hypothesis.

Q24. The effect of rejection of a hypothesis with decrease in sample size:

(a) Decreases (b) **Increases**

(c) Remains constant (d) Fluctuates

Hint: (b); If n decreases then the value of level of significance of each sample α decreases. Hence, 1-α increases which are called the rejection of test sample increases.

Q25. The independent values in a set of values of a test are called as?

(a) **Degrees of freedom** (b) Test Statistic

(c) Level of Significance (d) Level of confidence

Q26. The choice of on-tailed test and two-tailed test depends upon:

(a) Null hypothesis (b) Simple hypothesis

(c) **Alternative hypothesis** (d) Composite hypothesis

Q27. Which one is an example of two-tailed test?

(a) $H_1: \mu = 0$ (b) $H_1: \mu < 0$

(c) $H_1: \mu > 0$ (d) $\mathbf{H_1: \mu \neq 0}$

Q28. A statistic on the basis of which a decision is made about the hypothesis of interest is called:

(a) **Test statistics** (b) Significance level

(c) Statement of hypothesis (d) Critical region

Q29. A hypothesis may be classified as

(a) Composite (b) Simple

(c) Null (d) **All of them**

Q30. The dividing point between the region where the null hypothesis is rejected and the region where it is not rejected is said to be:

(a) Acceptance region (b) Critical region
(c) **Critical value** (d) Significant region

Q31. A statement that is accepted if the sample data provide sufficient evidence that the null hypothesis is falls is called:

(a) Null hypothesis (b) Simple hypothesis
(c) Statistical hypothesis (d) **Alternative hypothesis**

Q32. The level of significance is the risk of:

(a) **Reject H_0 when H_0 is true** (b) Reject H_0 when H_1 is true
(c) Reject H_1 when H_1 is correct (d) None of these

Q33. The region of rejection of H_0 is called:

(a) **Critical region** (b) Test statistics
(c) Type-I error (d) Acceptance region

Q34. A deserving player is not selected in National team, it is an example of:

(a) **Type-I error** (b) Type-II error
(c) Correct decision (d) Sampling error

Q35. When critical regicn is located on both side of the curve, it is called:

(a) Hypothesis testing (b) **Hypothesis**
(c) Test statistic (d) Level of significance

Q36. The probability of rejecting a true hypothesis is called:

(a) Critical region (b) **Level of significance**
(c) Test statistics (d) Statement of hypothesis

Q37. A statement about the value of a population parameter is called:

(a) **Null hypothesis** (b) Simple hypothesis
(c) Alternative hypothesis (d) Composite hypothesis

Q38. A hypothesis that specifies all the values of parameter is called

(a) Null hypothesis (b) **Simple hypothesis**
(c) Statistical hypothesis (d) Alternative hypothesis

Q39. Which of these distributions is used for a testing hypothesis?

(a) Normal distribution (b) **Chi-Squared distribution**

(c) Gamma distribution (d) Poisson distribution

Q40. The hypothesis $\mu \leq 10$ are a:

(a) **Composite hypothesis** (b) Simple hypothesis

(c) Statistical hypothesis (d) Alternative hypothesis

Q41. The region of accepting of H_0 is called:

(a) Critical region (b) Test statistics

(c) Type-I error (d) **Acceptance region**

Q42. The probability of rejection a false H_0 is:

(a) Critical region (b) Level of significance

(c) Level of confidence (d) **Power of the test**

Q43. Type-I error will occur if an innocent person is:

(a) Not arrested (b) **Arrested by Police**

(c) Police care of him (d) None of these

Q44. If a hypothesis specifies the population distribution is called

(a) Composite hypothesis (b) **Simple hypothesis**

(c) Alternative hypothesis (d) None of these

Q45. When critical region is located on both side of the curve, it is called:

(a) Left tailed test (b) Right tailed test

(c) **Two-tailed test** (d) One-tailed test

Q46. If population standard deviation is known and $n > 30$ then appropriate test statistics mean comparison is:

(a) **Z-test** (b) t-test

(c) Chi-square test (d) F -test

Q47. What is the assumption made for performing the hypothesis test with t -distribution?

(a) Distribution is non-symmetric

(b) Distribution has more than one modal class

(c) **Distribution follows a normal distribution**

(d) Distribution has a constant variance

Q48. Any statement whose validity is tested on the basis of a sample is called:

(a) Null hypothesis
(b) Simple hypothesis
(c) Alternative hypothesis
(d) **Statistical hypothesis**

Q49. A quantitative statement about a population is called:

(a) Null hypothesis
(b) Simple hypothesis
(c) **Statistical hypothesis**
(d) Research hypothesis

Q50. Which of the following distributions is used to compare two variances?

(a) Z - distribution
(b) t - distribution
(c) Chi-square distribution
(d) **F - distribution**

Q51. In a Z-test, the number of degree of freedom is:

(a) 1
(b) n-1
(c) n -2
(d) **None of these**

Q52. If $\alpha = 0.05\%$, the value of one-tailed Z test will be:

(a) **1.64**
(b) 1.96
(c) 2.38
(d) 2.58

Q53. Which hypothesis is always in an inequality form?

(a) Null hypothesis
(b) Simple hypothesis
(c) Composite hypothesis
(d) **Alternative hypothesis**

Q54. The degree of freedom for paired t-test based on n pairs of observations is:

(a) 2n-1
(b) 2n
(c) **n-1**
(d) n

Q55. Which of the following represents the Confidence coefficient?

(a) α
(b) **1-α**
(c) β
(d) $1 - \beta$

Q56. Test of hypothesis H_0: $\mu = 50$ against H_1: $\mu > 50$ leads to:

(a) Two-tailed test
(b) Left-tailed test
(c) **Right –tailed test**
(d) Difficult to tell

Q57. Test of hypothesis H_0: $\mu = 50$ against H_1: $\mu < 50$ leads to:

(a) Two-tailed test (b) **Left-tailed test**

(c) Right –tailed test (d) Difficult to tell

Q58. Test of hypothesis H_0: $\mu = 50$ against H_1: $\mu \neq 50$ leads to:

(a) **Two-tailed test** (b) Left-tailed test

(c) Right –tailed test (d) Difficult to tell

Q59. The range of test statistic-Z is:

(a) -1 to +1 (b) 0 to ∞

(c) **- ∞ to + ∞** (d) 0 to 1

Q60. The range of test statistic-t is:

(a) -1 to +1 (b) 0 to ∞

(c) **- ∞ to + ∞** (d) 0 to 1

Q61. A failing student is passed by an examiner, it is an example of:

(a) Type-I error (b) **Type-II error**

(c) Unbiased decision (d) Difficult to tell

Q62. A passing student is failed by an examiner, it is an example of:

(a) **Type-I error** (b) Type-II error

(c) Unbiased decision (d) Difficult to tell

Q63. $1 - \alpha$ is also called

(a) Power of the test (b) Size of the test

(c) Level of significance (d) **Confidence coefficient**

Q64. $1 - \alpha$ is the probability associated with:

(a) Power of the test (b) Size of the test

(c) Level of significance (d) **Level of confidence**

Q65. Area of the rejection region depends on:

(a) **Size of α** (b) Size of β

(c) Number of values (d) Test-statistic

Q66. Size of critical region is known as:

(a) β (b) 1-β

(c) Critical value (d) **Size of the test**

Q67. A null hypothesis is rejected if the value of a test statistic lies in:

(a) Acceptance region **(b) Rejection region**

(c) Significance level (d) None of these

Q68. Level of significance is also called:

(a) Size of the test (b) Power of the test

(c) Level of significance (d) Level of confidence

Q69. Level of significance α lies between:

(a) 0 to +1 (b) - 1 to +1

(c) $-\infty$ to $+\infty$ (d) 0 to $+\infty$

Q70. In testing hypothesis $\alpha + \beta$ is always equal to:

(a) Zero (b) One

(c) Two **(d) Difficult to tell**

Q71. If the magnitude of calculated value of t is less than the tabulated value of t has H_1 is two-tailed, we should:

(a) Reject H_0 (b) Accept H_1

(c) Not reject H_0 (d) Difficult to tell

Q72. The chance of rejecting a true hypothesis decreases when sample size is:

(a) Increased (b) Decreased

(c) Constant (d) None of these

Q73. The equality condition always appears in:

(a) Simple hypothesis (b) Null hypothesis

(c) Alternative hypothesis **(d) Both (a) & (b)**

Q74. Suppose that the null hypothesis is true and it is rejected, is known as:

(a) A type-I error and its probability is α

(b) A type-I error and its probability is β

(c) A type-II error and its probability is α

(d) A type-II error and its probability is β

Q75. Which of the following is composite hypothesis?

(a) $\mu = \mu_0$ (b) $\mu \geq \mu_0$

(c) $\mu \leq \mu_0$ (d) $\mu \neq \mu_0$

Q76. The power of the test is equal to:

(a) **1 - β** (b) β

(c) α (d) 1 - α

Q77. Student's t-test is applicable only when:

(a) n=30 and σ is known (b) n=30 and σ is unknown

(c) n >30 and σ is unknown (d) **n ≤ 30 and σ is known**

Q78. Student's t-statistic is applicable in case of:

(a) Small samples (b) Equal number of samples

(c) Unequal number of samples (d) **All of them**

Q79. Paired t-test is applicable when the observations in the two samples are:

(a) Correlation (b) Paired

(c) Equal in number (d) **All of them**

Q80. In an unpaired t-test with sample sizes n_1 and n_2, the degree of freedom is:

(a) n-1 (b) $n_1 + n_2$

(c) **$n_1 + n_2$ -2** (d) $n_1 + n_2$ -1

Q81. In an unpaired t-test with sample sizes n_1 = 14 and n_2 = 14, the value of tabulated t should be obtained for:

(a) 13 degree of freedom (b) 14 degree of freedom

(c) **26 degree of freedom** (d) 27 degree of freedom

Q82. In analyzing the results of an experiment involving eight paired samples, tabulated t should be obtained:

(a) **7 degree of freedom** (b) 8 degree of freedom

(c) 14 degree of freedom (d) 15 degree of freedom

Q83. The mean difference between 25 paired observations is 16 and the standard deviation of differences is 10. The value of statistic - t is:

(a) 24 (b) 15

(c) **8** (d) 1.6

Q84. Statistic - t is defined as deviation of sample mean from population mean μ expressed in terms of:

(a) Coefficient of standard deviation (b) Coefficient of variation

(c) Standard deviation (d) **Standard error**

Q85. Student's t-distribution has (n-1) d.f. when all the n observations in the sample are:

(a) **Independent** (b) Dependent

(c) Minimum (d) Maximum

Q86. The number of independent values in a set of values is called:

(a) **Degree of freedom** (b) Power of the test

(c) Level of significance (d) Level of confidence

Q87. When σ is known, the hypothesis about population mean is tested by:

(a) **Z-test** (b) t-test

(c) Chi-square test (d) F-test

Q88. A t-test sample has 7 pairs of samples. The distribution should contain:

(a) 5 degrees of freedom (b) **6 degrees of freedom**

(c) 13 degrees of freedom (d) 14 degrees of freedom

Q89. An advertising agency wants to test the hypothesis that the proportion of adults in India who read a Sunday magazine is 30%. The null hypothesis is that the proportion reading the Sunday magazine is:

(a) Less than 30% (b) More than 30%

(c) **Equal to 30%** (d) Different from 30%

Q90. Given $\mu_0 = 130$, $\mu = 150$, $\sigma = 25$ and $n = 4$. What test statistic is appropriate?

(a) **Z-test** (b) t-test

(c) Chi-square test (d) F-test

Q91. Given H_0: $\mu = \mu_0$, H_1: $\mu \neq \mu_0$, $\alpha = 0.01$ and we reject H_0; the absolute value of Z-statistic must have equaled or been beyond what value?

(a) 1.64 (b) 1.96

(c) 2.33 (d) **2.58**

Q92. What is the probability of a type II error when $\alpha = 0.05$?

(a) 0.05 (b) 0.01

(c) 0.001 (d) **Cannot be determine without more information**

Q93. The *t*- distributions are:

(a) Skewed
(b) Same as normal distribution
(c) **Symmetrical**
(d) None of these

Q94. The result of a statistical test, denoted p, shall be interpreted as follows:

(a) H_0 is rejected if $p > 0.05$
(b) **H_0 is rejected if $p < 0.05$**
(c) H_0 is accepted if $p < 0.05$
(d) H_1 is rejected if $p > 0.05$

Q95. For *t*-distribution, increasing the sample size, the effect will be on:

(a) The t-ratio
(b) Degree of freedom
(c) Standard error of means
(d) **All of them**

Q96. Chi-square is used to analyze:

(a) Rank
(b) **Frequencies**
(c) Score
(d) Correlation

Q97. Using a goodness of fit test, we can access whether a set of obtained frequencies differ from a set of ....frequencies:

(a) Actual
(b) Predicted
(c) **Expected**
(d) Mean

Q98. For the chi-square test to be effective, the expected value for each cell in the contingency table has to be at least:

(a) 3
(b) 5
(c) 8
(d) 10

Hint: (b); For the chi-square test to give meaningful results the expected frequency for each cell in the 2x2 contingency table is required to be at least 5.

Q99. What is the shape of Chi-square distribution?

(a) Nearly symmetrical
(b) **Skewed to the right**
(c) Skewed to the left
(d) More than one peak (Bimodal)

Q100.Analysis of Variance (ANOVA) is a test for equality of:

(a) Proportions
(b) Variances
(c) **Means**
(d) Sample mean with population mean

Q101.If the calculated value of t value is greater than the critical t value, then:

(a) Do not reject null hypothesis (b) **Reject null hypothesis**

(c) Again estimate (d) Can't say anything

Q102.A human resources manager is interested to know if the mean number of sick days taken per year differs between four specific jobs within a company. What test should the manager use to examine the problem:

(a) **One factor ANOVA**

(b) Two-factor ANOVA

(c) Goodness of fit for Chi-square test

(d) Comparison of two means for t-test

Q103. In t-distribution, if the sample size increases, it become close to:

(a) F-distribution (b) Chi-square distribution

(c) Binomial distribution (d) **Z- distribution**

Q104.The degree of freedom computed for testing the significance in the B-P test is equal to:

(a) 1 (b) 2

(c) k-1 (d) k-2

Q105.For testing of independence, the best suited test to apply is:

(a) Z-distribution (b) t-distribution

(c) F-distribution (d) **Chi-square distribution**

Q106.The expected value of F-distribution is equal to:

(a) k_2/k_2-1 (b) **k_2/k_2-2**

(c) $k_2 \times (k_2-1)$ (d) $k_2 \times (k_2-2)$

Q107.If the calculated value of Chi-square value is greater than the critical Chi-square value, then:

(a) **Reject null hypothesis** (b) Do not reject null hypothesis

(c) Again estimate (d) Can't say anything

Q108.In a distribution the variance is not defined and known, the best suited test to apply is said to be:

(a) **t-test** (b) F-test

(c) Chi-square test (d) Binomial test

Q109.While drawing a hypothesis, encountered errors can be classified into:

(a) One types (b) **Two types**

(c) Three types (d) Four types

Q110.Large confidence interval would lead to:

(a) Significant testing (b) Insignificant testing

(c) **Accept null hypothesis** (d) Reject null hypothesis

Q111.Skewness of Chi-square distribution depends on its:

(a) Estimated value (b) Expected value

(c) **Degree of freedom** (d) variance

Q112.In F-test, the mean sum of square for estimated sum of square is computer through:

(a) ESS/(n-1) (b) **ESS/(k-1)**

(c) ESS/(n-k) (d) ESS/(n-2)

Q113.The major applications of Chi-square distribution can be classified into:

(a) **Three ways** (b) Four ways

(c) Five ways (d) Six ways

Q114.The probability of making a type-I error when null hypothesis is true as equality is determined by:

(a) Level of acceptance (b) Level of rejection

(c) **Level of significance** (d) Level of insignificance

Q115.A statistic whose value helps determine whether a null hypothesis should be rejected is called:

(a) Level of significance (b) **Test statistic**

(c) Level of acceptance (d) Level of rejection

Q116.A method for comparing more than two study groups for equality of means is computed through:

(a) t-test (b) Chi-square test

(c) Z-test (d) **ANOVA**

Q117.The level of significance is denoted by:

(a) $\boldsymbol{\alpha}$ (b) β

(c) λ (d) μ

Q118. The decision value to reject the null hypothesis in case of a right tail test is said to be:

(a) Calculated t is less than negative of critical t-value

(b) **Calculated t must be less than critical t-value**

(c) Calculated t must be greater than critical t-value

(d) None of these

Q119.In F-test, the mean sum of square of total sum of square equivalent to:

(a) **TSS/(n-1)** (b) TSS/(k-1)

(c) TSS/(n-k) (d) TSS/(n-2)

Q120.The F-distribution reaches, but never touches the:

(a) Vertical axis (b) **Horizontal axis**

(c) Level of significance (d) Critical region

Q121.The Chi-square distribution is skewed towards:

(a) Left (b) **Right**

(c) Symmetrical (d) None of these

Q122.The F-distribution has minimum value defined to be zero and its maximum value is:

(a) 1 (b) 100

(c) 1000 (d) **Not defined**

Q123.The value range of Chi-square distribution falls between:

(a) 0 to 1 (b) -1 to +1

(c) **0 to ∞** (d) $-\infty$ to $+\infty$

Q124.If the number of observations is greater than 30, the best suited test is to apply:

(a) **Z-test** (b) t-test

(c) Chi-square test (d) F-test

Q125.In F-test, degree of freedom for total sum of square is taken as:

(a) 1 (b) k-1

(c) **n-1** (d) n-k

Q126.In F-test, degree of freedom for error sum of square is taken as:

(a) 1 (b) k-1

(c) n-1 (d) **n-k**

Q127.As the degree of freedom in F-distribution increases, it becomes closer to:

(a) Standard distribution (b) Chi-square distribution

(c) **Symmetrical** (d) Asymmetrical

Q128.The square of t-distribution with k degree of freedom is equal to F-distribution having degree of freedom equal to:

(a) 1, 1 (b) 1, k-1

(c) **1, k** (d) 1, k-2

Q129.The Z-test is measurable only when:

(a) Variance is known

(b) Number of observations are greater than 30

(c) Number of observations are less than 30

(d) **Both (a) & (b)**

Q130.When the distribution the variance is defined and known, the best suited test to apply is said to be:

(a) Z-test (b) t-test

(c) Chi-square test (d) F-test

Q131.For F-distribution application, the minimum amount of degree of freedom required is:

(a) 1 (b) **2**

(c) 3 (d) 4

Q132.The t-test assesses the means of two groups that are:

(a) Similar to each other (b) **Different for each other**

(c) Independent to each other (d) Dependent to each other

Q133.F-distribution has general application that can be categorized into:

(a) 2 types (b) **3 types**

(c) 4 types (d) 5 types

Q134.One of the crucial assumptions of ANOVA is:

(a) Homoscedascity

(b) Multicollinearity

(c) **Homogeneity of covariance**

(d) Estimation of partial coefficient

Q135.If the calculated F-value is greater than the critical F-value then:

(a) Accept null hypothesis (b) **Reject null hypothesis**

(c) Estimate again (d) None of these

Q136.The level of significance output lies in value ranging in:

(a) **0 to 1** (b) -1 to +1

(c) 0 to ∞ (d) -∞ to +∞

Q137.The generated output of F-distribution is always:

(a) Negative (b) **Positive**

(c) Zero (d) Infinite

Q138.A distribution of sum of k independent normal random variables and k degree of freedom is known to be:

(a) Z-distribution (b) t- distribution

(c) **Chi-square distribution** (d) F- distribution

Q139.If degree of freedom is greater than 15 in t-test and α is equal to 0.05 and calculated t-value is greater than 5 then:

(a) Accept null hypothesis (b) **Reject null hypothesis**

(c) Estimate again (d) None of these

Q140.If degree of freedom is greater than 15 in Chi-square test and α is equal to 0.05 and calculated Chi-square value is greater than 15 then:

(a) Accept null hypothesis (b) **Reject null hypothesis**

(c) Estimate again (d) None of these

Q141.If degree of freedom is 3 for numerator and 5 for denominator in F- test and α is equal to 0.05 and calculated F-value is 0.89 then:

(a) **Accept null hypothesis** (b) Reject null hypothesis

(c) Estimate again (d) None of these

Q142. When we reject the null hypothesis it indicates that the result is statistically:

(a) Biased (b) Unbiased

(c) **Significant** (d) Non-significant

Q143. For testing of independence, the parameters used in Chi-square distribution is observed value and:

(a) Estimated value (b) Error value

(c) **Expected value** (d) Mean value

Q144. For estimated sum of square for F-test, the degree of freedom provided is of:

(a) n (b) n-1

(c) **k-1** (d) n-k

Q145. The mean sum of square of residual sum of square in F-test is equivalent to:

(a) RSS/(k-1) (b) RSS/(n-1)

(c) **RSS/(n-k)** (d) RSS/(n-2)

Q146. For t-test calculation and analysis, the population mean is replaced with:

(a) Sample variance (b) Sample median

(c) **Sample mean** (d) Sample mode

Q147. Unpaired t-test is used when:

(a) Observations in the two samples are paired

(b) Observations in the two samples are dependent

(c) **Observations in the two samples are independent**

(d) Observations in the two samples are the same

Q148. In large sample size, for testing H_0: $\mu = \mu_0$ against H_1: $\mu < \mu_{0,}$ the critical region for $\alpha = 0.05$ is:

(a) $|Z| \leq 1.96$ (b) $|Z| > 1.96$

(c) $\mathbf{|Z| < -1.645}$ (d) $|Z| > 1.645$

Q149. A large sample is drawn from a population to test H_0: $\mu = \mu_0$ against H_1: $\mu \neq \mu_{0,}$ the critical region $\alpha = 0.10$ is:

(a) $Z > 1.645$ (b) $Z < -1.645$

(c) $\mathbf{|Z| > 1.645}$ (d) $|Z| < 1.645$

Q150.The test statistic $F = S_1^2/S_2^2$ is used for testing:

(a) $H_0: \mu_1 = \mu_2$ (b) $H_0: \mu_1 \neq \mu_2$

(c) **$H_0: \sigma_1^2 = \sigma_2^2$** (d) $H_0: \sigma^2 = \sigma_0^2$

Q151.The number of degree of freedom for a contingency table of order 4 x 5 is:

(a) **12** (b) 15

(c) 16 (d) 20

Q152.When sample size is small, the test statistic $t = |r| / S.E(r)$ is used for testing:

(a) **$H_0: \rho = 0$** (b) $H_0: \rho \neq 0$

(c) $H_0: \beta = 0$ (d) $H_0: \beta \neq 0$

Q153.When sample size is small, the test statistic $t = |b_{yx}| / S.E.(b_{yx})$ is used for testing:

(a) $H_0: \rho = 0$ (b) $H_0: \rho \neq 0$

(c) **$H_0: \beta = 0$** (d) $H_0: \beta \neq 0$

Q154.Student's t-distribution was discovered by:

(a) R. A Fisher (b) Karl Pearson

(c) **W.S. Gossett** (d) G.W. Snedecor

Q155.In a contingency table, degree of freedom is determined by:

(a) (n – 1) (b) **(r – 1)(c – 1)**

(c) (r + 1)(c – 1) (d) (r – 1)(c + 1)

Q156.Analysis of variance technique was developed by:

(a) **R. A Fisher** (b) Karl Pearson

(c) W.S. Gossett (d) G.W. Snedecor

Q157.The samples are of unequal sizes then analysis of variance is:

(a) Same (b) **To be used**

(c) Not used (d) Cant's say anything

Q158.In analysis of variance technique, the degree of freedom for sum of square within samples is:

(a) (k – 1) (b) (n – 1)

(c) **(n – k)** (d) None of these

Q159.In analysis of variance technique, the degree of freedom for sum of square between samples is:

(a) **(k – 1)** (b) (n – 1)

(c) (n – k) (d) None of these

Q160.In usual notations, in two-way analysis of variance, the degree of freedom for sum of square within rows and columns is:

(a) (k – 1) (b) (n – 1)

(c) (nk – k – m) (d) **(nk – k – m +1)**

Q161.In usual notations, in two-way analysis of variance, the degree of freedom for sum of square between rows is:

(a) **(k – 1)** (b) (n – 1)

(c) (nk – k – m) (d) (nk – k – m +1)

Q162.When sample size is large, the test statistic is $\chi^2 = \frac{\Sigma(x-\overline{x})^2}{\sigma_0^2}$ used for testing:

(a) $\mathbf{H_0: \sigma^2 = \sigma_0^2}$ (b) $H_0: \sigma^2 \neq \sigma_0^2$

(c) $H_0: \sigma_1^2 = \sigma_2^2$ (d) $H_0: \sigma_1^2 \neq \sigma_2^2$

Q163.In usual notations, in two-way analysis of variance, the degree of freedom for sum of square between columns is:

(a) (k – 1) (b) **(n – 1)**

(c) (nk – k – m) (d) (nk – k – m +1)

Q164.In usual notations, in three-way analysis of variance, the degree of freedom for sum of square between rows is:

(a) **(k – 1)** (b) (n – 1)

(c) (nk – k – m) (d) (nk – k – m +1)

Q165.What is the mean of a Chi-square distribution with 5 degrees of freedom?

(a) 4 (b) **5**

(c) 6 (d) 7

Hint: By the property of Chi Square distribution, the mean corresponds to the number of degrees of freedom. Therefore, Degree of freedom = 5. Hence, Mean = 5

Q166.The t-distribution is symmetric around it's:

(a) Error term (b) Variance

(c) **Mean value** (d) None of these

Q167.In ANOVA with 5 groups and a total sample of 45, the computer F-statistic is 2.32. In this case, the p-value is:

(a) Less than 0.05 (b) **Greater than 0.05**

(c) Equal to 0.05 (d) Cannot tell

Q168.In multicollinearity, due to high variance and standard error, t-test will become statistically:

(a) Very low (b) Very high

(c) Significant (d) **Insignificant**

Q169.The Student's t-test is:

(a) A test for comparing two variances

(b) **A test for comparing two means**

(c) A test for comparing more than two variances

(d) A test for comparing more than two means

Q170.The F-test is:

(a) **A test for comparing two variances**

(b) A test for comparing two means

(c) A test for comparing more than two variances

(d) A test for comparing more than two means

Q171.Variance of a standard normal distribution is:

(a) Complex (b) Undefined

(c) **Fixed** (d) Not fixed

Q172.To assess the reliability of mean differences between multiple groups, the best suited test to apply is:

(a) Z – distribution (b) **t – distribution**

(c) Chi-square distribution (d) F - distribution

Q173. Two samples of sizes 22 and 28 are independently drawn from two normal populations, where the unknown variances are assumed to be equal. The number of degrees of freedom for the equal variances t- test statistic is:

(a) 50 (b) 49

(c) **48** (d) 6

Q174. The shape of the normal curve is:

(a) Flat (b) Circular

(c) Skewed (d) **Bell shaped**

Q175. For a standard normal variate, the value of mean is:

(a) 1 (b) 0.5

(c) **0** (d) ∞

Q176. The area under a standard normal curve is:

(a) **1** (b) 0.5

(c) 0 (d) ∞

Q177. The standard normal curve is symmetric about the value:

(a) 1 (b) 0.5

(c) **0** (d) ∞

Q178. For a standard normal variate, the value of standard deviation is:

(a) **1** (b) 0.5

(c) 0 (d) ∞

Q179. Skewness of normal distribution is:

(a) Positive (b) Negative

(c) **Zero** (d) One

Q180. For a normal distribution, the relationship between mean, median and mode is:

(a) Mean < Median < Mode (b) Mean > Median > Mode

(c) **Mean = Median = Mode** (d) Mean = 3 Median – 2 Mode

Q181. The shape of the normal curve depends on its:

(a) Correlation (b) Regression

(c) Mean deviation (d) **Standard deviation**

Q182.The formula to calculate standard normal variable is:

(a) $(X - \sigma)/ \mu$ (b) $(\sigma - X)/ \mu$

(c) **$(X - \mu)/ \sigma$** (d) $(\mu - X)/ \sigma$

Q183.If the value of X is less than μ of standard normal probability distribution then the:

(a) Z-statistic is positive (b) **Z-statistic is negative**

(c) f(x) will be prime number (d) f(x) will be even number

Q184.Which of the following is true for the normal curve?

(a) Bi-model (b) Positive skewed

(c) **Symmetrical** (d) Flat

Q185.In a normal probability distribution of a continuous random variable, the value of standard deviation is:

(a) Less than zero (b) **Greater than zero**

(c) Zero (d) None of these

Q186.If $N \sim (\mu, \sigma^2)$, the points of inflation of normal distribution are:

(a) μ (b) σ

(c) $\mu \pm \sigma$ (d) σ^2

Q187.In a normal probability distribution, the value of mean deviation is approximately equal to:

(a) $1\sigma/5$ (b) $3\sigma/5$

(c) **$4\sigma/5$** (d) $2\sigma/3$

Q188.Chi-squared is computed by first squaring the difference between:

(a) Observed frequencies and the total sample size

(b) Observed frequencies and observed percentages

(c) **Observed frequencies and Expected frequencies**

(d) Expected values and observed percentages

Q189.The range of Chi-square is:

(a) -1 to $+1$ (b) 0 to 1

(c) **0 to ∞** (d) $-\infty$ to $+\infty$

Q190.For a 3 x 4 Contingency table, the degree of freedom is:

(a) 12 (b) 11

(c) 8 (d) **6**

Q191.Z –test will be used when:

(a) Sample size is greater than 30

(b) Population variance is not known

(c) Population variance is known

(d) **All of them**

Q192.t –test will be used when:

(a) Sample size is greater than 30 (b) **Sample size is less than 30**

(c) Population variance is known (d) None of these

Q193.Which one of the following is a parametric test?

(a) Z – test (b) t – test

(c) F – test (d) **All of them**

Q194.Power of a test is denoted by:

(a) α (b) $1 - \alpha$

(c) β (d) $\mathbf{1 - \beta}$

Q195.Critical value for testing H_0: $\mu = \mu_0$ against H_1: $\mu \neq \mu_0$ at 5% level when n = 48 is:

(a) 1.645 (b) **1.96**

(c) 2.145 (d) 2.58

Q196.Critical value for testing H_0: $\mu = \mu_0$ against H_1: $\mu \neq \mu_0$ at 5% level when n = 18 is:

(a) 1.645 (b) 1.96

(c) **2.145** (d) 2.58

Q197.Critical value for testing H_0: $\mu = \mu_0$ against H_1: $\mu > \mu_0$ at 5% level when n = 18 is:

(a) **1.761** (b) 1.96

(c) 2.145 (d) 2.58

Q198.In the test for proportion H_0: $p_1 = p_2$ against H_1: $p_1 \neq p_2$, the best critical region is given by:

(a) $Z < Z_\alpha$ (b) $Z > Z_\alpha$

(c) $Z > -Z_\alpha$ (d) **None of these**

Q199.To test if the given normal population has a specified variance we make use of:

(a) Z-test (b) t-test

(c) **Chi-square test** (d) F-test

Q200.Test statistic corresponding to large sample approximation of median test follows:

(a) Z - distribution (b) t- distribution

(c) **Chi-square distribution** (d) F- distribution

Q201.If a hypothesis is rejected at 5%:

(a) **It must be rejected at 10%** (b) It must be rejected at 2%

(c) Both (a) & (b) are correct (d) All the statements are wrong

Q202.What does ANOVA stands for:

(a) Assessment of variability (b) Assessment of vulnerability

(c) Analysis of value (d) **Analysis of variance**

Q203.What must you include when reporting an ANOVA?

(a) Means (b) F – statistic

(c) Degree of freedom (d) **All of them**

Q204.A Z-test is conducted for H 0: $\mu = 9.0$ vs. H a: $\mu \neq 9.0$. The test statistic is –1.44. Correct conclusions are:

(a) Reject at both 5% and 10% level of significance

(b) Reject at 5% but not at 10%

(c) **Reject 10% but not at 5%**

(d) None of the above

Q205.For the application of chi-square test, the expected frequency in each cell should be:

(a) At least Three (b) Four

(c) **At least five** (a) None of the above

Q206. Which p value indicates the most statistically significant results:

(a) 0.006 (b) 0.005

(c) 0.005 (d) **0.05**

Q207. If a hypothesis is rejected at 5 percent:

(a) **It must be rejected at 10%** (b) It must be rejected at 2%

(c) Both (a) & (b) are correct (d) All the statements are wrong

Q208. For a P value to be significant it must be

(a) **Less than 0.05** (b) More than 0.05

(c) Less than 50% (d) More than 50%

Q209. A test is conducted for H_0: $\mu = 34$, with $\sigma = 5$. A sample of size 100 is selected. The standard error of the sampling distribution is:

(a) 0.05 (b) 0.34

(c) **0.5** (d) 5

Q210. A test is conducted for H_0: $\mu = 20$, with $\sigma = 4$. A sample of size 36 has a mean of 21.40. The test statistic is:

(a) 0.21 (b) 0.35

(c) **2.1** (d) 12.6

Q211. If a 99% confidence interval is constructed to estimate μ with a known σ for two tail, the correct critical values for Z **are**:

(a) 1.65 (b) 1.96

(c) 2.33 (d) **2.58**

Q212. The critical value of z for a two tail 90% confidence interval is:

(a) **1.65** (b) 1.96

(c) 2.33 (d) 2.58

Q213. Two random samples have been selected and a two sample z□test for the difference in population means is conducted with H_0: $\mu_1 = \mu_2$ vs. H_1: $\mu_1 \neq \mu_2$. The results are $\sigma_1 = 1.4$, $x_1 = 6.8$, $n_1 = 40$, $\sigma_2 = 1.6$, $x_2 = 7$, and $n_2 = 44$. The standard error of the sampling distribution is:

(a) – 0.431 (b) – 0.462

(c) **– 0.611** (d) – 0.749

Q214.In testing a hypothesis about two population means, it the t distribution is used, which of the following assumptions is required?

(a) The standard deviations are not the same

(b) The sample sizes are equal

(c) **Both populations are normally distributed**

(d) Both population means are the same

Q215.In analysis of variance, which form of variance must be estimated?

(a) Between group variance

(b) Within group variance

(c) Difference of within group variance to between group variance

(d) **Both (a) and (b)**

Q216.If the true means of the samples are equal, then MS(between)/MS(within) should be:

(a) Close to – 1.00 (b) **Close to 1.00**

(c) Negative value between 0 and -1 (d) More than 1.00

Q217.Following information from ANOVA table is given to you. Using this information calculate the F-ratio: Degree of freedom for Between sample is 3, Total degree of freedom of Between and Within is 29, Total Sum of Squares of Between and Within sample is 57, Sum of Squares of Within sample is 35:

(a) 4.45 (b) **5.45**

(c) 6.45 (d) 7.45

Q218.Muscle strength was measured on the right and left hands of a random sample of 15 left-handed men and the difference (left - right) was found. The alternative hypothesis is one-sided (left hand stronger). The resulting t-statistic was 1.80:

(a) **Degree of freedom = 14, the null hypothesis can be rejected**

(b) Degree of freedom = 14, the null hypothesis cannot be rejected

(c) Degree of freedom = 28, the null hypothesis can be rejected

(d) Degree of freedom = 28, the null hypothesis cannot be rejected

Q219. What is listed in the first column of an ANOVA table?

(a) Degree of freedom (b) Sum of squares

(c) **Source of variation** (d) Mean sum of square

Q220. A summary of the experiment that includes weather hypothesis was right or wrong:

(a) Independent variable (b) Dependent variable

(c) Hypothesis (d) **Conclusion**

Q221. The part of the experiment that does not contain the independent variables used the comparison:

(a) Data (b) Experimental groups

(c) **Control group** (d) Result

Q222. What are all the things that are kept the same in an experiment?

(a) Data (b) Controls

(c) **Constant** (d) Independent variables

Q223. What is the part of the experiment that contains the independent variable?

(a) Independent variables (b) Dependent variables

(c) **Experimental group** (d) Control group

Q224. An organized procedure to study something under controlled conditions is:

(a) Field study (b) Model

(c) **Experiment** (d) Observation

Q225. A numbered step by step description that begins with a verb and tells exactly how the experiment is to be done:

(a) Problem (b) **Procedures**

(c) Results (d) Materials

13

Design of Experiment

In 1935, **Prof. Ronald A. Fisher** laid the foundation for the subject in his monumental work entitled "The Designs of Experiments". Experimental designs concern the arranging of treatments in such a manner that the inferences and conclusions regarding the effects of these treatments can be easily done and their reliability measured. Experiments are made with a view to find the validity of a particular hypothesis and to have an idea about the extent of the reliability that can be placed on a particular conclusion arrived. The selection of the design will have a very great bearing on the accuracy of the ultimate results. By a random selection of experimental units it is possible to remove the ambiguity about the casual interpretation of the observed associations. Random sampling is the most essential ingredient of all experimental designs. Besides, there are many devices for increasing the precision of the inference and the calculations.

To understand design of experiments, let us first get acquainted with some important terms and concepts associated with design of experiments:

- **Experiment**: Experiment is a scientifically planned method. The experiment is conducted draw a valid conclusion about a particular problem. The conclusion is based on statistically observation.
- **Treatment:** The object or the procedures under comparison in an experiment are known as **treatment. For Example:** Fertilizers, varieties, cultivation practices, irrigation level **in case of agricultural experiments** and drugs, breeds, farms, etc. **in case of animal experiments**.
- **Yield:** The response of the treatment is measured by some indicator such as crop production, milk production, body temperature, etc. Such an indicator is called **yield**. The treatments are applied to some units such as field plots, sample of cows, sample of patients, etc. and the effect on the yield is observed.
- **Experimental material:** The experimental material is nothing but a set of experimental units. In other words, the material which is used in experiment is known as experimental material. **For Example:** A piece of land *i.e.* Agricultural field in **agricultural experiments** and a group of cattle *i.e.* animals **in animal experiments.**

- **Experimental unit:** A unit to which one treatment applied is called **experimental unit**. In other words, the ultimate part of the experimental material is known as **experimental unit**. It is the smallest division of an experimental material to which the treatment applied and on which the variable under study is measured. **For Example:** A small plot of agricultural field in **agricultural experiment** and an animals or a bird **in animal experiment** or in the field of medicine it may be a patient.

- **Experimental Error:** The variation due to environmental factor or due to uncontrolled factor is called **experimental error**. In other words, it is the difference between a measurement and the true value or between two measured values. Experimental error, itself, is measured by its accuracy and precision. The precision of an estimate is inversely proportional to the variance of that estimate. Thus, precision can be increased by reducing the variance of that estimate. That means;

$$Precision \propto \frac{1}{Variance}$$

Whereas, variance is directly proportional to the standard error of that estimate

Variance $\propto$ Standard Error (S.E.)

Now, for the success our experiment, the standard error (S.E.) can be reduced by-

- Improving the sampling techniques; or
- Increasing the number of units in the experiment.
- **Block:** When the experimental material is heterogeneous then in order to increase the precision of our experiments. The heterogeneous experimental material is divided into different homogeneous sub-groups. These sub-groups are known as **block. For Example:** Division of animals on the basis of their age, body weight, breed, sex, etc.

Experimental Designs for Experiments

In agricultural and biological experiments, the most commonly used designs are:

(i) Completely Randomized Design (CRD)

(ii) Randomized Block Design (RBD)

(iii) Latin Square Design (LSD)

(iv) Switch Over Design (SOD)

(v) Factorial design

(vi) Incomplete Block Design (IBD)

(vii) Balanced Incomplete Block Design (BIBD)

(viii) Split plot design

(i) Completely Randomized Design (CRD): A completely randomized design, which is the simplest type of the basic designs, may be defined as a design in which the treatments are assigned to experimental units completely at random. That is, the randomization is done without any restrictions. The design is completely flexible, i.e., any number of treatments and any number of units per treatment may be used. Moreover, the number of units per treatment need not be equal. The two basic principles to be followed by completely randomized design (CRD) are randomization and replication. To apply the CRD, the experimental material should be homogeneous. So, in this design, the homogenous experimental material is divided into different experimental units according to the number of treatment and number of replication. Then, all the treatments are randomly allotted to these experimental units.

(ii) Randomized Block Design (RBD): When experimental material is not entirely homogeneous, that means, heterogeneous and fertility gradient is moving one direction, randomized block design (RBD) is most suitable than CRD. The RBD uses all the three basic principles enumerated by Prof. R. A. Fisher. In this case, the experimental material is grouped into homogenous sub-groups called blocks. As each block consists of the entire set of treatments a block is equivalent to a replication. If the fertility gradient runs in one direction say from north to south or east to west then the blocks are formed in the opposite direction. Such an arrangement of grouping the heterogeneous units into homogenous blocks is known as randomized blocks design. Each block consists of as many experimental units as the number of treatments. The treatments are allocated randomly to the experimental units within each block independently such that each treatment occurs once. The number of blocks is chosen to be equal to the number of replications for the treatments.

(iii) Latin Square Design (LSD): When experimental materials have heterogeneity due to one factor only, then we apply RBD. But, if experimental materials have heterogeneity in the two directions simultaneously, then RBD will not be applied and Latin square design (LSD) which controls the variations in the two directions will be applicable. In LSD, the number of treatments is equal to the number of replications. Thus in case of m treatments there has to $m \times m = m^2$ experimental units. The m^2 experimental units will be arranged in a square. The m treatments are then allocated at random to these rows and columns such that every treatment occurs once and only once in each row and each column. Such a lay out is known as $m \times m$ latin square.

(iv) **Switch Over Design (SOD):** In case of CRD, RBD and LSD, a treatment allotted to a experiment unit remains the same throughout the experiment or we can say, an experimental unit gets one treatment throughout the experiment period. In switch over design, all the treatments are allotted to each of the experimental units and this is done by dividing whole of the experiment period into different fractional periods equal in number to the treatments. That means, each experimental units will have fractional periods equal to the number of treatments, then all the treatments are randomly adapted to these fractional periods one after the other. Since, all the treatments are given in sequence therefore, this design is known as '**Switch over design**". In this design, each experimental unit acts as a full replicates.

(v) **Factorial Design:** An experiment in which the effects of two or more than two factors of variation are studied and investigated simultaneously. The treatments being all the combinations of different factors under study. In these experiments an attempt is made to estimate the effect of each of the factors and also the interaction effects, i.e., the variation in the effect of one factor as a result to different levels of other factors. A series of experiments in which only one factor is varied at a time would be both lengthy and unsatisfactorily because of systematic changes in the general background conditions. These simple experiments do not give us any information regarding the independency or dependency of one factor on the other. In that case, the only alternative is to try to investigate the variations in several factors simultaneously by conducting the experiment as a p x q factorial experiment, where p and q are the level of various factors under consideration. In general, if the levels of various factors are equal then s^n - factorial experiment means an experiments with n factors, each at s levels where n is any positive integer greater than or equal to 2. **For example**; 2^n - factorial experiment means that the effect of n factors (treatments) each at two levels has been considered in the experiment, where $n \geq 2$.

(vi) **Incomplete Block Design (IBD):** An incomplete block design is that design in which each block does not contain a complete set of treatments. So, an incomplete block design is one in which the block size is less than the number of treatments so that all the treatments cannot be accommodated in the same block. An incomplete block design is one having v treatment and b blocks each of size $K < V$, such that each of the treatment is replicated r times and each pair of treatments occurs once and only once in the same block, where v, b, r and k are known as the parameters of IBD.

(vii) **Balanced Incomplete Block Design (BIBD):** An arrangement of v treatments in b blocks each of size $K<V$ is known as BIBD, if

(a) Each block contains K distinct treatments

(b) Each treatment occurs in r blocks

(c) Each pair of treatments occurs together in λ block, where λ denotes the number of blocks in which two treatments occurs together or number of times any two treatments occurs together in the design.

(viii) Split plot design: If one has to study two type of factors say A and B in the same experiment where A requires larger plot and B requires smaller plot size then the structure of the experiment will be that of plots within blocks are known as split plots and blocks within replication are known as main plots. Such a layout of the experiment is known as **split plot design**. In this design, the main plot treatments are allotted at random to the main plots and the split plot treatments are allotted at random to the split plots. In split plot design randomization is done separately for the whole plot treatments and for the split plot treatments while in ordinary two factors experiment, all the treatment combinations of two factors are allotted at random to the plots of a block.

(ix) Strip Plot Design: There is a variant of the split plot design in which both factors A and B will be applied to the larger strips by dividing the experimental field into as many columns as the number of levels of second factor B. Then the two factors will be applied at random to the rows and columns. Such a layout is known as strip plot design. If both the treatments A and B require larger plots and the interaction AB is more important than the main effect A and B then strip plot design is most suitable design.

MCQ

Q.1. The design effect is the ratio of two:

(a) Standard deviations (b) Means

(c) **Variances** (d) None of the above

Q2. The design of the experiment is used to determine the variables which are ___________ affecting the state of the process:

(a) **The most** (b) The least

(c) Not (d) Not changing

Q3. In the experiment, the replication means:

(a) Total number of treatments

(b) Number of blocks

Bold means correct answer

(c) **Number of times a treatment occurs**

(d) None of them

Q4. The science of experimental designs is associated with the name:

(a) Block design (b) **Lain square**

(c) Latin cubes (d) None of these

Q5. The principle which assures that extraneous factors do not influence continuously one factor is:

(a) Replication (b) **Randomization**

(c) local control (d) Regularity

Q6. Experimental design methods are not used:

(a) **Evaluating the process capability**

(b) In process development

(c) In process troubleshooting to improve process performance

(d) To obtain a process that is robust and insensitive to external sources of variability

Q7. Which of these can be used to develop a new process?

(a) Histogram (b) **Design of experiments**

(c) Acceptance sampling (d) Control charts

Q8. Latin squares are most widely used in the field of:

(a) Astronomy (b) Industry

(c) Zoology (d) **Agriculture**

Q9. Two basic principles followed CRD are:

(a) Randomization and local control

(b) **Randomization and replication**

(c) Randomization, replication and local control

(d) None of these

Q10. Randomization in an experiment helps to eliminate:

(a) Systematic influences (b) **Human biases**

(c) Dependence among observation (d) All of the above

Q11. Use of designed experiments does not give:

(a) Reduced variability (b) Improved yield

(c) **Increased development time** (d) None of these

Q12. Which of these can be used in the process of engineering design?

(a) Control charting (b) Acceptance sampling

(c) **Design of experiment** (d) None of these

Q13. In RBD, the object is to keep the experimental error:

(a) Within each block as high as possible

(b) Within each block as small as possible

(c) Between two adjacent blocks as small as possible

(d) **Both (b) and (c)**

Q14. In RBD, the randomization is done:

(a) Separately within each block

(b) Independently within each block

(c) Not independently within each block

(d) **Both (a) and (b)**

Q15. In agricultural field experiment where fertility contour map is not known then a suitable design applicable is:

(a) CRD (b) RBD

(c) **LSD** (d) Factorial design

Q16. In LSD, the experimental material of heterogeneous nature is controlled simultaneously in:

(a) Two parallel directions (b) Two perpendicular directions

(c) Incomplete three way classification (d) **Both (b) and (c)**

Q17. Replication in an experiment eliminates:

(a) Heterogeneity among blocks

(b) Human bias

(c) Competition among neighboring plots

(d) **All of them**

Q18. The number of replications in an experiment is based on:

(a) Heterogeneity of experimental material

(b) Precision required

(c) Experimental material available

(d) **All of them**

Q19. In an experiment, the factors which are responsible for the number of replications are:

(a) **Minimum degree of freedom required for experimental error**

(b) Shape of the experimental units

(c) The desire of experimenter

(d) None of these

Q20. The decision about the number of replications is taken in view of:

(a) Fraction to be sampled

(b) Size of experimental units

(c) Competition among experimental units

(d) **All of them**

Q21. Replications are an experiment eliminates:

(a) Human bias

(b) Competition among neighboring plots

(c) Heterogeneity among blocks

(d) **None of these**

Q22. Local control in experimental design is meant to :

(a) Increase the efficiency of the design

(b) Reduce experimental error

(c) To form homogeneous blocks

(d) **All the above**

Q23. Local control is a device to maintain:

(a) Homogeneity among blocks

(b) **Homogeneity within blocks**

(c) Both (a) and (b)

(d) Neither (a) nor (b)

Q24. Replication in an experiment means:

(a) The number of blocks

(b) Total number of treatments

(c) **The number of times a treatment occurs in an experiment**

(d) None of the above

Q25. Replications are an experiment eliminates:

(a) Human bias

(b) Competition among neighboring plots

(c) Heterogeneity among blocks

(d) **None of these**

Q26. Experimental error is due to:

(a) Experimenter's mistakes

(b) **Extraneous factors**

(c) Variation in treatment effects

(d) None of the above

Q27. Local control is a device to maintain:

(a) Homogeneity among blocks

(b) **Homogeneity within blocks**

(c) Both (a) and (b)

(d) Neither (a) nor (b)

Q28. An experimental design is:

(a) A map

(b) **Plan of experiment**

(c) An architect

(d) All of the above

Q29. Experimental error is necessarily required for:

(a) Testing the significance of treatments effects

(b) Comparing treatment effects

(c) Calculating the information released from an experiment

(d) **All of the above**

Q30. Local control in experimental design is meant to:

(a) Increase the efficiency of the design

(b) Reduce experimental error

(c) To form homogeneous blocks

(d) **All the above**

Q31. In CRD, F-value equals:

(a) MSC x MSE (b) **MSC/MSE**

(c) MSE/MSC (d) MSE - MSC

Q32. In RBD, the F-value between rows equals:

(a) **MSR/MSE** (b) MSR/MSC

(c) MSR x MSE (d) MSR - MSE

Q33. In RBD, Error sum of squares is obtained by:

(a) TSS – SSC (b) TSS – SSR

(c) **TSS – SSC – SSR** (d) TSS – SSC + SSR

Q34. In RBD, degree of freedom is determined by:

(a) (n-1) (b) (n-r-c-1)

(c) (n-r-c) (d) **None of these**

Q35. Randomized block design is a:

(a) **One way classification** (b) Two way classification

(c) Three way classification (d) Four way classification

Q36. A randomized block design has:

(a) One way classification (b) **Two way classification**

(c) Three way classification (d) No classification

Q37. Error sum of squares in RBD as compared to CRD using the same material is:

(a) More (b) **Less**

(c) Equal (d) Not comparable

Q38. A completely randomized design is also known as:

(a) Unsystematic design (b) **Non-restrictional design**

(c) Single block design (d) All of theme

Q39. A Latin square design possesses:

(a) One way classification (b) Two way classification

(c) **Three way classification** (d) Four way classification

Q40. A Latin square design is a:

(a) One restrictional design
(b) **Two restrictional design**
(c) Three restrictional design
(d) Non-restrictional design

Q41. Completely randomized design is used when experimental material is:

(a) **Homogeneous**
(b) Heterogeneous
(c) Very large
(d) None of above

Q42. A specific combination of factor levels whose effect is to be compared with other treatments is called a(an):

(a) Effect
(b) Permutation
(c) **Treatment**
(d) Factor

Q43. Performing the same treatment combination more than once is called:

(a) Randomization
(b) **Replication**
(c) Repetition
(d) Local control

Q44. The Latin Square design allows for ______ blocking factors.

(a) One
(b) **Two**
(c) Three
(d) Four

Q45. A controlled independent variable whose levels are set by the experimenter is called:

(a) Treatment
(b) Level
(c) Response
(d) **Factor**

Q46. Which is true about CRD?

(a) CRD provides minimum number of degree of freedom
(b) **CRD provides maximum number of degree of freedom**
(c) It does not provide any degree of freedom
(d) None of the above

Q47. Completely randomized design is based on the principles of .... and randomization only:

(a) **Replication**
(b) Local control
(c) Compounding
(d) Divisibility

Q48. In which of the following designs, three principles of design are used:

(a) CRD
(b) RBD
(c) LSD
(d) **Both (b) and (c)**

Q49. In entire materials on which different treatment of comparative experiment are applied, is called:

(a) Experimental unit
(b) **Experimental material**
(c) Experimental error
(d) None of these

Q50. In randomized block design, the number of blocks is equal to:

(a) **Replication**
(b) Treatment
(c) Error
(d) None of these

Q51. Experimental error is due to:

(a) Variation in treatment effects
(b) **Extraneous factors**
(c) Experimenter's mistake
(d) None of these

Q52. The experiments in which the treatments are allocated to experimental units through a random process are categorized as:

(a) Randomized design
(b) Partially randomized design
(c) **Completely randomized design**
(d) None of these

Q53. When the interaction effect is confounded in all the replicates, then it is called:

(a) Complete confounding
(b) Incomplete confounding
(c) **Partial confounding**
(d) None of these

Q54. What term refers to the situation where two independent variables have an effect when they are in combination?

(a) **Interaction**
(b) Dependency
(c) Artifact
(d) Confounding variable

Q55. Local control in the field is maintained through:

(a) Natural factors
(b) Randomization
(c) **Uniformity trial**
(d) All of them

Q56. What does an A x B interaction mean in a two way ANOVA?

(a) **The effect of factor A depends on the level of factor B**
(b) There must be significant main effects for factors A and B

(c) The main effect for factors A and B must be short of significance

(d) If there are significant main effects, they must be interpreted first before interpreting the interaction

Q57. Factorial experiments:

(a) Focus on organismic factors

(b) Focus on unmeasured factors

(c) Include two or more dependent variables

(d) **Include two or more independent variables**

Q58. What is the major advantage of within subject's designs over between subjects design?

(a) **They are more sensitive**

(b) They are better tests of causality

(c) They require less time

(d) They require fewer participants

Q59. Factorial designs allow us to study both .... Effects of the independent variables on the dependent variables:

(a) Symbiotic and dichotomous

(b) Dependent and independent

(c) Rank order and correlational

(d) **Main and interactive**

Q60. Which of the following statements is correct about interactions?

(a) They are additive effects

(b) They are spurious effects

(c) They occur only in interaction with organimic dependent variables

(d) **They are enhancements of the effect**

Q61. Within-subjects factorials:

(a) Use statistics that take into account the correlated nature of the data

(b) Also called repeated measures factorials

(c) Must include independent groups of participants

(d) **Both(a) and (b)**

Q62. What are the factors in a factorial design?

(a) The dependent variables (b) **The independent variables**

(c) The organimic variables (d) The experimental variables

Q63. Factorial designs:

(a) **Contain more than one null hypothesis**

(b) Cannot test participants across more than one condition

(c) Are ineffective when matched participants are included

(d) Include no more than one research hypothesis

Q64. What information is given in the factorial design notation, 2 x 3 x 2?

(a) The design has 12 independent variables

(b) Interaction will be found

(c) **The design has three independent variables, two levels of A, three levels of B, and two levels of C**

(d) The design has two independent variables, three dependent variables, and two organimic variables

Q65. Which of the following is correct about a mixed design?

(a) It includes both ANOVA and Chi-square statistical analysis

(b) It includes both a manipulated and a non-manipulated variable

(c) It includes both a within-subjects and a between-subjects component

(d) **Either (a) or (b)**

Q66. Which of the following is correct?

(a) Whenever main effects are found, there must be at least one interaction

(b) **When both interactions and main effects are found, the interactions are interpreted first**

(c) Whenever interactions are found, there must be main effects

(d) If there are no main effects, there can be no interactions

Q67. What is the appropriate statistical test for a factorial design?

(a) Z –test (b) t – test

(c) Chi-square test (d) **ANOVA**

Q68. A 2 x 2 factorial:

(a) Contains four factors

(b) Does not have enough factors to show interactions

(c) **Is essentially two designs that have been combined into a single study**

(d) Is extremely difficult to interpret if interactions are found

Q69. Unlike an ANOVA, the MANOVA procedure:

(a) Must include at least three independent measures

(b) Must have no more than one independent measure

(c) **Has more than one independent measure**

(d) Has only one independent measure

Q70. Designs in which more than one variable are studied simultaneously are called.......... designs:

(a) Replicate
(b) Sum of squares
(c) **Factorial**
(d) None of these

Q71. For ANOVA models, the levels of a factor can be selected according to:

(a) Random effect model
(b) Mixed effect model
(c) Fixed effect model
(d) **None of these**

Q72. The .... models contain some fixed and some random effects:

(a) Random effects
(b) **Mixed effects**
(c) Fixed effects
(d) Factorial design

Q73. Use of experimental design techniques in the designs stages of a process is referred to:

(a) Factorial design
(b) Fractional factorial design
(c) On-line process control
(d) **Off-line process control**

Q74. An experimental design is:

(a) An architect
(b) A map
(c) **Plan of experiment**
(d) All of them

Q75. In which of the following designs, three principles of design are used:

(a) RBD (b) LSD

(c) **Both (a) and (b)** (d) None of these

Q76. Completely randomized design is suitable when:

(a) Heterogeneity of experimental is along one direction

(b) Heterogeneity of experimental is in erratic fashion

(c) **Experimental material are homogeneous**

(d) None of these

Q77. Which do you randomize first when setting up a Latin square design?

(a) Row (b) **Column**

(c) Treatments (d) Block

Q78. How are RCBD and LSD ANOVA similar?

(a) Column factors are of interest

(b) Row factors are of interest

(c) Row and column factors are of interest

(d) **Row and column factors are not of interest**

Q79. How is randomization different for RCBD and Latin square (LS) design?

(a) RCBD is randomized by column; LS is randomized by row

(b) RCBD is randomized by row; LS is randomized by row

(c) **RCBD is randomized by column; LS is randomized by row and column**

(d) RCBD is randomized by column and row; LS is randomized by row and column

Q80. Replication in a Latin square design is:

(a) Time consuming (b) Economical

(c) **Costly** (d) Time consuming and costly

Q81. An experimental design where the experimental units are appear once and only once in each direction:

(a) CRD (b) RBD

(c) **LSD** (d) Factorial design

Q82. In a factorial design, the independent variables are called:

(a) **Factors** (b) Types

(c) Levels (d) None of these

Q83. In a 2 x 2 factorial experiment, the possible interaction effects are:

(a) 0 (b) **1**

(c) 2 (d) 4

Q84. In a 2 x 2 factorial experiment, the degrees of freedom for the first main effect will be:

(a) 0 (b) **1**

(c) 2 (d) 4

Q85. In a 2 x 2 x 2 factorial experiment, the possible treatment combinations can be:

(a) 2 (b) 4

(c) 6 (d) **8**

Q86. In Latin square design, the number of rows, columns and treatments are:

(a) **Always equal** (b) Always different

(c) Not necessarily equal (d) None of these

Q87. A Latin square design is a .... Restriction design:

(a) One (b) **Two**

(c) Three (d) Four

Q88. A Latin square design is classified in:

(a) One way (b) Two way

(c) **Three way** (d) Four way

Q89. If in LSD the entries in rows are same as its columns, then LSD is known as:

(a) Orthogonal (b) Symmetric

(c) Conjugate (d) **Self-conjugate**

Q90. Two Latin squares are calculated such as one is obtained by interchanging the rows by the columns of other Latin square, then Latin squares are:

(a) Orthogonal (b) Symmetric

(c) **Conjugate** (d) Self-conjugate

Q91. In a k x k Latin square, the error degree of freedom is:

(a) (k – 1) (b) $(k^2 - 1)$

(c) **(k – 1)(k – 2)** (d) k(k – 1)(k – 2)

Q92. In a lower order LSD, the error d.f. can be obtained by taking:

(a) Latin square at two locations

(b) **Group of Latin squares at a time**

(c) Both(a) and (b)

(d) None of these

Q93. The cross-over design can also be named as:

(a) Change over design (b) Switch back design

(c) Reversal design (d) **All of them**

Q94. When the subjects are taken as replicates, the error d.f. for a cross-over design are:

(a) **Less** (b) More

(c) Zero (d) One

Q95. An experiment having several factors with equal number of level is called:

(a) Asymmetrical experiment (b) Complex experiment

(c) **Symmetrical factorial experiment** (d) All of them

Q96. The error d.f. in RBD when number of block is 4 and number of treatments is 5 with one missing value is:

(a) 8 (b) **11**

(c) 12 (d) 14

Q97. Mostly CRD are used in............... experiments:

(a) Agriculture field (b) Animal

(c) **Plot** (d) All of them

Q98. In CRD, ... observations are deleted:

(a) Fixed (b) Random

(c) **Missing** (d) None of them

Q99. For a Completely randomized design (CRD) with t treatments and n experimental units, the d.f. of error is:

(a) t – 1 (b) n – 1

(c) **n – t** (d) n – t + 1

Q100.For each experimental unit, the process of determining more than one observation pertaining the same character is called:

(a) Multiple sampling (b) **Sub-sampling**

(c) Repeated observations (d) None of them

Q101.The F-value of treatment is less than one because of:

(a) Selection of wrong statistical model

(b) Improper randomization

(c) Non-normality of response measure

(d) **All of them**

Q102.In a CRD, the treatment mean square of sub-sampling can be tested against:

(a) Sampling error mean square (b) Pooled error mean square

(c) **Experimental error mean square** (d) None of them

Q103.In an experiment, if each and every level of a factor is taken at every level of other factor, it is called:

(a) **Crossed classification** (b) Nested classification

(c) Hierarchical classification (d) All of them

Q104.In the field layout of a RBD, the blocks are formed in direction.............. to the fertility gradient:

(a) Diagonally (b) Parallel

(c) **Perpendicular** (d) None of them

Q105.An experimental design that is devoid of any randomization process is called:

(a) Systematic design (b) **Purposive design**

(c) Non-randomized design (d) All of them

Q106.When all the treatments are allocated to experimental units through a random process, this is called:

(a) Completely randomized design (b) **Randomized design**

(c) Partially randomized design (d) All of them

Q107.Factorial experiment may be conducted in:

(a) CRD
(b) RBD
(c) LSD
(d) **Any of all three**

Q108.In factorial experiment all possible combinations of two or more factors:

(a) **Are considered as treatment**
(b) Are not considered as treatment
(c) May or may not be considered as treatment
(d) Are treated as block size

Q109.If two factors A and B are considered in a factorial experiment, each at two level, then treatment combinations are denoted as:

(a) **1, a, b, ab**
(b) (1), (a), (b)
(c) (1), (b), (ab)
(d) None of these

Q110.In 2^2 factorial experiment with two factors A and B, the effect AB gives an estimate of:

(a) Interaction effect of A and B
(b) Inter dependence of A and B
(c) **Both (a) and (b)**
(d) None of these

Q111.In 2 x 2 factorial design:

(a) Cannot yield interactions
(b) Must include an organismic independent variable
(c) Called a one way ANOVA
(d) **Results in a four cell matrix**

Q112.Experimental design methods are not used:

(a) In process development
(b) In process troubleshooting to improve process performance
(c) **Evaluating the process capability**
(d) None of the above

Q113.Which of these can be obtained by using the experimental design?

(a) Increased cost
(b) Increased variability

(c) **Reduced distance from the nominal value**

(d) Reduced process capability

Q114.For two way ANOVA with 2 x 3 design, which of the given statements is true?

(a) The first independent has 2 levels

(b) The second independent has 3 levels

(c) There are two independent variables

(d) **All of them**

Q115.How many F-values are computed in one-way ANOVA?

(a) **1**

(b) 2

(c) 4

(d) 5

Q116.How many effects are possible in two independent variables in ANOVA?

(a) 1

(b) 2

(c) **3**

(d) 4

Q117.In a factorial design due to the response when the treatments of one factor interacts that of the other factor is called:

(a) Replication

(b) Main effect

(c) **Interaction**

(d) None of these

Q118.In a factorial design, the number of times each experimental condition is checked is called:

(a) **Replication**

(b) Partition

(c) Experimental condition

(d) Factor

Q119.In an experimental design the experimental units are randomly assigned to treatments. This process is known as:

(a) Random factor design

(b) **Completely randomized design**

(c) Factor block design

(d) None of these

Q120.Randomization is a process in which the treatments are allocated to the experimental units:

(a) At the will of investigator (b) In a sequence

(c) **With equal probability** (d) None of them

Q121.An experimental design which allows the statistical conclusions about two or more factors is known as:

(a) Random design

(b) Completely randomized design

(c) Randomized block design

(d) **Factorial design**

Q122.Randomization in an experiment provides:

(a) Impetus to treatments

(b) **Estimate of experimental error**

(c) Check to the variation in soil fertility

(d) None of these

Q123.Randomization is a process which enables the experimenter to:

(a) Treat errors independent (b) Apply mathematical theories

(c) Make probability statements (d) **All of them**

Q124.In an experimental design, the effect in blocks which is confounded is:

(a) **Not estimated** (b) Estimated more precisely

(c) Estimated less precisely (d) None of these

Q125.What are the factors in a factorial design?

(a) The dependent variables (b) **The independent variables**

(c) The experimental variables (d) The organismic variables

14

Non-parametric Tests

A **non parametric test** (sometimes called a *distribution free test*) is a statistical analysis method that does not assume the population data belongs to some prescribed distribution which is determined by some parameters. That means, a non-parametric test can be defined as a test that is used in statistical analysis when the data under consideration does not belong to a parameterized family of distributions. When the data does not meet the requirements to perform a parametric test, a non-parametric test is used to analyze it. The non-parametric test compared to parametric test, which makes assumptions about a population's parameters (for example, the mean or standard deviation); When the word "non parametric" is used in stats, it means that population data **does not have a normal distribution**.

If at all possible, we should us parametric tests, as they tend to be more accurate. Parametric tests have greater statistical power, which means they are likely to find a true significant effect. Nonparametric tests can perform well with non-normal continuous data if the data of the sample are sufficiently large. Non parametric do **not assume that the data is normally distributed. The only non parametric test we are likely to come across in elementary stats is the Chi-square test.** In nonparametric tests have several advantages as compared to parametric tests. It includes:

- More **statistical power when assumptions for the parametric tests have been violated.**
- Fewer assumptions, i.e. the assumption of normality don't apply.
- Small sample sizes are acceptable.
- It is used for all data types, including nominal data, rank data or ordinal data.
- Knowledge of the population is not required to conduct this test.
- Nonparametric tests are valid when our sample size is small and our data are potentially non-normal.

There are several non-parametric tests:

(i) **Sign test:** This test to estimate the median of a population and compare it to a reference value or target value.

(ii) Wilcoxon Signed Rank test: This test, we also estimate the population median and compare it to a reference/target value. However, the test assumes your data comes from a symmetric distribution.

(iii) Wilcoxon Rank Sum test (Mann-Whitney U test) : This test to compare differences between two independent groups when dependent variables are either ordinal or continuous.

(iv) Friedman test: This test is used to test for differences between groups with ordinal dependent variables. It can also be used for continuous data if the one-way ANOVA with repeated measures is inappropriate (i.e. some assumption has been violated).

(v) Goodman Kruska's Gamma: It is a test of association for ranked variables.

(vi) Kruskal Wallis test: This test instead of a one-way ANOVA to find out if two or more medians are different. Ranks of the data points are used for the calculations, rather than the data points them.

(vii) Mann-Kendall Trend test: This test, it looks for trends in time-series data.

(viii) Mood's Median test: Use this test instead of the sign test when you have two independent samples.

(ix) Spearman Rank Correlation Coefficient: When we want to find a correlation between two sets of data.

(x) Kalmogorov-Smirnov test: This test, the equality of continuous, one-dimensional probability distributions that can be used to compare a sample with a reference probability distribution, or to compare two samples.

(xi) Runs test: Running a test of randomness is a non-parametric method that is used in cases when the parametric test is not in use. In this test, two different random samples from different populations with different continuous cumulative distribution functions are obtained. Running a test for randomness is carried out in a random model in which the observations vary around a constant mean.

MCQ

Q.1. When testing for randomness, we can use:

(a) Sign test (b) Kruskal Wallis test

(c) Wilcoxon Rank Sum test (d) **Runs test**

Q2. The Mann-Whitney U test is preferred to a t-test when:

(a) Sample size are small

(b) Samples are dependent

Bold means correct answer

(c) **Assumption of normality is not met**

(d) Data are paired

Q3. When using the sign test, if two scores are tied, then we:

(a) **Discard them** (b) Depends upon the score

(c) Count them (d) None of these

Q4. In testing for the difference between two populations, it is possible to use :

(a) **Sign test**

(b) Mood's Median test

(c) Kalmogorov-Smirnov test

(d) Goodman Kruska's Gamma test

Q5. In a Wilcoxon Rank- Sum test:

(a) Tied within one sample may affect the decision

(b) Tied never affect the decision

(c) **Tied always affect the decision**

(d) Tied between two samples may affect the decision

Q6. The Wilcoxon Rank- Sum test can be:

(a) Lower tailed

(b) Upper tailed

(c) **Either of Upper tailed or Lower tailed**

(d) None of these

Q7. In the Kruskal-Wallis test of k samples, the appropriate number of degrees of freedom is:

(a) n - k (b) n / k

(c) n - 1 **(d) k - 1**

Q8. Which of the following test must be two sided?

(a) **Sign test** (b) Runs test

(c) Kruskal Wallis test (d) Wilcoxon Rank- Sum test

Q9. The non-parametric equivalent of an unpaired samples t-test is:

(a) Sign test (b) Kruskal Wallis test

(c) **Wilcoxon Rank Sum test** (d) Wilcoxon Signed Rank test

Q10. The sign test assumes that the samples are:

(a) Have the same mean | (b) Independent
(c) **Dependent** | (d) None of these

Q11. Which of the following test use rank sums?

(a) **Kruskal Wallis test** | (b) Sign test
(c) Runs test | (d) F-test

Q12. The Wilcoxon Rank Sum test compares:

(a) A sample mean to the population mean
(b) Two populations
(c) **Three or more populations**
(d) None of these

Q13. The Spearman rank-correlation test requires that the:

(a) Data must be from two independent samples
(b) Data must be measured on the same scale
(c) **Data at least ordinal scale**
(d) Data must be distributed at least approximately as a t-distribution

Q14. Which one of the following is a non-parametric test?

(a) Z – test | (b) t – test
(c) F – test | (d) **χ2 – test**

Q15. The Sign test is:

(a) More powerful than the Wilcoxon signed-rank test
(b) Equivalent to the Mann-Whitney test
(c) **Less powerful than that of the Wilcoxon signed-rank test**
(d) More powerful than the pared sample t-test

Q16. Three brands of tea are rated for taste on a scale of 1 to 10. Six persons are asked to rate each brand so that there is a total of 18 observations. The appropriate test to determine if three brands taste equally good is:

(a) Wilcoxon rank-sum test | (b) Spearman rank difference
(c) **Kruskal-Wallis test** | (d) One way analysis of variance

Q17. Comparing the times-to-failure of radar transponders made by firms A, B, and C, based on an airline's sample experience with the three types of instruments, one may use:

(a) **Kruskal-Wallis test**

(b) Spearman rank-correlation test

(c) Wilcoxon rank-sum test

(d) Kolmogorov-Smirnove test

Q18. The Sign test assumes that:

(a) Samples are independent (b) **Samples are dependent**

(c) Wilcoxon signed rank test is used (d) Samples have same mean

Q19. When testing for randomness, we can use:

(a) Mann-Whitney U test (b) Sign test

(c) **Runs test** (d) None of these

Q20. Which of the following test must be two sided?

(a) Runs test **(b) Sign test**

(c) Kruskal-Wallis test (d) Kolmogorov-Smirnove test

Q21. The best use of Wilcoxon Signed Rank test will be for comparison of which of the following types of data:

(a) Continuous, parametric paired data

(b) Continuous, parametric unpaired data

(c) Continuous, non-parametric paired data

(d) Continuous, non-parametric unpaired data

Q22. To perform a runs test for randomness the data must be:

(a) Divided into at least two classifications

(b) Divided into exactly two classifications

(c) **Quantitative**

(d) Qualitative

Q23. When using the Sign test, if two scores are tied, then we:

(a) Depend upon the scores (b) Count them

(c) **Discard them** (d) None of these

Q24. Compare to parametric methods, the non-parametric methods are:

(a) Computationally easier
(b) Less efficient
(c) Less accurate
(d) **Both (a) and (b)**

Q25. The Kruskal Wallis test is the non-parametric alternative to the:

(a) **One –way ANOVA**
(b) Two –way ANOVA
(c) Three –way ANOVA
(d) Factorial

Q26. The Friedman test is the non-parametric alternative to the:

(a) One –way ANOVA
(b) **Randomized block design**
(c) Chi-square test
(d) Factorial

Q27. For the ……………., when sample size is less than or equal to 15, it is treated as a small sample.

(a) Kruskal-Wallis test
(b) Sign test
(c) Runs test
(d) **Wilcoxon test**

Q28. Non-parametric test referred to as:

(a) Z – test
(b) t – test
(c) F - test
(d) **Distribution free test**

Q29. The non-parametric tests may be applied to ……… data or data can be ranked but not measured exactly.

(a) Chronological
(b) Geographical
(c) **Qualitative**
(d) Quantitative

Q30. Sign test is used to test:

(a) Critical region
(b) Acceptance region
(c) **Null hypothesis**
(d) Alternative hypothesis

Q31. When the mean more accurately the center of distribution of the data and sample size is large, then we use:

(a) Parametric test
(b) **Non-parametric test**
(c) Both (a) and (b)
(d) None of these

Q32. The best use of the Mann-Whitney U test will be for the comparison of which of the following types of data:

(a) Parametric paired data
(b) Non- parametric paired data
(c) **Non-parametric unpaired data**
(d) Parametric unpaired data

Q33. Which of the following statements you believe to be true when considering the Mann-Whitney U test?

(a) **It uses the ranking the data values**

(b) It is the equivalent of the paired t-test

(c) It is the equivalent of the unpaired t-test

(d) It assumes that the standard deviations of the two independent groups are equal

Q34. The null hypothesis for the Mann-Whitney U test is used to test that:

(a) Two samples are from different populations but have the same mean

(b) Two samples are from the same population and have the same mean

(c) **Two samples are from the same population and have the same median**

(d) Two samples are from different populations

Q35. The null hypothesis of the sign test is that:

(a) The lower half the ranks to have the same standard deviation as the upper half

(b) The lower half the ranks to have the same mean as the upper half

(c) **Half the ranks to be less than the median and half greater than the median**

(d) Half the ranks to be less than the mean and half greater than the mean

Q36. If there is no knowledge about the population or parameters then the test that will be used is called:

(a) Parametric test

(b) **Non-parametric test**

(c) Both (a) and (b)

(d) None of these

Q37. The test of significance for which of the following parameters is conducted by non-parametric test:

(a) Mean

(b) **Median**

(c) Mode

(d) Standard deviation

Q38. Which of the following tests cannot be used for testing the equality of two population medians?

(a) **Kruskal-Wallis test**

(b) Wilcoxon rank-sum test

(c) Mann-Whitney U test

(d) Median test

Q39. Non-parametric analogue of one - way ANOVA is:

(a) Kruskal-Wallis test
(b) Wilcoxon rank-sum test
(c) **Mann-Whitney U test**
(d) Median test

Q40. Kruskal-Wallis Test is used for testing the equality of:

(a) **Several population medians**
(b) Several population means
(c) Several population variances
(d) None of these

Q41. The Wilcoxon rank-sum test is a nonparametric alternative to the:

(a) One –way ANOVA
(b) Two –way ANOVA
(c) Factorial
(d) **Two sample t-test**

Q42. The Runs test results in rejecting the null hypothesis of randomness when:

(a) There is an unusually small number of runs
(b) There is an unusually large number of runs
(c) **Either of the above**
(d) None of these

Q43. Which of the following test use rank sums?

(a) Chi-square
(b) Sign tests
(c) Runs test
(d) **Kruskal-Wallis and Wilcoxon test**

Q44. Which of the following test is most likely assessing the null hypothesis of "the number of violations per apartment in the population of all city apartments is binomially distributed with a probability of success in any one trial of P = 0.4"

(a) Kruskal-Wallis test
(b) Mann-Whitney test
(c) **Kolmogorov-Smirnove test**
(d) Wilcoxon signed-rank test

Q45. Unlike the non-parametric tests, parametric test make certain assumptions about:

(a) The population size
(b) **The underlying distribution**
(c) The sample size
(d) None of these

Q46. The sign test is

(a) **More Powerful than the Wilcoxon signed-rank test**

(b) Less Powerful than that of the Wilcoxon signed-rank test

(c) More Powerful than the paired sample t-test

(d) Equivalent to Mann-Whitney test

Q47. Match the following parametric and its corresponding non-parametric test. Parametric -1: Independent t-test, 2. ANOVA, 3. Paired t-test. Non-parametric –(i) Wilcoxon, (ii) Kruskal-Wallis, (iii) Mann-Whitney:

(a) 1-(i), 2-(ii), 3-(iii)

(b) 1-(ii), 2-(iii), 3-(i)

(c) **1-(iii), 2-(ii), 3-(i)**

(d) 1-(i), 2-(iii), 3-(ii)

Q48. A pre post study gave the following results: Pre-study- 4, 4, 5, 4.6. Its corresponding Post study values are - 3.5, 4.5, 4, 3,9. Which non parametric test is suitable here?

(a) Kruskal Wallis test

(b) Mann Whitney test

(c) Paired t-test

(d) **Wilcoxon test**

15

Vital Statistics

Vital Statistics

It is defined as that branch of biometry which deals with data and the laws of human mortality, morbidity and demography. According to **N.B. Ryder**, they "provide cumulative summaries for successive time periods of population movements like birth, death, migration, marriage and marital dissolution as well as demographic and other relevant characteristics of the individuals involved in these events."

Measures of Vital Statistics of Population

The total number of vital events such as births and deaths are converted into rates and ratios for comparison of vital statistics from place to place or year to year. In order to determine the population at any time 't' after the census or between two censuses, a number of methods have been devised. Therefore, a suitable method which makes use of births, deaths and migration statistics, if we assume that:

(i) The census data gives us the total size of the population of a region or community together with age and sex distribution

(ii) The birth, death and migration statistics during different periods are obtained from registers.

Then the population P_t at any time is given by the relation

$$P_t = P_0 + (B - D) + (I - E)$$

Here, P_0 = Total population at the last census

B = Total number of births

D = Total number of deaths

I = Total number of immigrants

E = Total number of emigrants from the region during the period t.

Rates of vital event: It is defined as the ratio of the total number of cases of the vital to the total number of persons exposed to the risk of occurrence of that event and multiplied by thousand of population.

$$\text{Rates of vital event} = \frac{\text{Number of cases of the vital event}}{\text{Total number of persons exposed to the risk of the occurence of the event}} \text{ x K}$$

Where, K is a constant usually 1000. So, rates are generally expressed per thousand of population.

Ratios of vital events: It is a static quantity. It is defined as the ratio of the total number of cases of the vital event to the number of cases of another vital event (Which may or may not include the numerator or sub classes) and multiplied by a constant.

$$\text{Rates of vital event} = \frac{\text{Number of cases of the vital event}}{\text{Number of cases of another vital event}} \text{ x K}$$

Here, K is a constant usually 100.

Mortality Rates

The following are the principal rates used for mortality rates:

(i) **Crude Death Rate (C.D.R.):** It is defined as the number of deaths from all causes per K persons in the population of any given region during a given time period. It is given by:

$$\text{Crude Death Rate}\left(\text{C.D.R.}\right) = \frac{\text{Annual Deaths of any given region during a given period}}{\text{Annual mean population}} \text{ x K}$$

Where, K = 1000

We can also compute the crude death rate for males and females separately. For example, crude death rate for males in the given region during the given period is given by the formula

$$\text{C.D.R. for males} = \frac{\text{Number of male deaths in the given region during a given period}}{\text{Male population in the given region uring same period}} \text{x1000}$$

Similarly, crude death rate for females in the given region during the given period is given by the formula

$$\text{C.D.R. for females} = \frac{\text{Number of female deaths in the given region during a given period}}{\text{Female population in the given region during same period}} \text{x1000}$$

(ii) **Specific Death Rate (S.D.R.):** Death rate for a particular specified section of the population is termed as specific death rate (S.D.R.) and it is defined as:

$$\text{Specific Death Rate (S.D.R.)} = \frac{\text{Total number of deaths in the specific region during a given period}}{\text{Total population of the specific sregion in the same period}} \text{x K}$$

Where, K = 1000

(iii) **Age-Specific Death Rate (Age- S.D.R.):** It is the number of deaths per 1000 of the population for the age-group of a class in a given year is denoted as:

$$\text{Age} - \text{S.D.R.} = \frac{\text{Number of deaths in the specified age group}}{\text{Total population in the same age group}} \text{x}1000$$

(iv) **Infant Mortality Rate (I.M.R):** It is the number of infant deaths under the age under one year in a year per 1000 live births during the year. It is denoted as:

$$(\text{I.M.R.}) = \frac{\text{Number of deaths under one year in the region during a given period}}{\text{Number of live births during the same period}} \text{x}1000$$

(v) **Central Death Rate:** It is also known as central mortality rate and it is the probability that a person whose age is not known but lies in between a specify period will die within one year following the attainment of the age. It is denoted by:

$$\text{Central Death Rate} = \frac{\text{Number of deaths in the specified age group}}{\text{Mid} - \text{year total population in the same age group}} \text{x}1000$$

Fertility Rates

It is in relation to the actual production of children or specially live births. As a measure of the rate of growth of population various fertility rates are computed.

(i) **Crude Birth Rate (C.B.R.):** It is the number of live births in a year per thousand of the mid-year population. This provides the composite index of increase in population through child birth.

$$\text{Crude Birth Rate (C.B.R.)} = \frac{\text{Number of live births in the given region during a given period}}{\text{Total population of the given region during the same period}} \text{x K}$$

Where, K = 1000

(ii) **General Fertility Rate (G.F.R.):** It consists in relating the total number of live births to the number of females in the reproductive or child bearing ages and is given by:

$$(\text{G.F.R.}) = \frac{\text{Number of live births among the population in the given region during a given period}}{\text{Female population in the reproductive age of the given region during the same period}} \times K$$

Where, K = 1000

(iii) **Specific Fertility Rate (S.F.R.):** The fertility rate computed with respect to any specific factor is called specific fertility rate (S.F.R.) and is denoted by:

$$\text{S.F.R.} = \frac{\text{Number of births to the female population of the specified section in a given period}}{\text{Total female population in the specified section during the same period}} \times K$$

Where, K = 1000

(iv) **Age-specific Fertility Rate (Age-S.F.R.):** The fertility rate on the basis of specification with respect to age is called the age-specific fertility rate. Symbolically, the age-specific fertility rate for the specific age group is given by:

$$\text{Age} - \text{S.F.R.} = \frac{\text{Number of births to the females in the specified age group in a given period}}{\text{Average female population in the specified age group during the same period}} \times K$$

Where, K = 1000

(v) **Total Fertility Rate (T.F.R.):** It is a practically measure of the population growth, the age specific fertility rates for different groups have to be combined together to give a single quantity, known as total fertility rate (T.F.R.). Symbolically

$$T.F.R. = \sum_{\lambda_1}^{\lambda_2} \frac{B_x}{f_{Px}} x\, k$$

Where, k = 1000; B_x = Number of births to the females of age x during the given period;

${}^{f}P_x$ = Female population in the reproductive age x during the given period;

λ_1 and λ_2 is the lower and upper limits of the female child bearing age.

(vi) **Gross Reproductive Rate (G.R.R.):** It is defined as the sum of age-specific fertility rates calculated from female births for each year of reproductive period. Symbolically, if ${}^{f}B_x$ is the number of female births to the women of age x during the given period in the given region, then with usual notation, we have

$$G.R.R. = \sum_{\lambda_1}^{\lambda_2} \frac{f_{B_x}}{f_{Px}} x k$$

Where, k = 1000; ${}^{f}B_x$ = Number of female births to the women of age x during the given period;

${}^{f}P_x$ = Female population in the reproductive age x during the given period;

λ_1 and λ_2 is the lower and upper limits of the female child bearing age.

(vii) **Net Reproductive Rate (N.R.R.):** It measures the extent to which mothers produce female infants who survive to replace them. It measures the extent to which generations of girl babies arrive to reproduce themselves as they pass through the child-bearing age group.

$$\text{N.R.R.} = \frac{\text{Number of girls survived after the mortality experience}}{\text{Number of women survived at the end of reproductive period as per their mortality experience}} \text{ x K}$$

Where, k = 1000

Vital Index

It is the index of two vital events i.e. births and deaths. The formula is:

$$Vital\ Index = \frac{Total\ live\ births}{Total\ deaths} x1000$$

MCQ

Q.1. Vital statistics include the following except

(a) Mortality rate (b) Birth rate

(c) **Education rate** (d) Morbidity rate

Q2. The data on births, deaths, marriage and divorce called:

(a) Morbidity (b) Mortality

(c) **Vital statistics** (d) Survey

Q3. Population statistics mainly display the records pertaining to:

(a) Morbidity (b) Mortality

(c) Foetal deaths (d) **Population of regions**

Q4. Vital statistics is greatly utilized by:

(a) Social reformers (b) Planners

(c) Statistician (d) **All of them**

Bold means correct answer

Q5. Vital statistics is obtained through:

(a) Survey method
(b) Census operation
(c) Registration system
(d) **All of them**

Q6. Complete count of the heads of people of a country is known as:

(a) Vital statistics
(b) Demography
(c) **Census**
(d) None of these

Q7. Fertility rates mainly depend on:

(a) Total population
(b) Total male population
(c) Total female population
(d) Female population of child bearing age

Q8. Adequate basis for family planning is provided by:

(a) Total fertility rate
(b) General fertility rate
(c) Total marital fertility rate
(d) **Age specific marital fertility rate**

Q9. Fertility rates provide an adequate basis for:

(a) **Family planning**
(b) Female growth
(c) Population growth
(d) Checking the infant mortality

Q10. Population growth is mainly concerned with:\

(a) Number of male births
(b) **Number of female births**
(c) Total number of births
(d) All of them

Q11. Total fertility rates provides the basis for:

(a) Population projection
(b) Population increase in a desired period
(c) Expected maximum family size
(d) **All of them**

Q12. The death rate of women due to delivery of children is termed as:

(a) Foetal death rate
(b) Neonatal mortality rate
(c) Infant mortality rate
(d) **Maternal mortality rate**

Q13. The death rate of babies under one month is known as:

(a) Foetal death rate
(b) **Neonatal mortality rate**
(c) Infant mortality rate
(d) Maternal mortality rate

Q14. Age-specific mortality rates fail to reveal:

(a) Sex ratio
(b) Mortality conditions
(c) Age distribution of persons
(d) **All of them**

Q15. Standardized death rates are particularly useful for:

(a) **Comparing the death rates of two regions**
(b) Comparing the death rates in males and females
(c) Comparing the death rates in a desired period
(d) All of them

Q16. Vital rates are customarily expressed as:

(a) Per hundred
(b) **Per thousand**
(c) Per million
(d) Per trillion

Q17. The sampling registration system fails to record:

(a) Birth rates
(b) Death rates
(c) **Age and sex composition**
(d) None of these

Q18. Sex ratio is:

(a) Number of men per thousand of women
(b) **Number of women per thousand of men**
(c) Number of deaths per thousand live births of children
(d) None of these

Q19. Infant mortality average means:

(a) Number of deaths per thousand live births of children under 10 year of age
(b) Number of deaths per thousand live births of children under 5 year of age
(c) Number of deaths per thousand live births of children under 3 year of age
(d) **Number of deaths per thousand live births of children under 1 year of age**

Q20. Neonatal death is experienced in terms of:

(a) 0-5 days
(b) 0-10 days
(c) 0-15 days
(d) **0-27 days**

Q21. Number of live birth during a year per thousand estimated mid-year population is known as:

(a) General fertility rate
(b) Death rate
(c) **Birth rate**
(d) None of these

Q22. Number of live birth during a year per thousand women in reproductive age group is known as:

(a) **General fertility rate**
(b) Death rate
(c) Birth rate
(d) None of these

Q23. Gross reproduction rate is gross in the sense that:

(a) It is based on unrealistic assumptions
(b) Its basis is hypothetical
(c) It is based on clumsy grounds
(d) All of them

Q24. The main drawback of gross reproduction rate is:

(a) Non-consideration of divorces
(b) **Assumption of no mortality of female children till their fecundity age**
(c) Neglect of the number of unmarried women
(d) All of them

Q25. Demography is?

(a) **The study of human population**
(b) The study of human Health
(c) The study of human happiness
(d) The study of human education

Q26. Net reproduction rate (N.R.R.) is more viable than gross reproduction rate because:

(a) It utilizes survival rate
(b) It makes use of life tables
(c) It takes into account fertility rates as well as mortality rates
(d) **All of them**

Q27. The sampling registration system fails to record:

(a) Birth rates (b) Death rates

(c) **Age and sex composition** (d) All of them

Q28. The age-specific death rate for the babies of age less than one year is specifically called:

(a) **Infant mortality rate** (b) Neonatal death rate

(c) Foetal death rate (d) Maternal mortality rate

Q29. The death rate of babies under one month is known as:

(a) Infant mortality rate (b) **Neonatal death rate**

(c) Foetal death rate (d) Maternal mortality rate

Q30. The death rate of women due to delivery of children is termed as:

(a) Infant mortality rate (b) Neonatal death rate

(c) Foetal death rate (d) **Maternal mortality rate**

Q31. Vital statistics is mainly concerned with:

(a) Marriages (b) Deaths

(c) Births (d) **All of them**

Q32. Population statistics mainly display the records pertaining to:

(a) **Population of regions** (b) Morbidity

(c) Foetal deaths (d) Neonatal deaths

Q33. The registration of births, deaths and marriages are:

(a) **Legal document** (b) Part of medical research

(c) Part of political research (d) Fancy of society

Q34. The magnitude of population growth refers to:

(a) The number of females per thousand males

(b) The number of persons added each year

(c) **The total population of an area**

(d) The rate at which the population increases

Q35. Calculate crude marriage rate when the marriages registered were 1,84,568 and the estimated population in the mid year were 6,85,485:

(a) 0.72 (b) **0.27**

(c) 3.71 (d) 26.9

Q36. Study and measurement of facts related to health and growth of community is called:

(a) Demography (b) **Vital statistics**

(c) Morbidity (d) Prevalence

Q37. Natural increase in population is the equal to:

(a) Crude birth rate + Crude Death rate

(b) Crude birth rate - Crude Death rate

(c) Crude Death rate - Crude birth rate

(d) Total number of births in a region

Q38. What is replacement rate?

(a) Rate required to replace the father

(b) Rate required to replace the mother

(c) **Rate required to replace both the parents**

(d) None of these

Q39. Emigration of highly trained skilled people from the country is called:

(a) **Brain drain** (b) Mobility rate

(c) Immigration rate (d) None of these

Q40. Specify death rate is equal to:

(a) (Number of deaths/ Total population in specify class) x 1000

(b) **(Number of deaths in specify class / Total population in specify class) x 1000**

(c) (Number of deaths/ Total population) x 1000

(d) None of these

Q41. In a region, number of males = 8000, number of females = 12000 and the number of births are 300, and then the crude birth rate for the region is:

(a) 75 (b) 25

(c) **15** (d) 5

Q42. If gross reproductive rate is less than one:

(a) Population will remain stagnant

(b) Population will increase in spite of low birth rate

(c) **Population will decrease in spite of low birth rate**

(d) None of these

Q43. Morbidity is the increase in the rate of:

(a) **Sickness**

(b) Births

(c) Deaths

(d) None of these

Q44. In the broadest term, what do demographers study?

(a) Farming practices

(b) **Population statistics**

(c) Demolition

(d) Fertility rates of agricultural crops

Q45. Crude birth rate is a simplest measure of fertility because it includes:

(a) Births only

(b) Pre-term birth

(c) **Mid-year population**

(d) Total population

Q46. In the past, life expectancy was so low because the risk of death was very high during:

(a) **Infancy and childhood**

(b) Middle age

(c) Young adulthood

(d) Old age

Q47. The most important fertility reducing practice in our society is:

(a) Abortion

(b) Religious value

(c) **Contraception**

(d) Breast feeding

Q48. Crude birth rate not true is:

(a) It is independent of age of population

(b) It is a measure of fertility

(c) It is actually ratio not rate

(d) None of these

Q49. Birth rate – Death rate is equal to:

(a) Population density
(b) Reproduction rate
(c) **Population index**
(d) None of them

Q50. Rates is computed by adding the age specific fertility rates of various age groups of child bearing age:

(a) **Total fertility rate**
(b) Crude birth rate
(c) Net reproduction rate
(d) General fertility rate

Q51. Most significant indicator of fertility is:

(a) Family size
(b) Gross reproductive rate
(c) **Net reproductive rate**
(d) General fertility rate

Q52. The rate of natural increase in the human population is calculated by:

(a) CBR - CMR
(b) CDR - CBR
(c) **CBR - CDR**
(d) CBR + CDR

Q53. Population growth is said to be less than adequate requirement when NRR is:

(a) >1
(b) **< 1**
(c) = 1
(d) = 0

Q54. If TFR = 2000 per thousand and male and female ratio is 40:60, then GRR is:

(a) 0.4
(b) 0.8
(c) **1.2**
(d) 1.5

Hint: We know that, GRR = TFR x female birth/Total births = (2000/1000) x (60/100) = 1.2

Q55. If TFR = 2000 per thousand and male and female ratio is 40:60, then:

(a) No change
(b) The population of female is replacing itself
(c) **Population is increasing**
(d) Population is declining

Q56. If the female birth is 80 per thousand and survival rate is 500 per thousand then NRR?

(a) 0.04 (b) 0.16

(c) **0.4** (d) 0.8

Q57. Life expectancy means:

(a) The philosophical prediction that we are aware of our existence

(b) The exact number of years you will live

(c) People's personal expectations about the quality of their lives

(d) **The estimated number of years someone born in a certain year will live**

Q58. What do we create when we calculate age-specific death rates for an entire population?

(a) **Life expectancy** (b) Age-specific rates of death

(c) Crude death rate (d) Life table for that population

Q59. Which one of the following is not the measure of mortality?

(a) Infant mortality (b) Average life expectancy

(c) **Net reproduction rate** (d) Crude death rate

Q60. Direct standardization is used to compare the mortality rates between two countries. This is done because of the difference in:

(a) Denominators (b) Numerators

(c) Causes of death (d) **Age distributions**

Q61. Mortality rate means:

(a) The number of deaths annually worldwide

(b) The number of deaths in a specific region

(c) **The proportion of deaths to a population**

(d) All of them

Q62. The number of people who die per 1,00,000 population in a given year is called:

(a) **Mortality rate** (b) Morbidity

(c) Life expectancy (d) None of these

Q63. The simplest measure of fertility is:

(a) General fertility rate
(b) Total fertility rate
(c) Standardized birth rate
(d) **Crude birth rate**

Q64. The causes of infant deaths can be generally being separated in two broad group i.e. endogenous death and:

(a) Mature deaths
(b) Premature deaths
(c) Neonatal deaths
(d) **Exogenous deaths**

Q65. The main sources of information concerning fertility are population census and:

(a) **Birth registration system**
(b) Demographic surveys
(c) Questionnaire
(d) Stratified sampling

Q66. Components of population growth are fertility, mortality and:

(a) Immigration
(b) Emigration
(c) Wars
(d) **Migration**

Q67. Rate is computed by adding the age specific fertility rates of various age groups of child bearing age:

(a) General fertility rate
(b) Total fertility rate
(c) Crude birth rate
(d) Net reproduction rate

Q68. In case of demography errors can be divided into three categories i.e. sampling error, error of content and:

(a) Error of coverage
(b) Non-sampling error
(c) Standard error
(d) Percentage error

Q69. Population is one in which the rate of growth is zero:

(a) Stationery
(b) Increasing
(c) Unstable
(d) Stable

Q70. Any population with constant age distribution and which is increasing at a constant rate is called:

(a) Stationery population
(b) Increasing population
(c) Unstable population
(d) **Stable population**

Q71. The simplest measure of fertility, requiring only total births and total population:

(a) NRR (b) GFR

(c) SBR (d) **CBR**

Q72. Birth rate refers to the number of live births per thousand of the population in a region during a given period of time:

(a) **Crude** (b) Standardized

(c) Child (d) None of these

Q73. When death rates are computed for some specific class of people or specific age group of a population, they are called:

(a) Age specific crude death rate (b) Age specific mortality

(c) Age specific death schedules (d) **Age specific Death rates**

Q74. Three major sources of demographic data are categories, i.e., demographic survey, census and:

(a) **Vital registration system** (b) Vital index

(c) Questionnaire (d) Sampling

Q75. Gives a net biological status of the population:

(a) Mortality rate (b) Fertility rate

(c) Vital index (d) Health indicator

Q76. Shows weather the population is increasing, decreasing or stable:

(a) **Vital index** (b) Social index

(c) Health rate (d) Crude death rate

Q77. Population density will increase if:

(a) Number of births and number of deaths increase

(b) Number of births and number of immigrants is low

(c) **Number of births and number of immigrants is high**

(d) Number of deaths and number of immigrants is high

Q78. Which of the following is an important determinant of population change?

(a) Life expectancy (b) Net attendance ratio

(c) Human development report (d) **Migration**

Q79. Most useful single measure of mortality is:

(a) CDR (b) **ASDR**

(c) Proportional mortality rate (d) Case fertility rate

Q80. Demographic gap means:

(a) Differences in child and women ratio

(b) Differences in sex ratio

(c) **Differences between birth and death rates**

(d) Differences between age specific birth and death rates

Q81. Denominator of prenatal mortality is:

(a) Number of still births

(b) Number of live births

(c) Live births –still births

(d) **Number of still births plus the number of live births**

Q82. A person who is an international migrant entering the country from a place outside the country is:

(a) Out-migrant (b) In- migrant

(c) **Immigrant** (d) Emigrant

Q83. The probability of dying of a person of age between x and (x+1) years is known as:

(a) Infant mortality rate (b) **Central mortality rate**

(c) Age-specific death rate (d) None of these

Q84. A population maintaining a constant growth rate is said to be:

(a) Stationary population (b) Continuous population

(c) **Stable population** (d) Unstable population

Q85. A population has constant size and composition is called:

(a) **Stationary population** (b) Continuous population

(c) Stable population (d) Unstable population

Q86. Unemployment rates are similar to:

(a) Birth rates (b) **Death rates**

(c) Survival rates (d) Migration rates

Q87. A life table consists of:

(a) **Eight columns** (b) Nine columns

(c) Seven columns (d) Six columns

Q88. A life table is most utilized by:

(a) Employment exchange (b) **Life insurance companies**

(c) General insurance companies (d) All of them

Q89. Life tables are usually constructed:

(a) Only male populations

(b) Only female populations

(c) Jointly for male and female populations

(d) **Separately for male and female populations**

Q90. Construction of life tables is based on the assumption that:

(a) Mortality rates are same for male and female populations

(b) Age-specific death rates are constant at all ages

(c) **Death rates are uniformly distributed between two birth days**

(d) All of them

Q91. Life-table has also been named as:

(a) Life expectancy table (b) Survival table

(c) **Mortality table** (d) All of them

Q92. NRR =1 leads one to conclude that:

(a) Population remain constant

(b) Female population will exactly replace itself

(c) No mortality in female births till their child bearing age

(d) **All of them**

Q93. A value of NRR > 1 will result into:

(a) Decrease in population (b) **Increase in population**

(c) Zero increase in population (d) None of these

Q94. Vital index for ascertaining population growth is:

(a) **A rough measure** (b) A good measure

(c) Useless (d) None of these

Q95. The ratio of births to the total deaths in a year is called:

(a) Total fertility index
(b) Survival rate
(c) Population death rate
(d) **Vital index**

Q96. Normally a life-table is constructed for an age interval of:

(a) One month
(b) **One year**
(c) Five year
(d) 5-10 years

Q97. A life-table is a profile of human population which exhibits:

(a) Numbers living and dying
(b) Expectation of life at each age
(c) Probability of dying at each age
(d) **All of them**

Q98. Life-table is a mean of:

(a) Giving population projections
(b) Checking the census figures
(c) Monitoring the family planning programs
(d) All of them

Q99. Gross reproduction rate is gross in the sense that:

(a) It is based on clumsy grounds
(b) It is based on unrealistic assumptions
(c) Its basis is hypothetical
(d) **All of them**

Q100. The registration of vital statistics in India suffers from:

(a) Lack of accuracy
(b) Incomplete reporting
(c) Incomplete coverage
(d) **All of them**

Q101. The relation between N.R.R. and G.R.R. is:

(a) **N.R.R. never exceed G.R.R.**
(b) N.R.R. and G.R.R. are usually equal
(c) N.R.R. is generally greater than G.R.R.
(d) None of them

Q102.The central mortality rate m_x in terms of q_x is given by the formula:

(a) **$2q_x / (2 - q_x)$** (b) $2q_x / (2 + q_x)$

(c) $q_x / (2 - q_x)$ (d) $q_x / (2 + q_x)$

Q103.The ratio of the rate of decrease in initial population l_x at age x to l_x is known as:

(a) **Force of mortality** (b) Central mortality rate

(c) Nominal annual rate of mortality (d) None of them

Q104.The probability q_x of dying of a person between the age interval x and (x +1) and m_x, the central mortality rate are related as:

(a) $q_x = 2m_x /(2 - m_x)$ (b) **$q_x = 2m_x /(2 + m_x)$**

(c) $q_x = m_x /(2 - m_x)$ (d) $q_x = m_x /(2 + m_x)$

Q105.The probability of living of a person in the age group x to (x + n) can be obtained by the formula:

(a) $(l_x - l_{x+n}) / l_x$ (b) $(l_x - l_{x+n}) / l_{x+n}$

(c) **l_{x+n} / l_x** (d) l_x / l_{x+n}

Q106.The child bearing age in India is:

(a) 13 – 48 (b) **15 – 49**

(c) 20 – 24 (d) 25 – 29

Q107.The fertility of a woman in India is maximum in the age group:

(a) 13 – 48 (b) 15 – 49

(c) 20 – 24 (d) **25 – 29**

Q108.If P_1 and P_2 are the population at an interval of 10 years, the population just after five years will be:

(a) $P_1 \times P_2$ (b) $P_1 - P_2$

(c) $P_1 + P_2$ (d) **$(P_1 \times P_2)^{1/2}$**

Q109.In life table q_x is calculated by the formula:

(a) l_x/q_x (b) l_x/e_x

(c) l_x/d_x (d) **d_x/l_x**

Q110.For comparing the health conditions of two towns we have to calculate:

(a) Net reproduction rate (b) Crude birth rate

(c) **Standardized death rate** (d) Standardized birth rate

Q111.In life tables expectation of life at x is obtained as:

(a) $\mathbf{T_x/l_x}$

(b) l_x/T_x

(c) d_x/T_x

(d) T_x/d_x

Q112.When organisms move out of the population, this is known as:

(a) Succession

(b) **Emigration**

(c) Immigration

(d) Abandonment

Q113.If the number of births in a population is the same as the number of deaths in a population, what will happen to the growth of the population?

(a) It will fluctuate

(b) **It will stay the same**

(c) It will increase

(d) It will decrease

Q114.If the number of births is greater than the number of deaths in a population, what will happen to the growth of the population?

(a) It will fluctuate

(b) It will stay the same

(c) **It will increase**

(d) It will decrease